SPACE SCIENCE

T0345186

SPACE SCIENCE

editors

Louise K. Harra
Keith O. Mason
University College London

Imperial College Press

Published by

Imperial College Press
57 Shelton Street
Covent Garden
London WC2H 9HE

Distributed by

World Scientific Publishing Co. Pte. Ltd.
5 Toh Tuck Link, Singapore 596224
USA office: Suite 202, 1060 Main Street, River Edge, NJ 07661
UK office: 57 Shelton Street, Covent Garden, London WC2H 9HE

British Library Cataloguing-in-Publication Data
A catalogue record for this book is available from the British Library.

SPACE SCIENCE

ISBN   1-86094-346-2
ISBN   1-86094-361-6 (pbk)

Contents

Chapter 1

Introduction

Louise Harra and Keith Mason

Space research is one of the most evocative and challenging fields of human endea-vour. In the half-century that has elapsed since the beginning of the space age, we have been exposed to wonders beyond the imagination of even the most visionary of science fiction writers. We have peered deep into the Universe and studied physics that we can never duplicate on the Earth. We have begun to understand our own Sun, and the flow of energy outward through the Solar System that is fundamental to our existence. Robotic agents have explored the planets in the Solar System and the environment of our own planet, the Earth. Human beings have trodden on the surface of the Moon, and now have a constant presence in near-Earth orbit. We are also advancing on that most fundamental of questions: whether life has ever evolved elsewhere in the Solar System or elsewhere in the Universe.

Beyond the primeval urge to explore, this knowledge has a fundamental effect on the evolution of human society, its ethics and values (for example, how different would life be if we still believed the Earth to be flat?). Gaining that knowledge has also spawned more direct benefits. The challenge of successfully building machines to operate in the hostile environment of space has led to countless advances in technology and the way in which we organise and deliver complex systems. Earth-orbiting satellites, a novelty but a few short decades ago, are now taken for granted in areas such as communication, navigation and weather forecasting. Increasingly we have come to recognise the fragile nature of the Earth's ecosystem and how it can be influenced by external factors, from space weather, with its origins in the turbulent atmosphere of the Sun, to asteroid impact. By studying the Earth as a planet, we are also coming to appreciate the potential deleterious effects human society can have on its own environment. The tools that we have honed in the study of the distant Universe are now increasingly being deployed to monitor the health of our own backyard.

The purview of space science is nowadays very broad, covering a large number of subject areas that many would regard as distinct. Nevertheless there is a great deal of commonality in the physics that underlies these areas, not to mention the technology, data analysis techniques and management challenges. This book has its origins in a series of lectures given to graduate students at University College London's Mullard Space Science Laboratory (MSSL). These lectures were designed to broaden the perspective of students beyond their immediate field of study and encourage cross-fertilisation between fields. The subjects dealt with cover the core research themes of MSSL, the UK's oldest university space research laboratory. While certainly not a complete survey of space science disciplines, the chapters herein sample a breadth of activity that we hope will give anyone interested in space research, and particularly those starting out in the field, a better perspective of the subject as a whole. The topics included cover disciplines from the Earth's environment, through the interplanetary medium, to the dynamic and sometimes violent Sun, to the outer Universe looking at exotic regions such as black holes. To complement these areas there are chapters that cover the basic physics necessary for understanding these subjects, such as quantum physics, magnetohydrodynamics and relativity. There are also chapters that describe basic techniques necessary to progress in space science research. These include engineering skills, space instrumentation and data analysis techniques.

1.1   A Brief History of Discovery

The first artificial Earth satellite, *Sputnik 1*, was launched by the Soviet Union on October 4, 1957. While this was a momentous event in human history, the beginnings of space science preceded the launch of *Sputnik 1* by about a decade, and employed sub-orbital 'sounding rockets'. The earliest of these were captured German V2 rockets developed during World War II, and the sounding rocket programme achieved numerous successes, including mapping the upper atmosphere and discovering X-ray emission from the Sun.

The first major scientific breakthrough to be made using artificial satellites was the discovery of the Earth's radiation belts. This was achieved using *Explorer 1*, the first successful satellite launched by the United States on January 31, 1958. *Explorer 1* contained scientific detectors designed and built by James Van Allen that were intended to measure the flux of cosmic rays. In fact *Explorer 1* measured an anomalously low count rate, and Van Allen suggested that this was because the detectors were being saturated by a large flux of energetic particles. This was confirmed two months later using *Explorer 3* (number two failed to achieve orbit!), and established the existence of the 'Van Allen' belts, which we now know to be an important feature of the Earth's magnetosphere (Chapter 4).

Since those early days, the capabilities of spacecraft for scientific discovery have mushroomed. In general terms, scientific data from spacecraft can be obtained in two ways. We can fly a craft to the region of interest and take *in situ* measurements of its immediate surroundings, or we can equip the spacecraft with telescopes for *remote sensing* that allow it to examine distant objects, making use of the unique vantage point that space provides.

1.1.1   *Exploration of the Solar System*

The discovery of the Earth's radiation belts is a classic example of an *in situ* measurement, where detectors on the *Explorer 1* satellite were measuring the immediate environment through which the spacecraft was flying. This is the main way in which we have gathered information on the Earth's magnetosphere, the cocoon carved out of the solar wind by the Earth's magnetic field. An up-to-date example is the European Space Agency's *Cluster* mission (launched in 2000), in which four identical spacecraft equipped with an array of sensitive instruments fly in formation through the magnetosphere. Having measurements from more than one adjacent spacecraft means that we can determine how plasma and magnetic fields are moving in space, as well as their properties at any given point.

As the techniques of space flight were mastered, the range of *in situ* measurements was rapidly extended, first to the Moon, and then to other bodies in the Solar System. In September 1959, the Soviet *Lunik 2* spacecraft measured solar wind particles for the first time, confirming the existence of the solar wind, which until that time had been inferred only indirectly. A further advance in our knowledge of the solar wind came from the flight of the US *Mariner 2* spacecraft to Venus. *Mariner 2*, launched in August 1962, measured the velocity of the solar wind and identified both slow and fast components. In common with many planetary probes, it contained both *in situ* and remote sensing devices. The latter were in the form of microwave and infrared radiometers, which allowed us to measure the temperature and composition of Venus and its atmosphere as the spacecraft flew by the planet.

A common remote sensing device is some sort of camera, an early example of which was an instrument on *Lunik 3* that photographed the far side of the Moon for the first time in 1959. Close-up photographs of the Moon were obtained with the *Ranger* spacecraft in the last minutes before they crashed onto the surface, and the whole Moon was surveyed from the *Lunar Orbiter* spacecraft in 1966 and 1967. Several *Surveyor* spacecraft landed on the Moon between 1966 and 1968 and obtained photographs from its surface, as well as probing the terrain with mechanical scoops (Figure 1.1). The exploration of the Moon in the 1960's culminated in the landing of astronauts, who conducted a range of investigations including the deployment of seismic detectors and laser ranging targets, and collecting rock and dust samples that were returned to the Earth for analysis. The results of this exploration, the *Apollo* programme, have, over the years, contributed greatly to our understanding of the Moon's origin, and of the formation of the Solar System.

Turning further afield, spacecraft have now been sent to every planet in the Solar System bar Pluto/Charon. The *Mariner 4* spacecraft discovered craters on Mars during a flyby in July 1965, while a large part of Mars was mapped using

Figure 1.1   The *Surveyor 3* spacecraft landed on the Moon in April 1967. Astronauts Conrad and Bean later visited *Surveyor 3* in November 1969 during the *Apollo 12* mission. Here Conrad is pictured examining the camera on *Surveyor 3*. The *Apollo 12* lunar excursion module 'Intrepid' is visible in the background. (NASA)

Mariner 9, which was placed in orbit about the planet in 1971, and two *Viking* spacecraft, which arrived at Mars in 1976. The *Viking*'s also carried landers, which obtained the first pictures from the Martian surface. The *in situ* exploration of the planet was continued in 1997 from the *Mars Pathfinder* lander, which included a rover vehicle. Meanwhile surveys of the planet from orbit have continued using the *Mars Global Surveyor* (arr. 1997) and *Mars Odyssey* (arr. 2001) spacecraft.

   Mariner 10 visited both Venus and Mercury, obtaining the first detailed view of the latter and revealing Moon-like craters in three encounters with the planet between 1974 and 1975. As noted above, the first spacecraft to visit Venus was *Mariner 2* in 1962. It was followed by many others (more than 20 in all so far), including *Pioneer Venus*, which made the first detailed map of its surface, and the Soviet *Venera 7*, the first spacecraft to land on another planet. Another Soviet lander, *Venera 9*, returned the first photographs of the surface of Venus, while the lander *Venera 13* survived for over two hours in the hostile Venusian environment, returning colour pictures. In the early 1990's, the orbiting US spacecraft *Magellan* produced detailed maps of Venus' surface using radar.

Exploration of the outer Solar System began with NASA's *Pioneer 10* and *11* spacecraft, launched in 1972, followed by the pair of *Voyager* spacecraft, which took the first close-up photographs of the outer planets. *Galileo*, launched by NASA to Jupiter in 1989, included an atmospheric probe as well as an orbiter, while the NASA *Cassini* spacecraft, currently on its way to Saturn, also contains an atmospheric probe, known as *Huygens*, which was built by ESA. Spacecraft have explored minor Solar System bodies as well as the main planets. *Galileo* was the first mission to make a close flyby of an asteroid (Gaspra), and also the first mission to discover a satellite of an asteroid (Ida's satellite Dactyl; Figure 3.20). Another example is the ESA *Giotto* mission, which made a close approach to Comet Halley in 1986, and subsequently also encountered Comet Grigg-Skjellerup, in 1992.

There are future aspirations to return samples from a comet, which are expected to be representative of the primitive state of the Solar System, and to determine the composition of the Martian surface, first by *in situ* robotic methods (as planned for example with the UK *Beagle 2* lander which will shortly be launched on ESA's *Mars Express*) and later by returning samples to the Earth. A major goal is to identify how and where life originated in the Solar System by searching for evidence that primitive life once evolved on Mars. The ultimate aim is manned exploration of the planet.

1.1.2 *The Sun and beyond*

There are some regions of the Solar System where conditions are too hostile to send spacecraft, the Sun itself being a prime example. To study these regions, and anything beyond the Solar System, we must use telescopes of one form or another. The advantage of placing these telescopes in space is that we take advantage of the clarity afforded by the vacuum of space, and avoid the absorption effects of the Earth's atmosphere. The latter consideration means that we can access a much larger range of the electromagnetic spectrum.

Soon after the inception of NASA in the USA in 1958, programmes were started in three scientific areas that made use of specially designed spacecraft. These were the *Orbiting Geophysical Observatory* (*OGO*), *Orbiting Solar Observatory* (*OSO*) and *Orbiting Astronomical Observatory* (*OAO*) series. The *OGO* spacecraft provided *in situ* measurements of the Earth's environment. The *OSO* series was dedicated to studying the Sun, and designed to provide a continuous watch of its activity, concentrating on high-energy radiation from the Sun's outer atmosphere. The *OAO* programme was primarily aimed at ultraviolet observations of stars, but the last in the series, *OAO-3* or *Copernicus* (see Figure 1.2), also included an array of X-ray-sensitive telescopes. *Copernicus* was launched in 1972, and was operated

Figure 1.2   The *Copernicus* spacecraft (*OAO-3*), shown being prepared for launch. *Copernicus* carried an 80-cm UV telescope built by Princeton University, and a set of small X-ray telescopes built by MSSL/UCL. Launched in 1972, the last of the *OAO* series, it operated until it was switched off in 1981.

continuously until 1981, when it was switched off for financial reasons. The orbiting observatories were very sophisticated for their era, and provided a benchmark for work that was to follow.

An important milestone in studying the Sun from space was the programme conducted from the *Skylab* space station, launched in 1973. This was the first manned observatory in space, and the mission lasted for nine months. *Skylab* was placed in orbit 435 km above the Earth. It carried an array of instruments (the Apollo Telescope Mount — ATM), which included a white light coronagraph, a soft X-ray telescope, a soft X-ray spectroheliograph and an extreme ultraviolet spectroheliograph. The astronauts responded to events occurring on the Sun seen by means of $H\alpha$ telescopes. Their response to flares was fast — within a few minutes or so. *Skylab* results changed our vision of the Sun forever. The data showed that the corona was far from uniform — and summarised the coronal structure as consisting of coronal holes, coronal loops and X-ray bright points. This was the last mission to use photographic plates.

The next major solar observatory was the NASA *Solar Maximum Mission* (*SMM*), which was launched in 1980 and designed to study flares and solar activity. There were seven instruments on board, covering a wide wavelength range, from white light to gamma rays. The spacecraft was unmanned and hence the data were recorded electronically and telemetered to ground stations. There was unfortunately an early failure in the attitude control system. Astronauts on the space shuttle repaired this in 1984, and observations continued successfully until 1989. Re-entry occurred in 1989 (at solar maximum) and was ironically triggered by a storm from the Sun!

There were two other major missions launched around the same time as *SMM*. The first was a Japanese mission named *Hinotori* (meaning 'firebird'). *Hinotori* had a soft X-ray spectrometer and an imaging hard X-ray instrument on board with the purpose of studying solar flares. The second was a US satellite, *P78-1*, which was launched in 1979 and had soft X-ray spectrometers, and a white light coronagraph on board. These missions discovered and verified large shifts occurring in the early stages of a flare, which provided vital clues to understanding the flaring process.

In 1991, the highly successful Japanese spacecraft *Yohkoh* (meaning 'sunbeam') was launched. It had four instruments on board, including a soft X-ray imager and spectrometer, and a hard X-ray telescope. The data obtained with *Yohkoh* had the best resolution of X-ray images to date, and brought about huge leaps in understanding not only flares, as was its original goal, but also coronal mass ejections and coronal heating. The mission operated successfully for 10 years, finally ceasing in December 2001. *Yohkoh* provided the first-ever space dataset to cover a solar cycle, and will be studied for many years to come.

Another exciting mission launched in the 1990s was the joint ESA/NASA *Ulysses*. This mission has an extremely complex orbit that uses gravity assists from Jupiter in order to orbit over the solar poles. The first orbit over the poles took place in 1994/1995 during the solar minimum, and the second took place in 2000/2001 coinciding with the solar maximum. The results from this mission have allowed an accurate *in situ* measurement of the slow and fast solar wind.

The joint ESA/NASA *SoHO* spacecraft was launched in 1995 and contained 12 instruments. As discussed later, it is located at a Lagrangian point, and hence full 24-hour coverage of the Sun is achieved. The payload, consisting of a collection of imagers, spectrometers and particle instruments, continues to provide a wealth of information on the interior of the Sun, right out to 30 solar radii.

Looking beyond the Sun, early efforts naturally concentrated on those parts of the electromagnetic spectrum that were not visible from the Earth. The *Orbiting Astronomical Observatory* series, mentioned above, was the first venture into ultraviolet astronomy. Many important discoveries were made with the *OAO* spacecraft, including the first measurement of the cosmologically important abundance of deuterium, made using *Copernicus* (*OAO-3*). This was followed by the hugely successful NASA/ESA/UK *International Ultraviolet Explorer* (*IUE*) satellite, which operated between 1978 and 1996. *IUE* was the first general user space observatory, and was run much like a telescope on the Earth's surface. Because *IUE* was in geosynchronous orbit above the Atlantic Ocean, it could be operated continuously, via a real-time link from ground stations in the eastern USA and Spain. During its lifetime, *IUE* conducted many thousands of scientific programmes, ranging from observations of comets in the Solar System to quasars in the distant reaches of the Universe. *IUE*'s current successor is NASA's *Far-Ultraviolet Spectroscopic Explorer* (*FUSE*), launched in 1999.

Orbiting observatories have also opened up many other wavebands. X-ray astronomy is a prime example (see Chapter 7), and very many satellites devoted to X-ray observations have been built over the past three decades by many different countries. Observations in the X-ray band have revealed hitherto unsuspected phenomena, including exotic objects such as massive black holes in the centres of galaxies, accretion onto neutron stars, hot gas in galaxy clusters, and activity in nearby stars. The German/USA/UK *Röntgen Satellite* (*ROSAT*; 1990–1999) and the NASA *Extreme Ultraviolet Explorer* (*EUVE*; 1992–2001) have probed the region between X-rays and the ultraviolet, known as the extreme ultraviolet band, while missions such as ESA's *COS-B* (1975–1982) and NASA's *Compton Gamma-Ray Observatory* (*CGRO*; 1991–2000) have explored the gamma ray region of the spectrum, at wavelengths shorter than X-rays.

We have also probed the electromagnetic spectrum at wavelengths longward of the visible band. The *Infrared Astronomy Satellite* (*IRAS*), a joint venture between the USA, the UK and the Netherlands, surveyed the sky during 1983 and made a major impact on astronomy. Long-wavelength infrared radiation is able to penetrate dusty regions much more effectively than short wavelength light, and one of *IRAS*'s many achievements was to identify large numbers of powerful starburst galaxies, which are a key element in understanding how all galaxies formed. Many of these discoveries were followed up with ESA's *Infrared Space Observatory* (*ISO*), which operated between 1995 and 1998.

Another major advantage of flying a telescope in space is freedom from the distorting effects of the Earth's atmosphere. The *Hubble Space Telescope* (*HST*), a NASA/ESA spacecraft that was launched in 1990, has amply demonstrated this. Because it is in space, *HST* can operate over a wide wavelength range, from the ultraviolet through the visible band to the infrared. Telescopes on the Earth's surface can also measure visible light of course, but *HST* is able to resolve much finer detail in astronomical objects because it does not have to peer through the turbulent atmosphere. It can also achieve much higher sensitivity because the sky background in space is much darker. *HST* has produced spectacular images of a huge range of astronomical targets — from planets in our own Solar System to galaxies at the edge of the known Universe.

The lack of atmospheric twinkling is also a major advantage when it comes to measuring the position of stars. This was used to good effect by the ESA mission *Hipparchos*, which measured the distance to large numbers of nearby stars. This was done by accurately measuring the slight shift in the position of the stars that occurs when they are viewed from opposite sides of the Earth's orbit about the Sun, known as the parallax. There are plans for a future mission, *GAIA*, which will be capable of measuring the slight shift of stars due to the rotation of our Milky Way

Galaxy. It will also measure their line of sight velocity and distance, building up a 3-D picture of how every star that is visible in the Galaxy is moving. This will be used to reconstruct the history of the Galaxy, and the way it has grown by mergers with other galaxies.

HST continues to operate today, and because it was designed to be serviced by space shuttle crews, its instrumentation has been regularly maintained and updated. The 'Great Observatories' *Chandra* (NASA) and *XMM-Newton* (ESA), both of which operate in the X-ray band, have now joined it in orbit. Soon, the NASA *Space Infrared Telescope Facility* (*SIRTF*) will be launched, to be followed by the Japanese/USA *Astro E* X-ray observatory. Plans are being laid for a follow-up to *HST*, the *James Webb Space Telescope* (named for the former NASA administrator), which will concentrate on the infrared regime, and for larger X-ray telescopes capable of probing the early Universe.

The NASA *Microwave Anisotropy Probe* (*MAP*) satellite was launched in 2001 and is currently surveying the microwave background radiation, which is left over from the Big Bang which created the Universe. The microwave background was previously studied from space using the *Cosmic Background Explorer* (*COBE*; 1989–1993), but *MAP* has much better angular resolution and will search for fluctuations in the background that can be used to determine the fundamental parameters of the Universe. A further increase in sensitivity is planned with the ESA survey mission *Planck*. This is scheduled for launch in 2007, along with a large ESA space telescope for pointed observations in the microwave spectral region, *Herschel*. Further in the future, there are plans to build an array of telescopes that can search for Earth-like planets around other stars (for example the ESA mission *Darwin*), and to build a fleet of three spacecraft capable of detecting gravitational waves (*LISA*). To have the required sensitivity, the three *LISA* spacecraft will be separated by 5 million kilometres, but yet will need to determine their relative location to an accuracy comparable to the wavelength of light if they are to detect the tiny distortions in space caused by the orbits of binary stars scattered through our Galaxy and beyond.

1.2 Observing from Space

A great deal of experience has now been accumulated in the use of spacecraft for scientific investigations. Building a scientific spacecraft remains a challenging task, and many of the technical and design issues are discussed in Chapter 14. It is worth reflecting briefly here on how we go about using a space observatory, and how the technical issues affect what we can and cannot do.

A major consideration is the choice of orbit. This is important partly because of the concentrations of charged particles trapped in the radiation belts within the Earth's magnetosphere (see Figure 4.5). These high-energy particles interact with many types of astronomical detectors and greatly increase the background signal, rendering them insensitive. X-ray detectors are especially prone to these effects, for example, so it is wise to avoid these regions of charged particle concentration when making astronomical observations. To minimise the problem, one can choose to fly the observatory in a low Earth orbit (LEO), below the worst of the radiation belts, or in a high Earth orbit (HEO), above the radiation belts.

A disadvantage of a low Earth orbit, which will typically be at an altitude of 400–600 km, is that the Earth itself will block a substantial part of the sky. Moreover, the orbital period is only about 90–100 minutes, so the region of sky that is visible changes rapidly, making it difficult to observe a single target for an extended period. Another disadvantage for some applications is that the sunlit portion of the Earth is very bright. This can obviously lead to problems with stray light background for instruments that are sensitive to optical/infrared radiation, or problems with thermal control for instruments that need to be operated cold. A further disadvantage is that low Earth orbits decay relatively rapidly due to the drag caused by the tenuous outer atmosphere, limiting the lifetime of the satellite unless corrective action is taken. Nevertheless, many astronomical observatories have operated very successfully in a low Earth orbit, including *HST*. Satellites in a low Earth orbit can also in principle be serviced by the space shuttle, and repaired if they develop a fault. This was famously demonstrated in the recovery of the *Solar Maximum Mission* (*SMM*), while *HST* was designed from the outset with shuttle servicing in mind.

The detrimental effects of the Earth on observing conditions are much reduced in a high Earth orbit, which is ideal for applications where long observations of a target are required under stable conditions. However, a penalty for choosing a high Earth orbit is that more energy is required to reach it. Put another way, for a given launch vehicle (which equates to cost) you can place a heavier and more capable observatory into a low Earth orbit than a high Earth orbit. Similarly, it takes more energy to achieve a circular orbit at a particular altitude, than an eccentric orbit with an apogee (furthest distance from the Earth) at the same altitude. Thus a compromise that is often adopted is to launch into an eccentric orbit with a high apogee. This takes advantage of Kepler's third law, which states that an orbiting body moves more slowly at the apogee, and therefore spends most of its time at the furthest points in its orbit. ESA's *EXOSAT* X-ray observatory (launched in 1983) was the first to use an eccentric high Earth orbit of this kind, and similar orbits are employed by the latest generation of X-ray observatories, *XMM-Newton* (Figure 1.3) and *Chandra*. These satellites spend most of their time outside the Earth's radiation belts, but one of the penalties paid for straying outside the Earth's protective shield is that they are vulnerable to the occasional burst of energetic particles originating from the Sun (Chapter 5).

Communications are also an issue. A satellite in a low Earth orbit is typically visible to a given ground station for only a few minutes for each orbit, which means that some degree of autonomous operation is essential. NASA's *Tracking and Data*

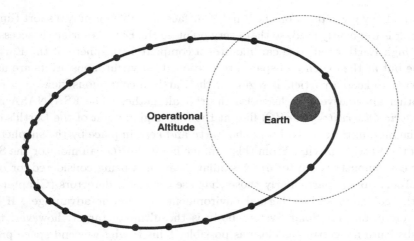

Figure 1.3 The orbit of *XMM-Newton* about the Earth, shown to scale. The solid circles mark the position of the spacecraft at two-hour intervals in its two–day orbit. The observatory collects scientific data only when it is above about 40,000 km altitude (shown as the dashed circle), but this amounts to about 90% of the time because of the fact that the spacecraft moves more slowly near the apogee than the perigee. The apogee is about 114,000 km above the Earth.

Relay Satellite System (*TDRSS*) provides an orbital relay through which continuous contact can be achieved, particularly for monitoring the state of the satellite, but this is expensive. The *IUE* satellite was placed into a geosynchronous orbit, which meant that it could be operated as a real-time observatory from a single ground station. (In fact the orbit was slightly eccentric, which meant that its apparent position in the sky oscillated every day to allow it to be operated from two ground stations, one in Europe and one in the USA). Geosynchronous orbits (altitude 40,000 km) are not completely free of the radiation belts, however, and the IUE detectors suffered from enhanced radiation background for part of each day. Instead of a fully geosynchronous orbit, X-ray observatories such as *EXOSAT*, *Chandra* and *XMM-Newton* were/are in eccentric orbits with periods that are an exact multiple of a day. These take them well above the radiation zones; in the case of *XMM-Newton*, which is in a two-day orbit, to an altitude of about 114,000 km at the apogee (Figure 1.3). Because the orbital period is a multiple of a day, continuous coverage can be achieved from a limited set of ground stations during the part of the orbit when the spacecraft are making observations, above the radiation belts. Set against the value of continuous coverage, the problem of transmitting the data collected to the ground becomes more acute the further the spacecraft is from the Earth.

Similar considerations apply in observing the Earth from space (Chapter 2). A low Earth orbit gives you the best spatial resolution, since the spacecraft is only a few hundred kilometres above the Earth's surface. However, the region being studied is moving quickly in the field of view, because of the satellite's orbital motion, and any given point on the Earth's surface is visible for only a short time. A polar orbit is frequently used, so that every point on the Earth's surface is accessible. From a high Earth orbit, one can monitor a complete hemisphere of the Earth at one time but at the expense of spatial resolution. Geosynchronous orbits are useful if you want to keep a particular region of the Earth in continuous view.

Another alternative is to leave Earth orbit altogether. The ESA/NASA *Solar Heliospheric Observatory* (*SoHO*) flies at the Lagrangian point of the Earth's orbit about the Sun, in the same orbit as the Earth and kept in place by the shepherding effect of the Earth's gravity. From this vantage point, *SoHO* can monitor the Sun's activity continuously (Chapter 6). A similar location is being considered for many future observatories, particularly those that use cryogenic detectors (Chapter 13), where the cold and stable thermal environment is a major advantage. If high spatial resolution of a Solar System body is the ultimate target, however, there is no substitute for getting as close as possible. This is why we send space probes to other planets. For the Sun also, we would like to get much closer in order to resolve the detail that we need to understand how it works. For this reason ESA are currently planning their *Solar Orbiter* spacecraft, which is intended to fly to within 40 solar radii of the Sun (for comparison, the Earth's orbit is at a radius of about 200 solar radii). Withstanding the thermal and radiation stresses of flying that close to our parent star will be a major challenge, but will provide a powerful addition to our investigative arsenal.

Chapter 2

Remote Sensing of the Earth's Climate System

Ian Mason

2.1 Remote Sensing of the Climate System

Observation from space has become an important tool for the scientific study of the Earth, and with increasing concern about anthropogenic global warming; much of this remote sensing is currently directed towards studying the Earth's climate. This chapter provides an overview of this field, outlining the basic principles involved in Earth observation from space (Section 2.2), and then providing one example of each of the three main types of remote sensing, as applied to climate research (Sections 2.3, 2.4 and 2.5). First, however, Section 2.1 provides an introduction to remote sensing and the climate system.

2.1.1 *Remote sensing and Earth system science*

The phrase 'remote sensing' can be used to describe almost any indirect measurement, but is used in this chapter in the context of measurement and observation of the Earth from satellites. Nevertheless, it should be noted that many of the methods for remote sensing of the Earth have been used in spacecraft missions to other planets, and are also deployed on aircraft, often so as to test new techniques and methods prior to their use in space-borne instruments. Remote sensing of the Earth from space goes back to the beginning of space exploration in the late 1950s/early 1960s, when it was initially applied to military surveillance and meteorology. Operational programmes in both of these areas were subsequently established by the USA, the USSR and other nations. Further socio-economic applications followed with the introduction of commercial high resolution visible/infrared imaging satellite programmes such as the USA's *Landsat* in 1972 and the French *Spot* in 1986, and these systems are now routinely used for applications such as geological prospecting and land use mapping. In addition, since the 1990s, imaging radar systems on non-military satellites have been used on a regular basis for economic purposes such as sea ice mapping and oil prospecting.

Alongside these developments, Earth scientists have also made extensive use of data from remote sensing satellites — not only the above operational and commercial systems, but also a number of dedicated scientific missions. Such data have been applied in disciplines such as atmospheric science, oceanography, glaciology, geology and biology. However, much Earth science is now interdisciplinary, treating the Earth as a single complex system, and such research is therefore often referred

Table 2.1   The first large *International Earth Observing System* (*IEOS*) satellites and their main instruments.

Satellite (in launch order)	Main Instruments (in alphabetical order)	
Terra (USA, 1999)	ASTER	Advanced Spaceborne Thermal Emission and Reflection Radiometer
	CERES	Clouds and the Earth's Radiant Energy System
	MISR	Multi-angle Imaging Spectro-Radiometer
	MODIS	MODerate-resolution Imaging Spectro-radiometer
	MOPITT	Measurements of Pollution in The Troposhere
Envisat (Europe, 2002)	ASAR	Advanced Synthetic Aperture Radar
	AATSR	Advanced Along Track Scanning Radiometer
	GOMOS	Global Ozone Monitoring by Occultation of Stars
	MIPAS	Michelson Interferometer for Passive Atmospheric Sounding
	MERIS	MEdium Resolution Imaging Spectrometer
	MWR	MicroWave Radiometer
	RA-2	Radar Altimeter 2
	SCIAMACHY	SCanning Imaging Absorption SpectroMeter for Atmospheric CHartographY
Aqua (USA, 2002)	AIRS	Atmospheric Infrared Sounder
	AMSR/E	Advanced Microwave Scanning Radiometer/EOS
	AMSU	Advanced Microwave Sounding Unit
	CERES	Clouds and the Earth's Radiant Energy System
	HSB	Humidity Sounder for Brazil
	MODIS	MODerate-resolution Imaging Spectro-radiometer
ADEOS-II (Japan, 2002?)	AMSR	Advanced Microwave Scanning Radiometer
	GLI	Global Imager
	ILAS-II	Improved Limb Atmospheric Spectrometer II
	POLDER	Polarization and Directionality of the Earth's Reflectances
	SeaWinds	Radar Scatterometer

to as Earth system science. For such studies, the Earth system is generally considered in terms of two major subsystems: the physical climate system and the biogeochemical system (IGBP, 1990). In view of possible anthropogenic changes to the climate and the environment, two international programmes were set up in the 1980s to promote and coordinate research into these two subsystems: the World Climate Research Programme (WCRP) and the International Geosphere–Biosphere Programme (IGBP), respectively. The IGBP also addresses Earth system modelling and past global change (IGBP, 1990). Research in these areas has subsequently received strong support, and an international series of large remote sensing satellites, conceived in the 1980s, is now being launched to provide high quality remotely sensed data for this research effort. The first, *Terra* (from the USA's NASA), was launched in 1999, followed by *Envisat* (from ESA) and *Aqua* (from NASA) in 2002. At the time of writing, *ADEOS-II* (from the Japanese NASDA) is about to be launched. Table 2.1 lists the main instruments on these satellites. Further scientific Earth observation missions are planned, including a range of dedicated smaller satellites such as ESA's *Cryosat* (designed to measure land ice and sea ice in polar regions). Improved versions of commercial and operational systems, such as *Landsat-7* (launched 1999) and various advanced meteorological satellites, are also expected to provide valuable additional data for Earth system science. The NASA missions are collectively known as the *Earth Observing System* (*EOS*); the whole set of missions in this '*EOS* era', including those from other agencies and nations, has been referred to as the *International Earth Observing System* (*IEOS*; King and Greenstone, 1999).

2.1.2 *The requirements for remote sensing of the climate system*

Much of the remote sensing for Earth system science is directed towards climate research, the main application considered in this chapter. Before discussing remote sensing methodology, therefore, it is useful to give an overview of the climate system, and to note how remote sensing can be used in the investigation of climate.

The Earth's climate system (Figure 2.1) is complex and dynamic, involving the mutual interaction of the atmosphere, ocean, cryosphere, biosphere and land surface, as they exchange energy, momentum, moisture, and other substances such as carbon dioxide or aerosols (suspensions of tiny solid or liquid particles in the atmosphere). Solar energy (mostly visible light) drives the whole system, mainly by heating the surface of the Earth, the majority of which is ocean. This heated surface then warms the atmosphere by various processes. Of particular importance is the greenhouse effect, whereby greenhouse gases such as water vapour and carbon dioxide absorb most of the radiant heat from the surface. These in turn radiate heat back and warm the surface further, the amount of warming depending on their concentration. The difference in solar heating between low and high latitudes causes the general circulation of the atmosphere, which in turn, by the action of wind on

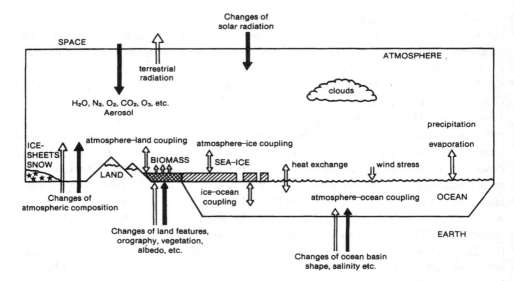

Figure 2.1   Schematic illustration of the climate system. The full arrows are examples of external processes, and the open arrows are examples of internal processes, in climatic change. From IPCC (1990, p. 75).

the ocean surface, is largely responsible for setting up the ocean circulation. Both of these circulations transport heat from low to high latitudes and play a major role in determining the weather and climate at any location. The weather is the instantaneous state of the atmosphere at any place, characterized by parameters such as temperature, precipitation, wind, cloud, pressure, humidity and visibility. Climate is the long-term description of the weather for a particular place or region, characterised by averages and other statistical measures of the weather.

Averaged over time, temperatures within the climate system are kept steady by the radiation of heat to space, mainly from the atmosphere, balancing the incoming solar radiation. Consequently, any changes to this radiation balance will, on average, tend to cause global warming or cooling. Because of the complexity of the system and its processes, there will also be many other alterations to the climate on a global and a regional scale, as well as environmental impacts of climate change such as a sea level rise. Climate change can therefore be induced by changes to any of the external constraints of the system (e.g. the full arrows in Figure 2.1), such as greenhouse gas concentrations, or the amount of incoming sunlight reflected by clouds or aerosols. Both of these effects are in fact occurring when we burn fossil fuels — one of the key human activities influencing climate. (Of the combustion products, carbon dioxide is a greenhouse gas, and sulphur dioxide creates sulphate aerosols that reflect sunlight and also help produce more highly reflecting clouds.)

There are also, however, natural fluctuations in the climate system's external constraints, particularly the release of sunlight-blocking aerosols into the

stratosphere by volcanic eruptions, and variations in the radiation output from
the Sun. In addition the climate displays a natural variability due to internal oscillations within the climate system itself. An important example is El Niño/Southern
Oscillation (ENSO). An interaction between the tropical ocean and the global atmosphere, ENSO is a quasi-periodic oscillation, every few years, between two different
climatic states (known as El Niño and La Niña), which involve alterations of climate
(e.g. average temperature, precipitation and wind) in many regions. The magni-
tudes of all these natural effects are such that it was only during the 1990s that an
anthropogenic climate change 'signal' (a net global warming) began to emerge from
this natural climatic 'noise' (Figure 2.2).

The Intergovernmental Panel on Climate Change (IPCC) was set up in 1988 to
assess the scientific information on climate change and its impacts, and to formulate
response strategies. It produced major reports in 1990, 1995 and 2001 (IPCC, 1990;
IPCC, 1996; IPCC 2001). The 2001 report showed that the Earth had warmed by
0.6°C ± 0.2°C since the late 19th century, with a strong increasing trend since the
mid-1970s that is still continuing (Figure 2.2). The IPCC concluded that 'most of
the warming observed over the last 50 years is attributable to human activities'.
The report summarised modelled predictions for the climate over the 21st century,
including a global warming of between 1.4°C and 5.8°C (depending on the greenhouse gas and aerosol emission scenario), mainly due to an enhanced greenhouse
effect, and a rise in sea level of between 0.09 and 0.88 m, mainly due to thermal
expansion of the oceans. Though such changes may seem small compared with,
say, daily fluctuations in temperature or tidal effects on sea level, they represent

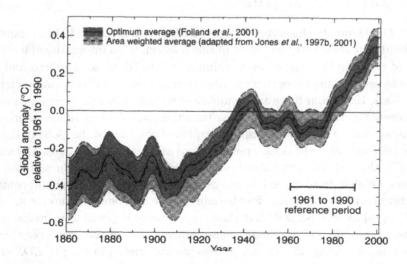

Figure 2.2   Smoothed annual anomalies of global combined land surface and sea surface temperatures (°C), 1861–2000, relative to 1961–1990, and twice their standard errors. From IPCC (2001,
p. 115).

exceptionally rapid changes in mean values compared with the past. These changes, together with associated changes in other climatic variables and oscillations such as ENSO, are therefore likely to have a profound impact on the biosphere and on human society.

Despite these results, there is still considerable uncertainty in establishing accurately, for both climate change and its impacts, the magnitude of current changes and predicted trends, particularly at the regional level. There are three ways in which research is helping to reduce these uncertainties:

- Predictive modelling

  — to predict more accurately future changes in climate and its impacts, by means of sophisticated atmosphere–ocean general circulation models (AOGCMs).

- Process studies

  — to understand, quantify and model the detailed processes involved in climate change and its impacts, so as to make the predictive models more accurate.

- Observation

  — to directly detect any changes in climate or its impacts;
  — to obtain input data for predictive models (e.g. data for initial or boundary conditions).
  — to obtain validation data for predictive models (to check model outputs, when they are tested by being run for periods ending at the present time).
  — to obtain data for process studies.

The Global Climate Observing System (GCOS) was set up in 1992 to coordinate and facilitate the observational aspect of this research at an international level, with the aim of ensuring that the necessary climate-related data are acquired and made available to users. Remote sensing can play an important role in making such observations (Karl, 1996), and there are a number of reasons why measurements made by remote sensing can in principle be better than the equivalent *in situ* measurements. The main advantage of remote sensing is improved coverage of the globe in time and space, particularly over the ocean, polar regions and remote land areas. Data access is generally also fast and guaranteed, via dedicated data centres. In addition, data quality can be high, in so far as it is consistent (from a single sensor) and controlled (via calibration and validation). Furthermore, there are some measurements (e.g. of ocean topography; see Section 2.5) that are virtually impossible to make *in situ*. Nevertheless, not all climate-related measurements (e.g. of the deep ocean) can be made by remote sensing, and in many cases remote sensing in the pre-*EOS* era has not matched the accuracy of *in situ* data. Therefore it is only with the advent of the *IEOS* that the potential of remote sensing as a tool for climate research is likely to be fully realised.

Table 2.2   Remote sensing of the climate system: main measurements.

Atmosphere	Land
Clouds and aerosols	Albedo
Lightning	Inland water (rivers, lakes, wetlands)
Precipitation	Soil moisture
Radiation budget	Surface composition
Temperature (vertical profiles)	Surface temperature
Trace gases	Surface topography
Water vapour (vertical profiles)	Vegetation
Winds	Volcanoes
Ocean	**Cryosphere**
Colour (phytoplankton)	Land ice thickness
Sea surface temperature	Sea ice extent, thickness, type
Sea surface topography circulation	Sea ice motion
Waves and surface wind speed	Snow extent and thickness

Table 2.2 lists most of the climate-related measurements that can be made by remote sensing. It is based on a number of such lists in the literature (Butler *et al.*, 1984; Butler *et al.*, 1987; IPCC, 1990; Harries, 1995). For each of the listed or associated variables, there are particular observational requirements for climate research (e.g. accuracy, spatial resolution, temporal resolution). A comprehensive set of these requirements was listed, for example, in one of the early planning documents for the *EOS* (Butler *et al.*, 1984). For most of the listed measurements, the atlas edited by Gurney *et al.* (1993) gives examples of remotely sensed observations in the pre-*EOS* era.

2.2   Remote Sensing Methodology

Satellite remote sensing from space, particularly for climate research, involves a particular methodology. In this section we begin by describing the main constraints that result from observing from space, and then describe the basic concepts involved in making the measurements. Finally, we describe some of the main factors affecting the measurement process.

2.2.1   *Constraints due to observing from space*

One of the main constraints imposed on space-based Earth observation is the nature of the satellite's orbit. There are two main orbits used for remote sensing: the geostationary orbit (GEO), used for the international set of geostationary meteorological satellites, and the near-polar circular low Earth orbit (LEO), used for most other remote sensing satellites. In addition, other LEOs are occasionally used for specific purposes.

Assuming a spherically symmetrical Earth, the period P of an Earth satellite in a circular orbit of altitude h is given by

$$P = 2\pi\sqrt{\frac{(R_E + h)^3}{GM_E}}\,,$$

where G is the universal gravitational constant and M_E and R_E are the Earth's mass and radius respectively. If P is given in days, the number of orbits per day $n_{\text{orb}} = 1/P$.

The near-polar circular LEO is shown in Figure 2.3. Typically the *orbital inclination* i (the angle between the orbital plane and the equatorial plane) is slightly greater than $90°$ (usually to provide a *Sun-synchronous* orbit; see below). In this

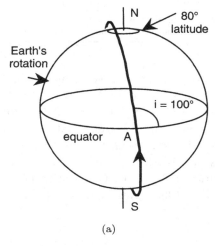

(a)

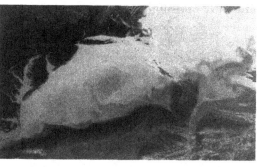

(b)

Figure 2.3   (a) Schematic diagram of a typical near-polar LEO for remote sensing. Inclination angle i and ascending node A are shown (see text). (b) Four minutes of thermal infrared data from AVHRR in such a LEO, showing its wide swath. The satellite is travelling approximately northward across the warm (dark) Gulf Stream (*bottom*) and east coast of the USA (*top*). The cool coastal waters and cold cloud appear light. Image from Cracknell and Hayes (1991).

case the satellite is *retrograde* (travelling from east to west), with the sub-satellite point reaching a maximum latitude of $(180-i)°$. The altitude h is typically between 500 km and 1000 km, with $P \approx 100$ minutes. Observations of the Earth are built up along the ground track (the track of the sub-satellite point on the Earth's surface) in the form of strips or *swaths* (Figures 2.3, 2.13). As the satellite travels round its orbit, the Earth rotates from west to east while the orbital plane remains fixed in space. Thus each successive revolution of the satellite around its orbit provides a new swath. Approximating the Earth as a sphere and ignoring Earth rotation, the speed with which the sub-satellite point moves along the ground track is given by $v_{\text{track}} = (2\pi R_{\text{E}}/P)$; note that this is slower than the actual satellite speed. For $h = 800$ km, for example, $v_{\text{track}} \approx 6.6$ km s^{-1}. The separation S_{eq} between successive swath centres, measured along the equator, is given by $S_{\text{eq}} = (2\pi R_{\text{E}}/n_{\text{orb}})$. Again using the example of $h = 800$ km, we find that $n_{\text{orb}} \approx 14.3$, so $S_{\text{eq}} \approx 2800$ km. However, the separation reduces dramatically as the ground tracks converge at high latitudes.

The frequency or time resolution for observing any location is therefore better for wide swaths and higher latitudes. For a typical near-polar LEO, the maximum realistic swath width (avoiding excessively oblique viewing) is approximately the same as the spacing of successive ground tracks at the equator. For example, for a satellite with $h = 800$ km and $S_{\text{eq}} \approx 2800$ km at the equator, an on-board instrument with a swath width of 2800 km centred at the *nadir* (the direction of its sub-satellite point) will be viewing at an angle of $\approx 17.5°$ to the horizontal at the edge of the swath. An instrument with such a swath width (e.g. the Advanced Very High Resolution Radiometer, AVHRR, on the NOAA meteorological satellites; see Figure 2.3 and Section 2.3) can therefore make observations of any location on the equator twice per day (approximately 12 hours apart) near the *ascending and descending nodes* of the orbit (the points where the sub-satellite point crosses the equator travelling northward and southward, respectively). Locations at higher latitudes can be observed on several successive swaths. If the view angle needs to be less oblique, or if the swath width of the instrument is intrinsically narrower, there will inevitably be gaps between successive swaths at low latitudes. On successive days, however, assuming n_{orb} is not an integer, the orbital tracks will lie in different positions on the surface, and this allows any viewing gaps between successive swaths gradually to be filled in. Thus full coverage of the Earth's surface is possible from a satellite in near-polar LEO, except perhaps at the very highest latitudes (depending on swath width and orbital inclination).

The horizon-to-horizon view for a satellite in LEO represents a small fraction of the circumference of the Earth (e.g. $\approx 15\%$ at 800 km altitude; Figure 2.4). By the same token, a satellite in LEO can only be observed by any one ground station for a small fraction of the orbital period. Therefore data are usually stored on board and downloaded to ground stations at high latitudes, where the convergence of the ground tracks means that many different orbital passes will be within range of a

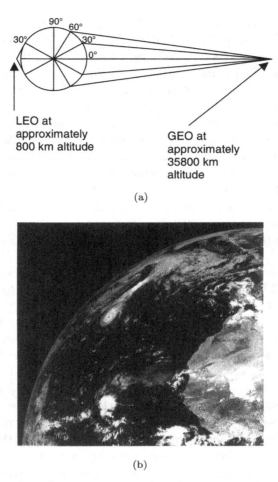

(a)

(b)

Figure 2.4   (a) Comparison of imaging geometry from GEO and LEO (approximately to scale). The view is either looking down on the North Pole, or looking perpendicular to the Earth's axis. (b) A quarter of a full-Earth image from Meteosat, in GEO. Note the hurricane Gloria. Image from ESA.

single ground station. Some imaging instruments, however, have such a high data rate that on-board storage is not realistic. For such instruments, e.g. the synthetic aperture radar (SAR) on *ERS-1* and *ERS-2*, the satellite can only acquire useful data when it is actually passing over a ground station, and can transmit the data in real time.

A special type of LEO is one whose altitude, and therefore value of n_{orb}, is adjusted such that after a *repeat period* of N days (where N is an integer), the ground tracks start to repeat their geographical positions again exactly (usually within ≈ 1 km). The criterion for such an orbit is therefore that the product $(N\ n_{\mathrm{orb}})$ is an integer. Sometimes known as an *Earth-synchronous* orbit, this is particularly valuable for radar altimeters, which have a very narrow swath (typically < 10 km) and are used to look for time-varying heights on the ocean or ice sheets (see Section 2.5).

As mentioned earlier, a near polar LEO may also be Sun-synchronous. For this type of orbit, the Earth's gravitational asymmetry is used as a natural way of slowly rotating the orbital plane around the Earth's axis, so that the orbital plane retains the same orientation with respect to the Sun rather than to the 'fixed stars'. Thus the *local* time of the ascending node does not change with the seasons, but remains fixed. Sun-synchronous orbits are often used for remote sensing satellites, so as to lock them into the diurnal cycle (e.g. with regard to solar illumination and meteorological conditions). For the natural rate of rotation of the orbital plane to be the required 360° in 365 days, the inclination has to be ≈ 100° for the typical LEO altitudes considered here.

Other LEOs are occasionally used for specific remote sensing purposes. For example, the *Tropical Rainfall Measuring Mission* (*TRMM*, launched in 1997) used a low inclination LEO ($i = 35°$, with the sub-satellite point reaching a maximum latitude of 35°) to give more frequent coverage of the tropics. Another example is *Cryosat*, due for launch in 2004, which will use a high inclination LEO (92°) to monitor land ice and sea ice near the poles with a narrow swath radar altimeter.

The geostationary orbit (GEO) has an inclination of 0° and an orbital period equal to the rotation period of the Earth, so that a satellite in GEO is always situated over a single longitude point on the equator. This requires an orbital altitude of ≈ 35800 km. The viewing geometry is shown in Figure 2.4. The horizon is at approximately ±80° in latitude or longitude, so the poles are not visible, and at least three geostationary satellites are required, located at different longitudes, in order to image the whole Earth. To avoid extremely oblique viewing at intermediate longitudes, five satellites are in fact used, forming the current operational set for meteorology: *GOES-W* (USA; 135°W), *GOES-E* (USA; 75°W), *Meteosat* (Europe; 0°), *GOMS* (Russia; 70°E), *GMS* (Japan; 140°E).

Nevertheless, viewing from satellites in GEO is still very oblique at high latitudes (Figure 2.4), producing image distortion with degraded spatial resolution. The main advantage of the GEO for remote sensing is high time resolution. Images can be produced continuously (in practice, a few times an hour), allowing rapidly changing phenomena, particularly clouds, to be monitored. For climate research and meteorology, such cloud data are valuable for estimates of rainfall and wind speed, and enable storms such as hurricanes to be tracked accurately. All the data can also be continuously transmitted in real time to a ground station, avoiding the necessity for on-board data storage. One disadvantage of remote sensing from GEO is the fact that the diffraction limit on spatial resolution becomes more important (see Section 2.2.5), mainly because of the high altitude but also because it is difficult and costly to place massive large aperture telescopes in GEO. In practice this currently precludes any useful radar or passive microwave observations from GEO.

(In addition, radar observations generally require specific viewing angles, which are clearly unavailable for the majority of locations from GEO.) The diffraction limit also currently restricts infrared observations from GEO to a moderate spatial resolution (~ 1 km), though such a resolution is acceptable for most climate-related applications. A summary of the advantages and disadvantages of the GEO and near-polar LEO for climate-related remote sensing is given in Table 2.3.

As well as these orbital constraints on observing the Earth, the usual spacecraft constraints (e.g. the need to survive the space environment, restrictions on mass and power, etc.) clearly also apply to remote sensing satellites. Attitude control (i.e. stabilisation of the satellite's orientation) is also an important consideration. Gravity gradient stabilisation, involving a mass on a boom that automatically points directly away from the Earth, seems an attractive method for Earth remote sensing. However, residual pointing oscillations mean that such systems cannot generally be used, and most LEO remote sensing satellites require three-axis stabilisation to provide sufficiently accurate pointing. An exception was the *GEOSAT* mission in the late 1980s. This LEO satellite was able to use gravity gradient stabilisation because

Table 2.3   Comparison of main satellite orbits for Earth remote sensing.

Geostationary Orbit	Near-Polar Low Earth Orbit
Advantages	*Disadvantages*
High time resolution (each location observed several times per hour). Good for cloud and storm monitoring	Time resolution as poor as every 12 h (broad swath) or many days (narrow swath). However, this is adequate for many climate applications.
Continuous ground communications.	Discontinuous ground communications. Need data recording, or acquisition over ground stations only (if high data rate).
Disadvantages	*Advantages*
Large payload mass in GEO is difficult/ expensive	Large payload mass more feasible.
Spatial resolution worse (large telescope aperture less feasible, and large viewing distance). Microwave observations unrealistic, but visible/infrared spatial resolution adequate for many climate applications.	Reasonable spatial resolution possible even at microwave wavelengths.
Limited latitude range (poles invisible).	All latitudes can in principle be observed.
Fixed viewing angles for each location. Prevents radar measurements. High latitudes viewed obliquely	Nadir viewing possible at all latitudes.
In-orbit recovery or repair currently impossible.	In-orbit recovery or repair possible in principle.

its only instrument was a pulse-limited radar altimeter, which could tolerate small amounts of mispointing (see Section 2.5). For GEO, a useful attitude control option, used by most geostationary meteorological satellites to date, is spin stabilisation, with a north–south spin axis. The spin is then also used for east–west scanning of the Earth by a narrow field of view telescope, so as to build up the image (see Section 2.2.5).

2.2.2 *Measurement by remote sensing: basic concepts*

Remote sensing instruments are often classified as either *passive* instruments, observing natural electromagnetic radiation reflected or emitted by the Earth, or *active* systems, where the instrument provides its own artificial source of electromagnetic radiation and observes the backscatter from the Earth. Thus we can distinguish three basic techniques for remote sensing of the Earth:

- Passive sensing of reflected sunlight (using visible and near-infrared wavelengths)
- Passive remote sensing of thermal emission (using thermal infrared and microwave wavelengths)
- Active remote sensing using laser or microwave radar systems

The wavelengths utilised by these systems are constrained by the transmission of the atmosphere and, for the passive systems, by the spectra of the radiation sources (Figure 2.5). It can be seen that there are only certain spectral bands or 'windows' that are available for observation of the Earth's surface from space, as follows. (Note that the nomenclature and boundaries for regions of the electromagnetic spectrum are ill-defined; here we adopt those of Stephens, 1994.) Over the whole of the *visible* region (0.4–0.7 μm) the atmosphere is highly transparent, but in the *near-infrared* region (0.7–3 μm) it has a number of absorption features, mainly due to water vapour and carbon dioxide. In the *thermal infrared region* (3–15 μm) the atmosphere is fairly opaque due to absorption by water vapour, carbon dioxide and other trace gases, except in the three main transmission 'windows' from 3.5–4.0 μm, 8.0–9.5 μm and 10.5–12.5 μm. At longer wavelengths the atmosphere is highly opaque until a window opens up in the *microwave* region (from 1 mm to 1 m, equivalent to 300 GHz to 0.3 GHz, as the product of the wavelength and frequency for electromagnetic waves is equal to the speed of light). The main absorption peaks here are due to oxygen at approximately 2.5 mm (120 GHz) and 5.0 mm (60 GHz), and water vapour at approximately 1.5 mm (200 GHz) and 15 mm (20 GHz).

If we wish to observe the atmosphere rather than the Earth's surface, the absorption features are not always a hindrance but can be put to positive advantage so as to carry out what is usually called 'remote sounding' of the atmosphere. In this special type of passive remote sensing of thermal emission, thermal infrared or microwave observations are made simultaneously in several narrow wavelength bands, each having a different value of atmospheric absorption. For each wavelength, the source

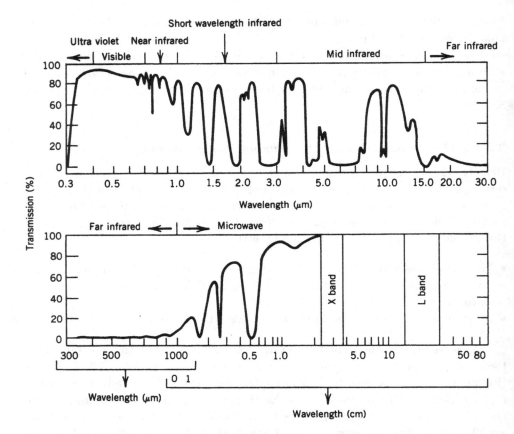

Figure 2.5   Generalised absorption spectrum of the Earth's atmosphere along a vertical path. The curve shows the total atmospheric transmission. (Note the slightly different nomenclature and divisions compared with the text.) From Elachi (1987). Copyright 1987. This material is used by permission of John Wiley & Sons, Inc.

of the detected radiation has a different height distribution in the atmosphere, with a peak at a different height. These differences allow vertical atmospheric profiles of temperature and water vapour to be obtained.

In discussing the natural radiation sources for passive remote sensing, it is useful to note the definitions of some of the radiometric quantities used in remote sensing (Table 2.4). The Sun and the Earth can, to a first approximation, be considered as blackbody radiation sources, whose spectrum is therefore given by the Planck function, which in terms of spectral radiance R_{BB} is given by

$$R_{\mathrm{BB}}(\lambda, T) = \frac{2hc^2}{\lambda^5} \frac{1}{e^{ch/\lambda kT} - 1},$$

where h is Planck's constant, k is Boltzmann's constant, c is the speed of light, λ is the wavelength, and T is the absolute temperature of the blackbody. Integrating

Table 2.4   Radiometric quantities.

Quantity	Definition	SI Units
Radiant flux	Total radiant power.	W
Radiant flux density	Radiant flux intercepted by a unit area of a plane surface.	$W\ m^{-2}$
Exitance	Radiant flux density emitted in all directions (i.e. into a hemisphere).	$W\ m^{-2}$
Irradiance	Radiant flux density incident from all directions (i.e. from a hemisphere).	$W\ m^{-2}$
Radiance	In a particular direction, the radiant flux per unit solid angle, per unit projected area (i.e. perpendicular to that direction).	$W\ m^{-2}\ sr^{-1}$
Spectral Q (Q is any of the above quantities)	Any of the above quantities defined at a particular wavelength, per unit wavelength.	As above, including m^{-1}

over all angles (i.e. a complete hemisphere) gives the spectral exitance $E_{\mathrm{BB}} = \pi\, R_{\mathrm{BB}}$. Integrating over all wavelengths then provides the well-known Stefan–Boltzmann law for blackbody radiation. The maximum of the Planck function occurs at a wavelength λ_{max} given by Wien's displacement law:

$$\lambda_{\mathrm{max}} = \frac{a}{T},$$

where the constant $a = 2898\ \mu\mathrm{m\ K}$. Values for λ_{max} for the Sun ($T \approx 5780$ K) and the Earth's surface ($T \approx 288$ K) are therefore $\approx 0.5\ \mu\mathrm{m}$ and $\approx 10\ \mu\mathrm{m}$ respectively. Figure 2.6 shows blackbody spectra approximating to the irradiance from incoming sunlight and the exitance of outgoing infrared radiation at the top of the Earth's atmosphere. (The former is greatly reduced in magnitude because of the Earth's distance from the Sun.)

As a framework for describing individual remote sensing instruments and their application to climate research, it is useful to generalise the remote sensing measurement process, as outlined in Figure 2.7. Following the arrows, we start (bottom left) with the aim of measuring a particular climate-related geoscience variable (e.g. sea surface temperature, SST). There are various factors that affect the measurement. They are categorised here as surface effects (e.g. emission), atmospheric effects (e.g. absorption) and instrumental effects (e.g. radiometric calibration); examples are discussed in the next section. The influence of each

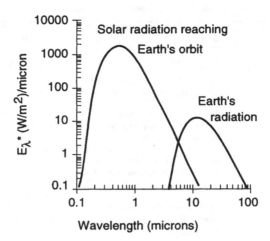

Figure 2.6   Approximate spectral radiant flux densities at the top of the Earth's atmosphere, assuming the Sun to be a blackbody at 5780 K, and the Earth/atmosphere system to be a blackbody at 255 K. ('Micron' is short for 'micrometre'.) From Stull (2000). Reprinted with permission of Brooks/Cole, a division of Thomson Learning (www.thomsonrights.com; fax 800 730-2215).

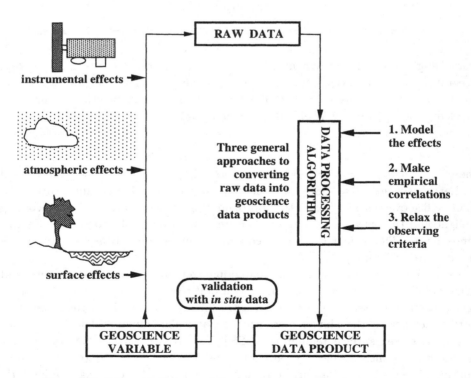

Figure 2.7   Schematic diagram of the remote sensing measurement process.

of these factors on the measurement needs to be considered when analysing the data. Some of the effects must be taken into account in data processing algorithms that convert the raw data into a geoscience data product (e.g. a map of SST). For remote sensing it is convenient to identify three general ways of producing these algorithms, as discussed below: modelling the effects, making empirical correlations, and relaxing the observing criteria. (In practice the development of the algorithms is also intimately associated with the design of the instrument itself, e.g. through choice of wavelengths, viewing geometry, etc.) Finally, the algorithm can then often be checked or *validated* using *in situ* measurements of the same geoscience variable, coincident in time and space with the remote sensing observations.

Let us briefly consider the three ways of producing the algorithms. The first method, namely physical and mathematical modelling of the effects, is perhaps the ideal method, and is commonly used in other fields such as astronomy. The resulting algorithms are generally of two types. For 'best fit' algorithms, the circumstances of the measurement (e.g. surface and atmospheric properties) are varied in a simulation of the measurement so as to produce the best match to the actual raw data value. For 'inverse' algorithms, the modelling leads to a relatively simple formula for the geoscience variable as a function of the raw data. An example of this type from climate research is the sea surface temperature (SST) algorithm used by the Along Track Scanning Radiometer (ATSR) series of instruments (see Section 2.4). In practice it is also common to carry out empirical 'tuning' of modelled algorithms (e.g. adjusting the algorithm coefficients) using *in situ* data, as part of the validation process noted above.

In some cases, however, the processes involved in the remote measurement are too complex to model. Fortunately, in the case of remote sensing of the Earth a second method for producing data processing algorithms is available, because it is possible to acquire coincident *in situ* measurements, as noted above. This method involves making empirical correlations between the raw data and the *in situ* measurements. A climate-related example is the measurement of wind speed using radar altimeters (see Section 2.5). The functional form for such correlations may sometimes be based on theoretical or modelled concepts. However, difficulties may arise in this method if other circumstances arise which have not been present in the initial correlation.

The third method for processing the raw data noted in Figure 2.7 involves relaxing the observing criteria, particularly the need for measurement accuracy. The measurement then becomes essentially qualitative rather than quantitative. An example is vegetation remote sensing in the pre-*EOS* era using AVHRR data, where it has proved difficult to take into account various surface, atmospheric and instrumental effects, and so climate research applications, using a relatively simple vegetation index, have of necessity tended to be rather qualitative (see Section 2.3).

In the next three subsections we consider in general terms some of the main surface, atmospheric and instrumental factors that affect the measurement process.

(We concentrate on passive remote sensing, because the factors affecting active remote sensing tend to be very specific to the instruments and so a general discussion is not appropriate.) Later we will provide examples of how such factors are taken into account in the design of particular instruments and their data processing algorithms.

2.2.3 *Some surface factors affecting the remote sensing process*

Thermal emission from a surface

The Planck function (see above) gives the spectral radiance R_{BB} emitted by a blackbody, which is defined as a theoretical object that absorbs all radiation that falls on it. Real surfaces, however, are not blackbodies, and the observed spectral radiance R_{obs} from any surface at temperature T at wavelength λ is given by

$$R_{obs} = \varepsilon R_{BB}(T, \lambda),$$

where ε is the emissivity of the surface, a fraction between 0 and 1. Blackbodies therefore have an emissivity of 1. The emissivity of a surface depends on its chemical makeup and its detailed structure. For any particular surface, the emissivity generally varies with wavelength and with the direction of emission. For land surfaces, there will often be variations in the surface type, and therefore in the emissivity, within a single picture cell, or *pixel* of the image.

For convenience, the thermal emission from a surface is often described in terms of *brightness temperature* rather than spectral radiance. The brightness temperature T_b is the temperature a blackbody would need to have, in order to emit the observed spectral radiance R_{obs}. Thus, for a surface with an actual temperature T, at wavelength λ we have the equalities

$$R_{obs} = \varepsilon R_{BB}(T, \lambda) = R_{BB}(T_b, \lambda).$$

Illumination of a surface

Sources of illumination in remote sensing can be classified as either approximately unidirectional (direct sunlight, lasers and radar) or omnidirectional (thermal or scattered radiation from the atmosphere). A special case of omnidirectional radiation is isotropic radiation, having equal radiance R_{iso} in all directions, where the total irradiance I onto a surface from a hemisphere is given by $I = \pi R_{iso}$.

For a unidirectional source, the illumination per unit area of a surface reduces with angle from the perpendicular, as the beam spreads out on the surface. For example, if F is the radiant power from the Sun per unit area (measured perpendicular to the beam), the irradiance I onto the surface is given by $I = F \cos \theta$, where θ is the angle of the Sun from the perpendicular. (For a horizontal surface, θ is the angle from the zenith.)

Shadowing effects can be important in remote sensing. With solar illumination, not only can a surface be directly shadowed from the Sun on a large scale (e.g. by a nearby mountain or cloud), but shadowing on a small scale (e.g. shadowing of leaves by other leaves and branches within a tree canopy) can change the average illumination value for the observed surface. With sideways-looking imaging radar (e.g. SAR), steep mountains can produce a 'radar shadow' area where the radar pulse is blocked by the mountain.

Reflection from a surface

There are two extreme cases of surface reflection. Smooth surfaces, whose surface roughness is much smaller than the illuminating wavelength, will exhibit specular (mirror-like) reflection. Approximations to this in remote sensing are sunlight reflecting from very calm water and radar reflecting from calm water or very smooth land. The reflected image of the Sun is referred to as *sunglint*, which will only be seen over water, at the appropriate view angle. The term 'sunglint' is used even if wave roughening substantially blurs the solar image. Images containing sunglint are difficult to interpret, and orbits are usually planned to avoid excessive amounts of sunglint. For a radar altimeter, which observes at the nadir, specular reflection would give a very bright return signal. For a sideways-looking radar instrument such as an imaging SAR, however, specular reflection is away from the instrument so the resulting image would show zero backscatter from the surface.

The other extreme form of reflection is diffuse reflection from rough surfaces (i.e. with a roughness much greater than the illuminating wavelength). The reflection is said to be *Lambertian* if the reflected spectral radiance R_{Lam} is equal in all directions, for unidirectional illumination. The total spectral exitance E from the surface is then given by $E = \pi R_{\text{Lam}}$. Some surfaces (e.g. snow) are close to being Lambertian reflectors. Many surfaces, however, fall somewhere between the specular and diffuse extremes, and the nature of the reflection also depends on the illumination direction.

The reflection from a surface is quantified as the *reflectivity* or *reflectance*, a fraction between 0 and 1, usually defined as the ratio of the output and input spectral radiances. However, there are various different types of reflectivity, depending on the illumination and observation geometry. For directional input and directional output, we can define the bidirectional reflectance ρ_{bidir}. For solar spectral irradiance $I = F \cos \theta$ onto a horizontal surface (see above), it is convenient to use the *bidirectional reflection function* (BDRF), which can be defined as the ratio of the observed output spectral radiance to the output spectral radiance I/π which would be provided by a perfect Lambertian reflector. If measured for all illumination directions and all output directions, ρ_{bidir} or the BDRF fully characterises the reflection properties of a surface. In practice, for remote sensing, both the illumination and the observation directions will be limited, but even so, for most surfaces, ρ_{bidir} or the BDRF will not be known for the whole range of angles involved. In the case of radar measurements, the viewing is in the same direction as the illumination. However, imaging radar systems have a range of illumination directions.

Another type of reflectivity, $\rho_{\text{hem-dir}}$, involves an isotropic hemispherical input (e.g. isotropic illumination of a horizontal surface) and a directional output (e.g. a single observation direction). In some cases this situation approximates reasonably well to the observation of 'sky' radiation (i.e. radiation that is either scattered or thermally emitted downwards by the atmosphere) reflected from the surface, although the sky radiance is generally greater near the horizon due to the increased optical thickness of the atmosphere. For isotropic illumination of a surface by thermally emitted sky radiation, a useful relationship for any particular wavelength is

$$\varepsilon_{\text{dir}} + \rho_{\text{hem-dir}} = 1\,,$$

where ε_{dir} is the emissivity of the surface in the same direction as the reflected output radiation.

2.2.4   *Some atmospheric factors affecting the remote sensing process*

Absorption and emission of radiation by gases

As noted earlier, for remote sensing of the Earth's surface, we need to use the 'window' regions of the spectrum, where the transmission of the atmosphere is higher. Even in these regions, however, there is a certain amount of absorption (e.g. by water vapour), and in the thermal infrared and microwave parts of the spectrum the atmosphere is also emitting its own thermal radiation, which adds to the signal. For remote sounding of the atmosphere, both of these effects are made use of by deliberately choosing regions of the spectrum with substantial absorption and emission. The atmospheric and emission processes are quantified as follows.

Let us consider a thin layer of the atmosphere of thickness dz with perpendicular illumination (Figure 2.8). For any wavelength, the atmospheric gases absorb a fraction α of the input spectral radiance R_{in}, where α is the *absorptivity* or *absorptance* of the layer. The reduction in spectral radiance dR_{abs} due to the layer is therefore given by

$$dR_{\text{abs}} = \alpha R_{\text{in}}\,.$$

Since dR_{abs} is proportional to both R_{in} and dz, we can more conveniently write

$$dR_{\text{abs}} = \alpha_{\text{abs}} dz R_{\text{in}}\,,$$

where the constant of proportionality α_{abs} is known as the *absorption coefficient*, with units of m^{-1}.

If we are considering wavelengths in the thermal infrared or microwave parts of the spectrum, the same gases in the layer also emit thermal radiation. At any particular wavelength, the spectral radiance dR_{emit} emitted by a thin layer at temperature

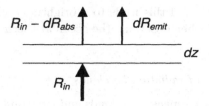

Figure 2.8   Absorption and emission in an atmospheric layer of thickness dz.

T_{layer} is given by

$$dR_{\text{emit}} = \varepsilon R_{\text{BB}}(T_{\text{layer}}),$$

where R_{BB} is the Planck function and ε is the emissivity of the layer. However, it can be shown that for radiation of any given wavelength, the emissivity ε of an object is equal to its absorptivity α. Thus

$$dR_{\text{emit}} = \alpha R_{\text{BB}}(T_{\text{layer}})$$

or

$$dR_{\text{emit}} = \alpha_{\text{abs}} dz R_{\text{BB}}(T_{\text{layer}}).$$

These basic concepts are used in numerical radiative transfer models of the atmosphere so as to simulate the spectral radiance observed by a radiometer in space. The atmosphere is simulated as a set of layers, each with its appropriate pressure, gas concentrations and (for thermal remote sensing) temperature, and then radiation is followed upwards from the surface. Downward radiation is also considered, and reflected from the surface to add to the upward beam. For completeness the presence of aerosols, and scattering (both out of and into the beam), must also be considered (see below).

For thermal remote sensing, note that if the Earth's surface were a blackbody (which is a reasonable approximation for the ocean at thermal infrared wavelengths), and the surface and the whole atmosphere were at the same temperature T, the output spectral radiance R_{out} after each layer would just be the same as the input spectral radiance R_{in}, namely the Planck function $R_{\text{BB}}(T)$. This can easily be shown from the above equations, starting with the first layer above the surface:

$$
\begin{aligned}
R_{\text{out}} &= R_{\text{in}} - dR_{\text{abs}} + dR_{\text{emit}} \\
&= R_{\text{BB}}(T) - \alpha R_{\text{BB}}(T) + \alpha R_{\text{BB}}(T) \\
&= R_{\text{BB}}(T).
\end{aligned}
$$

Thus the atmospheric emission would perfectly compensate for the absorption, and the brightness temperature observed in space would just be T. However, in reality the atmosphere is colder than the surface (and generally reducing in temperature with height), and this leads to a brightness temperature deficit compared with the true SST when observing the ocean at thermal infrared wavelengths (see Section 2.4).

Absorption and emission of radiation by clouds

In the microwave part of the spectrum, clouds and light rain are highly transparent, particularly at longer wavelengths. This is a great advantage when remote sensing with passive microwave imagers rather than visible or infrared imagers, particularly in areas with persistent cloud cover, such as the high latitude regions where most of the sea ice occurs (see later). Radar systems, which also use the microwave region of the spectrum, can also observe under cloudy conditions, and this again gives them a considerable advantage over the equivalent visible or infrared instruments, such as high-resolution imagers or laser altimeters.

In the visible and infrared regions of the spectrum, thick clouds are completely opaque, and as they are colder than the surface, with high clouds being colder than low clouds, their thermal infrared brightness temperatures are reduced accordingly. This provides a method of determining the height of the cloud top, as long as the vertical atmospheric temperature profile is already known. Thin clouds are partially transparent at visible and infrared wavelengths. In the thermal infrared region they absorb some surface radiation but also emit their own. However, as with the atmosphere, the fact that the clouds are colder than the surface means that overall effect over the ocean is an observed brightness temperature deficit compared with the SST. A reduction in brightness temperature at thermal infrared wavelengths will also result from the presence of a sub-pixel cloud (a cloud that is smaller than the image pixel size).

Scattering of radiation

When electromagnetic radiation is scattered by molecules, aerosols or water droplets (cloud or rain), the loss of spectral radiance dR_{scat} from a unidirectional beam of spectral radiance R_{in} in a layer of thickness dz can be considered in the same way as absorption (see earlier), with the scattering coefficient α_{scat} taking the place of the absorption coefficient α_{abs}. Thus

$$dR_{\text{scat}} = \alpha_{\text{scat}} dz R_{\text{in}}.$$

Furthermore, radiation initially travelling in other directions can be scattered into the beam, acting as a 'scattering source' in a similar way to emission.

The amount and angular distribution of scattered radiation depend largely on the size of the scattering object and the wavelength of the radiation. If the scattering object is very much smaller than the wavelength, then the process approximates to Rayleigh scattering, with the scattering coefficient being inversely proportional to the fourth power of the wavelength and the scattering intensity being peaked and equal in the forward and backward directions. An example is the scattering of visible solar radiation by atmospheric molecules or small aerosol particles, where the enhanced scattering at short wavelengths produces the blue appearance of the sky. A similar enhancement of the short wavelength end of the spectrum occurs when observing the surface from space, particularly if the atmosphere is hazy (containing small aerosol particles).

The other extreme case is where the wavelength of the radiation is much smaller than the scattering object. In this case the scattering coefficient is independent of wavelength, so this is known as non-selective scattering, which is isotropic. An example is the scattering of visible solar radiation by cloud droplets, giving clouds their diffuse, white appearance. If the wavelength is comparable to the scatterer dimensions, detailed scattering models are required. Lorenz–Mie theory gives a complete description of scattering by dielectric spheres, and can therefore be used, for example, to describe the scattering of thermal infrared radiation by cloud droplets. For non-spherical particles such as ice crystals (in high cloud), however, alternative methods have to be used.

2.2.5 *Some instrumental factors affecting the remote sensing process*

Spatial resolution

The instruments used for remote sensing of the Earth are generally telescope systems. The telescope's instantaneous field of view (IFOV) is usually defined as the smallest angular element of the observation (Figure 2.9). For an imaging instrument, the IFOV therefore usually represents the angular resolution. This angular IFOV can be mapped as an instantaneously observed region on the Earth's surface, where, for an imaging instrument, it therefore represents the spatial resolution. (Note that, strictly, the terms 'angular resolution' and 'spatial resolution' refer to the ability of the telescope to resolve two close point sources. In remote sensing, however, they are commonly used to refer to the IFOV and the image pixel size, and it is in this looser sense that they are used in this chapter.)

Because the IFOV's spatial dimensions on the surface at nadir are very small compared with the altitude h of the telescope, they are well approximated by simply multiplying the IFOV angular dimensions (in radians) by h (i.e. we can ignore the curvature of the surface of the sphere of radius h when defining the IFOV in radians, and we can also ignore the curvature of the Earth). Likewise the area, a, of the IFOV on the surface at nadir can be approximated as $a = (\Omega h^2)$, where Ω is the solid angle of the IFOV in steradians. There are two main limiting factors that determine how small a telescope's IFOV can be: the diffraction limit, and the limit set by geometric optics.

The diffraction limit to the IFOV is important for long wavelengths and/or a small telescope collecting aperture. If a telescope with a circular collecting aperture

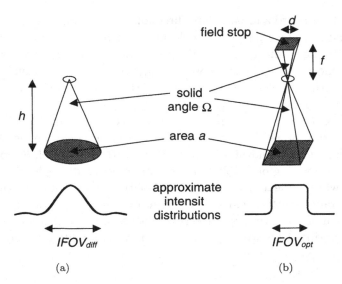

Figure 2.9   Instantaneous fields of view (IFOVs) as defined by (a) the diffraction limit and (b) a square field stop. The small circle represents the collecting and focussing optics.

of diameter D is observing a uniform scene, a point in the focal plane receives radiation of wavelength λ that has entered the telescope from a range of angles in the form of a circular diffraction pattern of intensity (the Airy pattern). This has a central peak of maximum intensity reducing to zero at the first diffraction minimum (Figure 2.9), whose full angular width we may define as the diffraction-limited instantaneous field of view IFOV$_{\text{diff}}$, given (in radians) by

$$\text{IFOV}_{\text{diff}} = 2.44\lambda/D.$$

In practice this diffraction limit to the IFOV is only important for microwave instruments, and for thermal infrared imagers requiring high spatial resolution (of order 10 m) from LEO or moderate spatial resolution (of order 1 km) from GEO.

If the wavelength is short and/or the collecting aperture is large, the diffraction effects are less important and the angular IFOV is largely determined by geometric optics, generally by means of a *field stop* (a small aperture) or a detector element in the focal plane (Figure 2.9). In this case the IFOV can be any shape; it is often square, and provided that the detector efficiency is constant over its area, the IFOV will have a flat top in intensity for a uniform scene (Figure 2.9). Generally the edges of the IFOV will be blurred due to optical aberrations, residual diffraction and manufacturing tolerances. For a telescope with a focal length f and an angular field of view IFOV$_{\text{opt}}$ (in radians) determined by a field stop of width d, then

$$\text{IFOV}_{\text{opt}} \approx d/f.$$

Even if the angular IFOV is (as in these examples) circular or square, it will represent a circle or square on the Earth's surface only at nadir. In other directions the oblique viewing and the curvature of the Earth will change its shape as mapped onto the surface.

Images from remote sensing instruments are generally digital, and as such are made up of square pixels. Like the IFOV, pixels represent solid angle elements that can be mapped onto the Earth's surface as area elements. However, the pixel size is determined independently of the IFOV in a way that (a) depends on the instrument and (b) is often different for the two dimensions of the image. The latter point means that square image pixels may map onto the nadir point as rectangles rather than squares. (In other directions, as with the IFOV, their surface shape will be distorted.) As an example, let us consider the imager on one of the spinning satellites in GEO, the original *Meteosat* (Figure 2.10). The satellite is located over a fixed point on the equator with its spin axis aligned north-south. Considering one of the wavelength channels that uses a single detector (e.g. the thermal infrared channel on *Meteosat*), the image is built up one pixel at a time, with the small square IFOV being scanned across the Earth by means of the east-west spin of the satellite, so as to build up an individual scan line. A north-south tilt of the telescope after each rotation of the satellite then allows the next scan line in the image to be produced. In the scan direction, the pixel size is therefore determined by electronic sampling of the continuous detector output. In the pixel sample time t_p the satellite (with its telescope) has rotated through an angle representing the pixel's angular size in the scan direction. In the perpendicular direction the pixel's angular size is twice the angle of tilt of the mirror between each scan line. As in this example, instrument designers generally try to make the pixel angular size equal to the IFOV angular size in both dimensions; otherwise the scene is either over-sampled or under-sampled. However, in the case of a circular IFOV (as with a

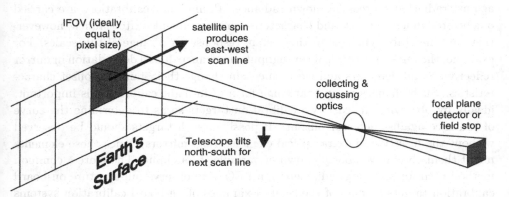

Figure 2.10 Image formation with a *Meteosat*-type geostationary imager. Schematic imaging is shown for one of the wavelength bands that uses a single detector.

diffraction-limited instrument such as a passive microwave radiometer) it is clearly impossible to match the pixel and IFOV sizes exactly.

Spectral resolution

Visible and infrared sensors observe in particular wavelength bands or 'channels'. The choice of bands depends on the source spectrum, the atmospheric transmission and the particular objectives of the instrument (Figure 2.5). Often, filters define the bands (e.g. for AVHRR, Landsat, ATSR, MODIS). There is usually a separate filter and detector (or set of detectors) for each band. In some instruments (e.g. ATSR) there is a single field stop, after which beam-splitting mirrors, which reflect and transmit in different regions of the spectrum, deflect the relevant wavelength range to the various filtered detectors, via extra focussing mirrors or lenses. In other instruments (e.g. *Landsat*'s Thematic Mapper) the filtered detectors are laid out within the focal plane, usually in the along-scan direction, enabling all pixels to be observed directly by each band, though not simultaneously. With such a system, however, accurate co-alignment of pixels for different spectral channels is more difficult. To select very narrow bands, especially for ocean colour applications, diffraction gratings are sometimes used (e.g. on *Envisat*'s MERIS) and special techniques (e.g. using Michelson interferometers) have also been employed for atmospheric sounders. For remote sounding the bands are matched to the appropriate atmospheric absorption features. For surface remote sensing the bands are matched to the surface reflectance features and the atmospheric windows.

Radiometric calibration

Most climate-related applications of passive remote sensing require accurate measurements of radiance, and the instruments therefore require radiometric calibration against radiation sources of known radiance. Preliminary calibrations are carried out before launch to check and characterise the instrument. After launch, however, there are inevitably changes to the system response on a range of timescales. For example, the telescope's optical throughput can change due to degradation in mirror reflectivity, and detectors and electronics can change their gain and offset characteristics. Such changes mean that some form of in-flight calibration is important, ideally by observing at least two calibration targets so as to determine the curve of radiance against detector output. If possible, such targets should be observed without the insertion of extra optical elements (e.g. mirrors), since those elements might themselves have poorly known or variable throughput. Therefore a common method used for scanning radiometers in LEO is to observe full-aperture on-board calibration targets as part of the scan. Examples of on-board calibration systems for the passive remote sensing of reflected sunlight and thermal infrared emission are described in Sections 2.3 and 2.4 respectively.

Radiometric resolution

The radiometric resolution of an imaging sensor represents the smallest brightness variation that can be observed, usually defined per pixel of the image. Though we will consider here the case of thermal infrared sensors, many of the same general principles can also be applied to visible and near-infrared sensors.

For thermal infrared radiometers the radiometric resolution is directly related to the *noise equivalent delta temperature* (NEΔT), representing the *root mean square* (rms) variation in the measured brightness temperature values from pixel to pixel in the image of a uniform scene of brightness temperature T_b. The NEΔT depends on the noise of the detector and on how much signal it produces for a given value of T_b. In discussing the nature of this dependence, however, it is more convenient to define the *noise equivalent delta spectral radiance* (NEΔR), representing the rms variation in the measured spectral radiance values from pixel to pixel in the image of a uniform scene of spectral radiance equal to the Planck function $R_{BB}(T_b, \lambda)$. For a narrow wavelength band this can readily be converted to NEΔT using the fact that the ratio (NEΔR/NEΔT) is equal to the derivative dR_{BB}/dT, evaluated at T_b.

Let us first consider the detector noise. Infrared radiometers for remote sensing use semiconductor photon detectors that are either photovoltaic (i.e. photodiodes) or photoconductive. Different semiconductors have different band gaps E_g. Thus they have different long wavelength cut-offs λ_{max}, because photons with energies less than $E_g = hc/\lambda_{max}$ (where h is Planck's constant and c is the speed of light) will be unable to generate an electron–hole pair. Typical detector materials, e.g. used by ATSR, are indium antimonide (InSb, having a 5.5 μm cut-off, for its 3.7 μm channel), and cadmium–mercury–telluride (CMT, having a cut-off depending on the component concentrations, for its 11 μm and 12 μm channels). To avoid the thermal creation of electron–hole pairs, which would create an internal background 'signal' with an equivalent of 'photon noise' (see below), such detectors must be cooled to a temperature T such that $kT \ll E_g$, where k is Boltzmann's constant. The rms noise in such detectors has a variety of sources. For any particular level of input radiant power, however, there is a lower limit to the detector noise per pixel, i_{rms}, known as 'photon noise', where random fluctuations in the photons' arrival time lead to varying numbers of photons detected within each pixel sample time. It is common to characterise *all* detector noise as if it were caused by fluctuations in input radiant power. One measure of this is the *noise equivalent power* (NEP), being the rms variation, from pixel to pixel, of the radiant power onto the detectors that would be needed, to give the observed noise level per pixel. For many noise sources, including photon noise, NEP is inversely proportional to $\sqrt{t_p}$, where t_p is the pixel sample time.

Let us now consider the detector signal. When observing a scene of brightness temperature T_b, the output signal from the detector, i_{det}, is proportional to the power incident on the detector P_{det}, which can be approximated as

$$P_{det} \approx \eta_{opt} R_{BB}(T_b, \lambda) A\Omega\Delta\lambda + P_{stray} ,$$

where η_{opt} is the optical throughput of the telescope (i.e. the fraction of the radiation onto the telescope aperture that arrives at the detector), $R_{BB}(T_b, \lambda)$ is the Planck function spectral radiance for brightness temperature T_b, A is the area of the telescope collecting aperture, Ω is the solid angle of the telescope IFOV (assumed to be narrow, flat-topped and equal to a single pixel), $\Delta\lambda$ is the width of a narrow wavelength band centred at wavelength λ (assuming the Planck function is approximately linear with wavelength over the band), and P_{stray} is the contribution from stray light and any internal detector background.

Using the definitions given above we can write down the following equalities for the signal-to-noise ratio for the detector. (For simplicity, so as to show the variables involved, an approximation is made for the second equality, namely that P_{stray} is negligible.)

$$\frac{i_{det}}{i_{rms}} = \frac{P_{det}}{NEP} \approx \frac{R_{BB}}{NE\Delta R} ,$$

i.e.

$$NE\Delta R \approx \frac{R_{BB}NEP}{P_{det}} \approx \frac{NEP}{\eta_{opt}\Omega A\Delta\lambda} .$$

Thus, to improve the radiometric resolution by decreasing $NE\Delta R$ (and hence decreasing $NE\Delta T$), we can either increase η_{opt}, Ω, A or $\Delta\lambda$, or decrease NEP (e.g. by increasing t_p). Normally η_{opt} is fixed by the choice of optical system, and Ω and $\Delta\lambda$ are fixed by the choice of spatial and spectral resolution respectively. Increasing A will not affect any such observational parameters (except where the resulting decrease in the diffraction contribution to the spatial resolution is significant) but it will generally have a major impact on the mechanical design of the instrument. Decreasing NEP by increasing, for each wavelength band, the pixel sample time t_p, however, is relatively easy to achieve by increasing the number of detectors in the focal plane, as shown in Figure 2.11 for the previously discussed geostationary scanning instrument.

The simple case of a single detector is shown again in Figure 2.11(a). If the scan speed were decreased, for example, by a factor of 3, it would indeed be necessary to increase t_p by a factor of 3 and decrease the NEP by (typically) a factor of $\sqrt{3}$. However, retaining a single detector would then be undesirable for a GEO scanner (because the overall image time would be increased by a factor of 3) and impossible for a LEO scanner (because the distance between scan lines is determined by the ratio of the scan frequency to the satellite speed, which is fixed). In both cases the solution is as shown in Figure 2.11(b), where three detectors are placed in the focal plane so as to produce three scan lines simultaneously, but with the scan

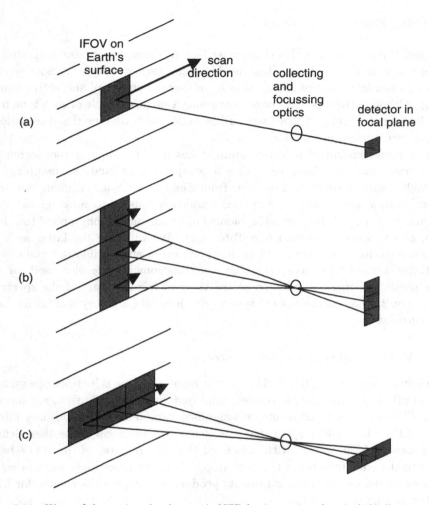

Figure 2.11   Ways of decreasing the detector's NEP by increasing the pixel dwell time t_p. (a) Initial situation, as in Figure 2.10. (b) Reduced scan speed, with multiple detectors producing multiple scan lines, to retain the same average time per scan line. (a) Unchanged scan speed, with time delay integration of multiple along-scan detectors.

speed three times slower. Such multiple scan-line systems have often been used in GEO (e.g. in the GOES imagers) and in LEO (e.g. in MODIS on *Terra* and *Aqua*).

An alternative method, also used in MODIS, which (if used on its own) avoids reducing the scan speed, is to place multiple detectors in the along-scan direction, as in Figure 2.11(c). Here the detectors' signals are added (with time delays) for the same ground pixel. With three detectors, such *time-delay integration* again increases t_p by a factor of 3 and reduces the detector noise per pixel, and hence NEP, by a factor of $\sqrt{3}$. The same principle could also be applied to a pushbroom system (see later) by using three detector arrays instead of one.

2.3   Using Reflected Sunlight

In the last three sections of this chapter we look at examples of remote sensing for climate research, in the three categories identified earlier: passive remote sensing of reflected sunlight, passive remote sensing of thermal emission, and active remote sensing. Although there is only space to examine a single example of each type from the wide range in Table 2.2, they will suffice to demonstrate how the observational principles are applied in practice.

The remote sensing of reflected sunlight has a wide range of socio-economic applications, many of which, e.g. mineral prospecting or land use mapping, require high spatial resolution data, e.g. from *Landsat* or *Spot*. Climate research, however, makes use of moderate spatial resolution sensors to make global scale observations of, e.g. clouds, aerosols, blooms of ocean surface phytoplankton, land albedo, and vegetation on land (see Table 2.2). We will take the latter as a detailed example for this section. First, however, we briefly outline the main types of instrument used. In most cases the same instruments are also used for the remote sensing of thermal emission in the thermal infrared part of the spectrum (see Section 2.4), since the optical system requirements are very similar for both wavelength ranges.

2.3.1   *Visible/infrared imaging systems*

At the heart of any visible/infrared imager for remote sensing is its telescope system. Various optical arrangements have been employed; one of the simplest, e.g. as used by the ATSR series of instruments, has an offset paraboloid as the primary mirror (Figure 2.12). The field stop or detector element in the focal plane then defines the angular IFOV as shown, with rays from the edge (centre) of the IFOV being focused to the edge (centre) of the field stop. There are three main ways in which telescopes on remote sensing instruments produce an image — as shown, for LEO satellites, in Figure 2.13.

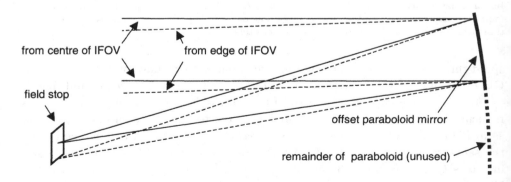

Figure 2.12   Optical arrangement for the telescope on the ATSR series of instruments.

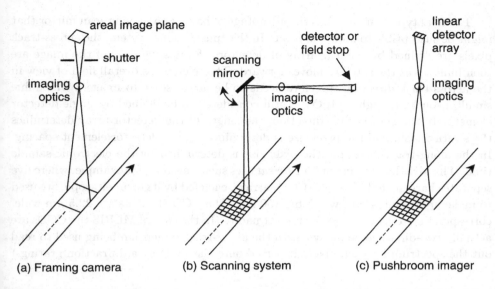

Figure 2.13 Methods of imaging for passive remote sensing instruments in low Earth orbit. After Elachi (1987). Copyright 1987. This material is used by permission of John Wiley & Sons, Inc.

Framing cameras take 'snapshots' of regions of the surface using a two-dimensional detector such as film. An example is the German *Metric Camera*, a modified Zeiss aerial survey camera used on the first ESA/NASA Spacelab manned mission in 1983. The Metric Camera Experiment produced about a thousand 190 km × 190 km images with ≈ 20 m spatial resolution, on 23 cm × 23 cm format film.

Scanning (or 'whiskbroom') systems, by contrast, build up the image over a period of time, pixel by pixel. A GEO example was discussed earlier, with the scanning being provided by the satellite rotation. Such imagers were employed, for example, on all the early geostationary meteorological satellites, such as ESA's *Meteosat*. For an instrument in LEO, a rotating or oscillating mirror scans the IFOV across the surface to build up scan lines of the image, ideally in an 'across-track' direction (perpendicular to the ground track). The satellite motion ensures that successive scan lines occur further along the track. In the across-track direction, the angular range of the scan therefore determines the swath width, and the angular size of a pixel is determined by the choice of electronic sample time. In the along-track direction, the pixel size is just the ratio of the speed of the satellite's ground track (see Section 2.2.1) and the scan frequency. As with some GEO sensors, it is common for LEO imagers to produce more than one scan line simultaneously, by arranging several adjacent detector elements in the along-track direction in the focal plane, with a separation equivalent to 1 pixel (see Section 2.2.5). Most of the current moderate resolution imagers for climate research (e.g. AVHRR, AATSR, MODIS, GLI) are scanning systems.

The final type of imager has the advantage of having no moving scan mirror that might be susceptible to failure in orbit. In this 'pushbroom' system, the across-track pixels are defined by a linear array of detectors. Successive lines of the image are then built up as the satellite moves forward. The telescope's overall field of view in the across-track direction therefore needs to be larger so as to image a whole line simultaneously, though its IFOV is still considered to be defined by single detector element. In the across-track direction, the length of the detector array determines the swath width, and the pixel size is determined by the detector element spacing. In the along-track direction, the pixel size is determined by the electronic sample time. The MERIS instrument on *Envisat* uses such a system, for example, where five separate charge-coupled device (CCD) arrays, each fed by its own telescope, are used to make up the full swath width of 1150 km. The CCD elements are 22.5 μm wide, corresponding to a 300 m pixel width at nadir. (In the case of MERIS the CCDs are actually two-dimensional arrays, with the along-track dimension being used to read out the spectrum for each pixel, dispersed onto the CCD by a diffraction grating.)

2.3.2   *Global vegetation remote sensing*

The presence of vegetation on land has a number of important climatic effects. Vegetation has a much lower albedo than bare ground, affecting the radiation balance at the Earth's surface. Vegetation also strongly affects the exchange of heat and moisture between land and atmosphere, and slows the wind near the surface because of friction and shielding. The destruction by burning of vegetation, particularly trees, is not only a major source of atmospheric aerosols, but also increases atmospheric carbon dioxide (if the trees are not replaced), and thus enhances the greenhouse effect and global warming. Changes in vegetation cover can also affect the production of methane, another greenhouse gas, in rice paddies and other wetlands. In addition, vegetation plays a crucial role in various biogeochemical cycles (e.g. the carbon cycle), and is of direct socio-economic importance through agriculture, forestry and the consequences of wildfires. There is therefore great interest in monitoring vegetation on a global scale at moderate ($\sim$ 1 km) spatial resolution. The main instruments used for this purpose to date, which are discussed below, are AVHRR (until 1999) and MODIS (since 1999). (Note that the two sensors' names make them sound very different, but the word 'resolution' refers to radiometric and spatial resolution, respectively.) Other sensors used for vegetation remote sensing at moderate spatial resolution include ATSR-2 on *ERS-2* (from 1995), AATSR and MERIS on *Envisat* (from 2002), and GLI on *ADEOS-II* (due for launch in 2002).

AVHRR (NOAA, 2000) is a scanning radiometer whose rotating scan mirror produces an across-track scan from horizon to horizon, with a useable swath width of about 2800 km. The IFOV at nadir is $\approx$ 1.1 km square, and the image is built up pixel by pixel, one scan line at a time, with individual filtered detectors for each wavelength channel. In the along-track direction the pixel size approximately equals the IFOV size, but in the across-track direction the pixel size is ≈ 0.8 km at nadir, slightly over-sampling the IFOV. The latest version of the instrument (AVHRR/3) has six channels, ranging from the visible to the thermal infrared. There are two bands that are used for vegetation remote sensing, namely channel 1 (0.58–0.68 μm, red) and channel 2 (0.725–1.00 μm, near-infrared), and these two bands have been included on all versions of the instrument since the first was launched on *TIROS-N* in 1978. Since 1979, two AVHRR instruments have been in operation continuously on pairs of NOAA operational meteorological satellites in Sun-synchronous LEOs (see Section 2.2.1), having different local times for their ascending nodes (typically 19:30 and 13:40).

MODIS (Barnes *et al.*, 1998) is also a scanning radiometer with a rotating scan mirror, having a 2330 km across-track swath width and 36 wavelength channels. The size of the square IFOV at nadir is 250 m for channels 1 and 2 (the red and near-infrared vegetation channels at 0.620–0.670 μm and 0.841–0.876 μm), 500 m for channels 3–7, and 1 km for channels 8–36. For all channels, each scan consists of multiple scan lines making up an along-track distance of 10 km on the ground. Thus for each of the 250 m channels there is a 40-element linear detector array, for each of the 500 m channels a 20-element array, and for all except two of the 1 km channels a 10-element array. Each of the other two bands (channels 13 and 14) has a dual 10-element detector array with time delay integration, so as to improve the radiometric resolution (see Section 2.2.5). Currently there are two MODIS instruments in operation on *EOS* satellites *Terra* and *Aqua* (see Table 2.1). *Terra* and *Aqua* are both in Sun-synchronous LEOs, again with different local times for their ascending nodes (22:30 and 13:30 respectively).

2.3.2.1   *Surface effects*

The remote sensing of vegetation is based on the detection of reflection features in the visible and near infrared part of the spectrum, so the main surface effects to consider (see Section 2.2.3) are surface illumination and reflection. The spectral signature of healthy leaves (Figure 2.14) shows a dramatic step change in reflectance at about 0.7 μm, whereas background soils show a relatively smooth spectrum over that region. Below 0.7 μm vegetation absorbs photosynthetically active radiation (PAR), and so appears dark in the visible part of the spectrum, though with a slight reflectance peak in the green. In the near infrared from ≈ 0.7 μm to ≈ 1.2 μm the reflectance is very much higher, though above this range there are water absorption features. Therefore a comparison of the reflectances in two spectral bands (red and near-infrared) on either side of the 0.7 μm step can be used to indicate canopy cover within a pixel, as seen in Figure 2.14. In fact, differences and ratios between the band radiances have been related not only to percentage vegetation cover, but to other useful vegetation properties such as green leaf biomass (Figure 2.14), the fraction of absorbed PAR (FPAR) and leaf area index (LAI) (Townshend *et al.*, 1993).

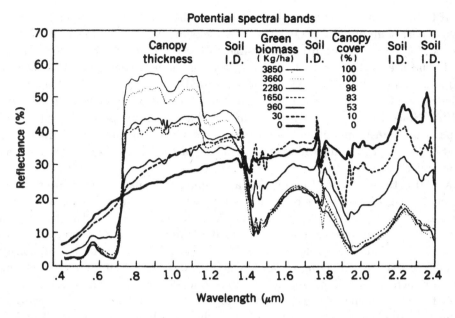

Figure 2.14   Variations in spectral reflectance as functions of amounts of green biomass and per cent canopy cover, for vegetation over a soil background. Potentially useful bands in the near-infrared region are also shown. From Short (1982). Copyright 1987. This material is used by permission of John Wiley & Sons, Inc.

One of the simplest ways to compare the two reflectances is by means of a ratio. The *Ratio Vegetation Index* (RVI) is given by

$$\text{RVI} = \frac{\rho_{\text{NIR}}}{\rho_{\text{R}}},$$

where $\rho_{\text{NIR}}(\rho_{\text{R}})$ is the pixel reflectance in the near-infrared (red) channel. Rather than the RVI, however, the most commonly used index is the *Normalised Difference Vegetation Index* (NDVI):

$$\text{NDVI} = \frac{\rho_{\text{NIR}} - \rho_{\text{R}}}{\rho_{\text{NIR}} + \rho_{\text{R}}}.$$

The NDVI has been widely employed with AVHRR data since 1982, using the channels noted earlier. It is a simple function of the RVI, namely NDVI $= (\text{RVI}-1)/(\text{RVI}+1)$, and is thus still a ratio-based index. This means it has the advantage that its value is largely independent of illumination variations, not only uncorrected large scale effects such as Sun angle changes and shadowing (e.g. by a mountain or a nearby cloud), but also small-scale shadowing of one leaf by another within the canopy. The NDVI is generally preferred to the RVI, however, as it is a more linear function of pixel canopy cover, ranging from ~ 0 for bare soil to ~ 1 for 100% canopy cover. The NDVI has been shown to have a close relationship with the photosynthetic capacity of specific vegetation types (Townshend *et al.*, 1993), though it has the disadvantage of displaying asymptotic (saturated) behaviour for high leaf biomass (Justice *et al.*, 1998).

Another major difficulty with the NDVI is that its value changes with variations in soil reflectance, especially where vegetation is sparse. The alternative *Perpendicular Vegetation Index* (PVI; Richardson and Wiegand, 1977) was an attempt to address this problem, but the *Soil Adjusted Vegetation Index* (SAVI; Huete, 1988) and *Modified SAVI* (MSAVI; Qi *et al.*, 1994) are able to improve on both the NDVI and PVI by compensating for a further residual effect, namely the differential transmittance of the two wavelengths through the canopy. This makes the SAVI and MSAVI highly resistant to soil reflectance variations. The same canopy background correction is also provided by some other indices that also incorporate atmospheric correction; they are discussed in Section 2.3.2.3.

An important surface effect is the variation of vegetation reflectance with the view angle and illumination angle. In principle this can be accounted for with some accuracy in the *EOS* era. For MODIS, for example, a BDRF model of the surface can be produced using reflectance observations over 16 days at multiple angles (also incorporating MISR data), and this can then be used to obtain nadir-equivalent reflectances for a representative solar zenith angle (Justice *et al.*, 1998).

2.3.2.2 *Instrumental effects*

For AVHRR, there are two main instrumental effects affecting global vegetation monitoring. First, the efficiency of the optical/detector system has been found to reduce with time for the visible/near-infrared channels, requiring a re-calibration of the signal vs. radiance curve. In the absence of an on-board calibration system, this has been attempted by viewing a 'standard' desert ground target. Second, limitations to the on-board data storage capacity mean that full spatial resolution or *Local Area Coverage* (LAC) data are obtainable for only part of the stored data, or for data acquired whilst passing over a ground station. Instead, most of the global data are averaged and sub-sampled as *Global Area Coverage* (GAC) data (Tarpley, 1991).

In the case of the *IEOS* instruments, however, the data are available globally at full resolution, and are calibrated. The MODIS vegetation channels, for example, are calibrated using three on-board modules (Barnes *et al.*, 1998). The first is the *solar diffuser* (SD), an on-board diffusely reflecting target that is illuminated by the Sun and observed regularly by the telescope. The reflectance of the diffuse target is monitored in turn by the *SD stability monitor* (SDSM), an optical system that alternately views the target, a dark region and the Sun itself. Finally there is a *spectral radiometric calibration assembly* (SRCA) that monitors spectral, spatial and radiometric performance. The resulting calibrated top-of-the-atmosphere radiances can then be used for calculating the nadir-equivalent surface reflectances that are used for the vegetation indices and other products (Justice *et al.*, 1998).

2.3.2.3   *Atmospheric effects*

For vegetation remote sensing the main atmospheric effects are the scattering of sunlight by molecules and aerosols, and molecular absorption (both of the incoming sunlight and the outgoing reflected radiation). As molecules and many aerosols are smaller in size than the wavelength of the light, the scattering can largely be approximated as Rayleigh scattering, whose magnitude is inversely proportional to the fourth power of the wavelength. The main effect is therefore an enhancement of the signal, especially at shorter wavelengths, due to scattered sunlight adding to the radiation entering the telescope. (The sunlight involved may reflect from the surface before scattering.) The effects of molecular scattering are fairly straightforward to calculate, but correction for the effect of aerosols or thin cirrus clouds, which are highly variable, is more difficult. As regards molecular absorption, the main difficulty is water vapour, which is highly variable in concentration and has a number of absorption features in the near-infrared. Absorption by the variable trace gas ozone also needs to be accounted for.

In the case of AVHRR, the near-infrared channel is broad, and encompasses a water absorption feature (see Figure 2.5), and the limited number of channels means that effects of aerosols and thin cirrus are difficult to estimate. These effects therefore cause a relatively large residual uncertainty in AVHRR's NDVI values.

For MODIS, the near-infrared channel is narrower, minimising the effect of water vapour. In addition there are extra channels and other *EOS* instruments that allow the estimation of aerosol properties, as well as cloud information and ozone and water vapour concentrations. One approach, therefore, is full correction of the top-of-the-atmosphere radiances for the effects of atmospheric gases, aerosols and thin cirrus clouds, e.g. using radiative transfer modelling and incorporating BDRF information for surface-reflected radiation (Justice *et al.*, 1998). Alternatively, partial (but usually more accurate) correction for atmospheric effects can be applied (e.g. for molecular scattering and ozone absorption only; Miura *et al.*, 2001). Nadir-equivalent surface reflectances are then estimated from the corrected radiances, as noted earlier. Finally, various vegetation indices and other products are produced (Justice *et al.*, 1998).

For either full or partial atmospheric correction, there will be aerosol and thin cirrus effects that are not accurately estimated, and the vegetation indices devised for use with MODIS generally aim to minimise the resulting 'atmosphere noise' by incorporating the reflectance ρ_B for the blue channel (channel 3; 0.459–0.449 μm). The first of these indices (Kaufman and Tanré, 1992) were the *Atmospherically Resistant Vegetation Index* (ARVI) and the *Soil Adjusted and Atmospherically Resistant Vegetation Index* (SARVI). The latter incorporates the resistance to 'canopy background noise' of the SAVI (see above). Liu and Huete (1995) devised the *Modified NDVI* (MNDVI), which used feedback techniques to minimise the effects of both canopy background noise and atmosphere noise. A simplified version of the MNDVI, the SARVI2, was then proposed (Huete *et al.*, 1997), and subsequently re-named the *Enhanced Vegetation Index* (EVI; Justice *et al.*, 1998; Miura *et al.*, 2000; Miura *et al.*, 2001). It is given by

$$\text{EVI} = \frac{G(\rho_{\text{NIR}} - \rho_{\text{R}})}{L + \rho_{\text{NIR}} + C_1 \rho_{\text{R}} - C_2 \rho_{\text{B}}},$$

where L is a canopy background and snow correction factor, C_1 and C_2 are to correct for aerosol effects, and G is a gain factor. Initial values of these coefficients are $G = 2.5$, $L = 1$, $C_1 = 6$ and $C_2 = 7.5$.

2.3.2.4   *Vegetation monitoring with MODIS*

As noted above, the NDVI has residual uncorrected surface effects, and with AVHRR's version of the NDVI, variations also occur due to residual atmospheric effects, and poorly quantified instrumental effects. Despite these difficulties, AVHRR's NDVI (often using uncorrected channel values rather than reflectances) has been remarkably successful in monitoring vegetation on a large scale in the pre-*EOS* era (Tarpley, 1991; Townshend *et al.*, 1993). Kaufman and Tanré (1992) suggested that this is because the normalisation of the NDVI actually makes the index relatively insensitive to calibration effects and angular effects of the surface reflectance and the atmosphere. Nevertheless, most of the applications of AVHRR NDVI data, e.g. phenological studies of the growing season, and land cover classification, have been somewhat qualitative (Miura *et al.*, 2000).

With the advent of MODIS, however, the use of calibrated radiances, and vegetation indices such as the EVI that are resistant to canopy background noise and atmosphere noise, means that more quantitative remote sensing of vegetation is now feasible. The main vegetation index products from MODIS involve the use of the NDVI and the EVI. The NDVI products are included partly for continuity with the AVHRR archive, but partly because the NDVI is complementary to the EVI, with the NDVI being more responsive to chlorophyll-related parameters such as FPAR and fractional cover, and the EVI being more responsive to canopy structural parameters such as LAI (Justice *et al.*, 1998). Figure 2.15 shows global EVI data for two two-month periods around the summer and winter solstices. The large seasonal variations in vegetation amount at mid-latitudes are very apparent.

The approach to producing the vegetation indices described so far, though based on theoretical understanding, is fairly empirical. Where the required products are specific vegetation variables rather than general indices, however, approaches have been adopted that involve more detailed numerical modelling of the surface and atmospheric radiative transfer processes. Examples are the production of LAI and FPAR products for MODIS (Justice *et al.*, 1998) and the devising of vegetation indices that are optimised to specific sensors and vegetation variables (e.g. MERIS and FPAR respectively; Gobrom *et al.*, 2000).

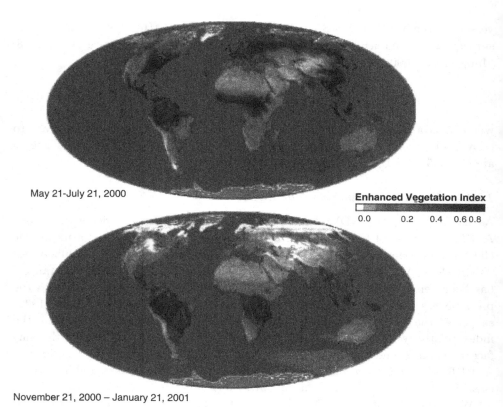

May 21-July 21, 2000

Enhanced Vegetation Index

0.0   0.2   0.4   0.6 0.8

November 21, 2000 – January 21, 2001

Figure 2.15   Averaged global EVI data from MODIS for Northern Hemisphere summer and winter. EVI ranges from 0, indicating no vegetation, to nearly 1, indicating densest vegetation. Images from NASA-GSFC/University of Arizona.

2.4   Using Thermal Emission

Thermal emission from the Earth's surface or atmosphere is used for a number of climate-related remote sensing observations.   One rather specialised type of measurement, as briefly noted earlier, is atmospheric sounding. For example, the High Resolution Infrared Radiation Sounder (HIRS) and the Advanced Microwave Sounding Unit (AMSU) on the polar-orbiting NOAA meteorological satellites have been used routinely to determine vertical profiles of temperature and water vapour concentration.   In the microwave part of the spectrum the Scanning Multichannel Microwave Radiometer (SMMR; 1978–1987) and the Special Sensor Microwave/Imager (SSM/I; since 1987) have been highly successful in monitoring sea ice concentration, making use of the large emissivity difference between water and sea ice at microwave wavelengths, as well as the instruments' ability to observe through persistent cloud cover and during the dark polar winter. Thermal infrared radiometers in LEO and GEO are used to measure the temperature of cloud-tops, and the Earth's surface. Land surface temperatures are difficult to measure accurately because surface emissivities are usually poorly known. However, sea surface temperatures (SSTs) can now be measured with high accuracy from space, and it is this application that will be considered here.

2.4.1 *Global sea surface temperature measurement*

The ocean is a major component of the climate system (see Section 2.1.2), and ocean heat has an important role in the global climate. Most of the solar energy falling onto the Earth is absorbed in the ocean (which covers the majority of the surface) and the ocean is, therefore, the main source of heat, as well as moisture, for the atmosphere. Latitudinal variations in solar heating of the surface (and hence of the atmosphere) give rise to the general circulation of the atmosphere, and indirectly, via surface winds and thermohaline effects, of the ocean. By means of this circulation the ocean also acts as a major transporter of heat from low to high latitudes before release into the atmosphere. (In fact, roughly as much heat is transported polewards by the ocean as by the atmosphere itself.) This is because the solar energy, mainly in the form of visible light, is absorbed down to considerable depths in the ocean, and its transport back to the surface and release into the atmosphere are relatively slow. The ocean heat transport produces a big effect on regional climates. A famous example is the Gulf Stream/North Atlantic drift, where vast quantities of heat are carried from low latitudes and released into the atmosphere in the North Atlantic, providing the western fringe of Europe with its mild winter climate.

Sea surface temperature (SST) turns out to be of crucial importance for these ocean-related climatic processes, in that it is a major determining factor for fluxes of heat and moisture from ocean to atmosphere, with SST regional anomalies affecting atmospheric circulation details. SST is also an indicator of ocean circulation details (e.g. eddies). Finally SST, together with surface air temperature, is one of the key measures of global warming and its regional variations.

Therefore there is a need for global monitoring of SST, e.g. for the validation of the AOGCMs that are used to predict future climatic change, for making heat flux estimates in process studies, for studying the growth and decay of SST anomalies (e.g. in ENSO), and for helping to monitor global warming. All these applications call for SST measurement accuracy of a few tenths of a degree or better. In the open ocean SST typically varies by less than this amount over 1 km or 1 day, so these are roughly the values of spatial and temporal resolution that are required for regional SST monitoring, as well for the monitoring of circulation details. Finally, as with most climate-related applications, there is a clear need for the coverage of SST by remote sensing to be global and long term.

To make SST measurements from space, we need to observe thermal emission from the ocean surface through either the thermal infrared windows or the microwave window (Figure 2.5). In the late 1970s and early 1980s two pioneering instruments, already mentioned in other contexts, were used for SST remote sensing: AVHRR, using its thermal infrared channels, and SMMR, a passive microwave imager. Microwave sensors have the advantage of observing through cloud, and though their spatial resolution is limited by diffraction to typically tens or even hundreds of kilometres, this is not necessarily a major problem for large-scale, average global SST measurements (though clearly it is a limitation for monitoring coastal SST or small-scale features such as ocean currents). However, SMMR was unable to achieve adequate SST accuracy for climate research. Using a channel at 6.6 GHz, with a conical scan of the surface to avoid the emissivity variations that occur at different view angles, and the use of other channels for correction of various surface and atmospheric effects, SMMR was only able to retrieve SST to an accuracy of 1–2 K (e.g. Milman and Wilheit, 1985; Pandey and Kniffen, 1991). Calibration difficulties were a major source of error for SMMR (Milman and Wilheit, 1985). Improved accuracy may be achievable by the next multichannel microwave radiometers to include a suitable low frequency channel (6.925 GHz) for global SST monitoring, namely AMSR-E and AMSR on the *IEOS* satellites *Aqua* and *ADEOS-II* respectively.

AVHRR was described in Section 2.3.2. The original version of the instrument had two thermal infrared wavelength bands, but most of the units flown (i.e. versions AVHRR/2 and AVHRR/3) included three such bands, each with the same ≈ 1 km IFOV and pixel size as the visible/near-infrared bands. For AVHRR/3 these channels are numbered 3b, 4 and 5, with wavelength ranges 3.55–3.93 μm, 10.30–11.30 μm and 11.50–12.50 μm respectively. AVHRR has been used since the early 1980s to produce global SST data, particularly for meteorology. However, though more accurate than SMMR, with an accuracy that has improved with changes to the retrieval algorithm, its current SST retrieval accuracy of ≈ 0.5 K (Walton *et al.*, 1998; Li *et al.*, 2001) is still not ideal for climate research. The residual uncertainty is due to a combination of surface, atmospheric and calibration effects (see below), and is considerably worse in the presence of stratospheric aerosols from volcanic eruptions such as that of Mount Pinatubo in 1991 (Walton *et al.*, 1998). MODIS is designed to measure SST in a way that is based on the AVHRR method, though MODIS has improved features such as accurate aerosol estimates (using extra visible and near-infrared channels), extra thermal infrared channels, improved radiometric resolution and improved calibration (Esaias *et al.*, 1998). Therefore its accuracy is expected to be better than that of AVHRR. However, the ATSR series of instruments, though based on the same general concept, incorporate a number of new features specifically designed to improve the accuracy of SST retrieval. They are discussed in detail below.

For completeness it is necessary to mention the other main sensors incorporating thermal infrared channels, namely the Thematic Mapper (TM) on *Landsat* and the imagers on the geostationary meteorological satellites. To date, however, these have had limitations with regard to correcting for atmospheric effects and instrumental effects (particularly calibration), and so have been unable to produce accurate enough SST measurements for climate research.

2.4.2   The Along Track Scanning Radiometer (ATSR)

To overcome the limitations of existing sensors, ATSR was designed with the primary goal of measuring SST, globally, to 0.25 K (rms) or better (Delderfield *et al.*, 1986). Versions of the instrument flown so far are ATSR (on ESA's *ERS-1*, 1991–96), ATSR-2 (on *ERS-2*, 1995–present) and the Advanced ATSR (AATSR, on *Envisat*, launched 2002). The SST measurement aspect of these three instruments is essentially identical, and is based on the AVHRR methodology. However, ATSR incorporates various novel features that enable it to achieve its improved SST accuracy, namely a dual-angle view of the surface through two different atmospheric path lengths to improve the atmospheric correction, an active cooler for the detectors to improve the radiometric accuracy, and a high accuracy on-board blackbody calibration system to improve the radiometric calibration (Delderfield *et al.*, 1986; Tinkler *et al.*, 1986; Edwards *et al.*, 1990; Mason, 1991; Mason *et al.*, 1996). Before discussing how these new features affect SST retrieval, however, it is useful to outline ATSR's optical design.

ATSR's dual-angle view of the surface is achieved by using a flat rotating scan mirror to provide a conical scan (Figure 2.16). Thus two separate images of the surface are built up, one obliquely through a forward view-port, and one approximately vertically through a nadir view-port, with one complete scan providing a single 'across-track' scan line of each image. The resulting curvature of the scan lines can be corrected when the data are processed, but the choice of forward angle at 55° to the vertical (see Section 2.4.2.2) means that the swath width is limited to

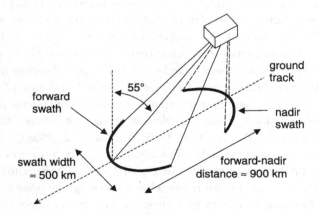

Figure 2.16   ATSR viewing geometry, showing a single conical scan producing forward and nadir scan lines for the forward and nadir images of the surface (not to scale).

≈ 500 km. An offset paraboloid mirror (see Section 2.3.1) focuses the beam onto a square field stop in the focal plane. After the field stop, beam-splitting mirrors within the cooled focal plane assembly enable the appropriate wavelength ranges to be focussed onto the filtered detectors. Thus the field stop is common to all wavelength bands, ensuring that the region viewed by the IFOV is the same for all channels. The IFOV and pixel size are equal, and equivalent to a 1 km × 1 km square at nadir (i.e. similar to the equivalent AVHRR values). In the scan direction the pixel size is determined by electronic sampling; in the along-track direction it is equal to the scan line separation, which is just the speed of the sub-satellite point along the ground track divided by the scan frequency.

2.4.2.1 *Instrumental effects*

For SST retrieval, the main instrumental effects to consider are radiometric resolution and radiometric calibration. For both AVHRR and ATSR, the collecting mirror diameter (203 and 106 mm respectively) is chosen to ensure a large enough signal, and the detectors are cooled to reduce the thermal production of electron–hole pairs, with its associated noise (see Section 2.2.5). AVHRR uses passive cooling to 105 K by radiating to deep space, whereas ATSR uses an active refrigeration system (a Stirling cycle cooler), which was able to reduce the detector temperature to between 90 and 95 K for the early part of the mission (Merchant *et al.*, 1999). For AVHRR, the NEΔT is designed to be $\leq$ 0.12 K. For ATSR, however, with detectors at 91 K, pre-launch NEΔT values per pixel were significantly lower, being measured at between 0.03 and 0.05 K for a 285 K scene brightness temperature (Mason, 1991; p. 165). These values enable detailed observations of regional SST patterns to be made, and in principle would allow single-pixel SST retrieval to extremely high accuracy ($\approx$ 0.1 K), provided that other uncertainties could be reduced to that level.

AVHRR and ATSR both employ on-board calibration systems to provide regular calibration of the detector/optics system. For AVHRR, the telescope views a single simulated blackbody source at the instrument temperature (10–30°C, i.e. within the range of observed brightness temperatures), and also views deep space to provide a 'zero' point. However, there have been significant difficulties in obtaining a sufficiently accurate calibration with this system, not only because of errors due to the blackbody emissivity, temperature uniformity and temperature readout accuracy, but also because of non-linearities when interpolating between the two widely spaced calibration points (Brown *et al.*, 1985; Walton *et al.*, 1998).

For ATSR it was decided to use two simulated blackbody sources of high intrinsic accuracy, with temperatures at about −10°C and 30°C, i.e. approximately at either end of the brightness temperature range when observing the ocean. Mason *et al.* (1996) describe the system and its performance. The two cavity-style blackbodies are situated in the gaps between the two view-ports, and so are observed by the full aperture beam as part of each conical scan (Figure 2.16). They use a number of technical innovations that give them a very high emissivity, very small temperature gradients and a very high absolute accuracy for the measurement of their temperature. The maximum (three standard deviations) value for the uncertainty of the calibration was specified to be 0.1 K, and the measured performance is at least as good as this (Mason *et al.*, 1996). Such a value would again, in principle, allow SST retrieval accuracies to ≈ 0.1 K, if other uncertainties could be reduced sufficiently.

2.4.2.2   *Surface effects*

ATSR uses essentially the same wavelength channels as AVHRR, situated in two atmospheric windows (Figure 2.17). In this part of the spectrum the emissivity of the ocean surface is well known and very high, varying between 0.97 and 0.99. Therefore the reduction in emission compared with a blackbody, and the amount of reflected 'sky' radiation, are both small (see Section 2.2.3), and can easily be corrected for a part of the atmospheric correction algorithm (see below). In the case of ATSR, where an oblique along-track forward view is required, the large reduction

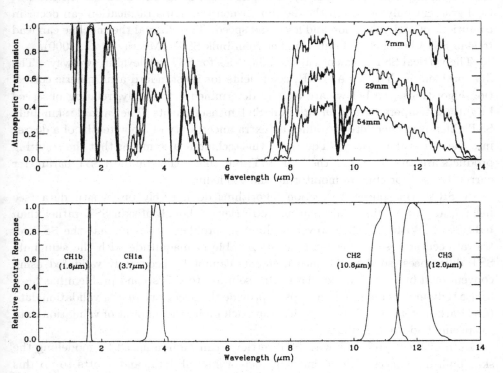

Figure 2.17   The upper graph shows atmospheric transmission for three different amounts of precipitable water vapour: 7 mm (polar), 29 mm (temperate), 54 mm (tropical). The lower graph shows the ATSR spectral channels. From Mason (1991, p. 11).

in emissivity that occurs at large view angles from the vertical led to a choice of 55° from the vertical for the forward view to ensure a high enough value of emissivity. In the 3.7 μm channel during the daytime, however, for both view angles there is a significant amount of reflected sunlight as well as thermal emission, so this channel is only used for SST retrieval at night (Figure 2.6). (Note that the 3.7 μm channel on the first ATSR, on *ERS-1*, failed in May 1992.)

The vertical temperature gradient, which occurs at the top of the ocean, is the main surface effect that causes difficulty for SST remote sensing. The surface-leaving radiance is related to the skin SST at the surface of the ocean, whereas *in situ* measurements are of the bulk SST further below (e.g. at a 1 m depth for the Tropical Atmosphere Ocean array of oceanographic drifting buoys; Merchant and Harris, 1999). One contributor to the gradient is the 'skin effect', whereby the top millimetre of the ocean surface has a temperature gradient across it due to the (largely upward) flux of heat through the ocean surface. At lower levels the temperature is relatively uniform with depth because of vertical mixing. The skin SST varies considerably with factors such as wind speed, but is typically 0.1–0.5 K lower than the bulk SST (Murray *et al.*, 2000). The second influence on the vertical gradient only occurs in the daytime, when thermal stratification can occur in conditions of high insolation and low wind speed. This diurnal thermocline can lead to skin SSTs being considerably higher than bulk SSTs (Murray *et al.*, 2000).

The vertical SST gradient causes difficulties for SST retrieval in two ways. The first problem is that, for AVHRR, coefficients for its standard atmospheric correction algorithms (see below) are usually determined empirically, relating observed brightness temperatures to *in situ* bulk SST measurements. The variable skin–bulk SST differences therefore introduce an extra uncertainty of a few tenths of a degree into the retrieval accuracy. (The use of this technique also means that the resulting products are an estimate of the bulk SST, but this is the standard SST parameter currently used for climate monitoring and modelling.)

ATSR's atmospheric correction algorithms do not rely on in situ data (see below), and so provide a more accurate retrieval (though of skin SST rather than bulk SST). With ATSR, however, a different problem arises in that the SST retrieval accuracy (see below) is now comparable in magnitude with the skin–bulk SST differences, so that the measurements cannot be accurately validated using conventional bulk SST *in situ* data. One solution to this second problem has been to use radiometers mounted on ships to provide the necessary *in situ* validation data (e.g. Parkes *et al.*, 2000), though this approach limits the amount of validation data to specific regions and times.

Both of these problems with SST retrieval can be minimised by modelling the skin–bulk differences using coincident meteorological data, and efforts to do this for ATSR validation have met with some success (e.g. Merchant and Harris, 1999). Such modelling could also, in principle, be used to convert skin SST data products to the bulk SST equivalents if required.

2.4.2.3   Atmospheric effects

Clouds

At any particular time, about 50% of the ocean is covered by cloud (Duggin and Saunders, 1984), which causes a major difficulty for thermal infrared observations of SST (Section 2.2.4). The first problem is that the cloud will potentially cause gaps in spatial coverage. However, the relatively slow variation in SST with time means that over a period of several days fairly complete SST coverage can usually still be achieved from LEO. Nevertheless there are still difficulties with obtaining regular SST data in areas of persistent cloud. The second problem is how to identify the pixels containing cloud, so as to ignore them when applying the SST retrieval algorithm.

Various cloud-screening schemes were initially developed for AVHRR (e.g. McClain et al., 1985; Saunders and Kreibel, 1988), and the initial cloud-clearing scheme for ATSR data used similar techniques (Zavody et al., 1994; Zavody et al., 2000). These schemes apply a series of tests to the data, so as to ensure that only cloud-free pixels are used for SST retrieval. The schemes involve three main methodologies for cloud clearance (Duggin and Saunders, 1984).

The first methodology involves the use of simple brightness thresholds, and is based on the fact that clouds are brighter than the ocean in visible/near-infrared channels, and have lower brightness temperatures than the ocean in thermal infrared channels. Visible/near-infrared thresholds can only be used during the day. They need to be adjusted to take into account illumination conditions (i.e. sun angle), and also have difficulties in the presence of sunglint, thick aerosols and cloud shadows. To avoid these difficulties the ATSR scheme uses a threshold that is dynamically determined from a brightness histogram of a small sub-scene. Thermal infrared thresholds also need to be variable, depending on the local SST value itself. In the ATSR scheme the threshold depends on the time of year, viewing angle and pixel location.

The second methodology involves image processing. The presence of clouds broadens the distribution of brightness temperatures within an image. Methods have therefore been developed that examine small regions of an image over which SST is expected to be uniform, and either reject the whole region if the brightness distribution is too broad, or use the shape of the distribution to estimate the temperature of the cloud-free pixels in the region, based on the fact that the highest brightness temperatures must correspond to the cloud-free pixels. In the ATSR scheme, the rms variation of the brightness temperatures in 3 km × 3 km regions is used to determine the presence or absence of cloud.

The third methodology involves the use of thresholds on channel combinations. A threshold on the ratio of visible to near-infrared radiances can identify cloud over the ocean, as cloud-free scenes include a large molecular and aerosol scattering component in the visible channel (Saunders and Kreibel, 1988). At thermal wavelengths, thresholds on brightness differences can distinguish clouds by their emissivity variations with wavelength. In the ATSR scheme, a 10.8–12 μm threshold detects thin cirrus, a 10.8–3.7 μm threshold detects fog and low stratus, and a 12–3.7 μm threshold detects medium/high level cloud. Such tests also help to detect sub-pixel cloud.

Even with such schemes, undetected cloud remains a significant source of residual error in SST retrieval. More sophisticated multichannel image processing techniques have therefore been developed for AVHRR (e.g. Gallaudet and Simpson, 1991; Simpson *et al.*, 2001) and ATSR (the ATSR Split-and-Merge Clustering (ATSR/SMC) algorithm; Simpson *et al.*, 1998). The ATSR/SMC algorithm, which also incorporates some of the earlier techniques, was found to be superior to the standard ATSR scheme (Simpson *et al.*, 1998), retaining more valid pixels, and being more effective in detecting multilayer cloud structures, sunglint, subpixel cloud near cloud boundaries, and low-lying stratiform cloud.

Molecules and aerosols

Even in the thermal infrared atmospheric 'windows', there is significant molecular absorption, mainly due to water vapour. The atmospheric water vapour content (and hence the absorption) varies widely with time and space. In particular there is a large latitude effect (Figure 2.17) with tropical atmospheres receiving more evaporation from the warm tropical ocean and, because of their high temperature, being able to contain a higher concentration of water vapour before saturation. The absorbing molecules also emit their own radiation, but because they are colder there is a net brightness temperature deficit at the top of the atmosphere (see Section 2.2.4). This temperature deficit is different for each channel, but can be up to ≈ 10 K in the tropics.

The difference in absorption between channels is exploited to produce a multichannel atmospheric correction algorithm. It can be shown theoretically that for optically thin atmospheres (and making a number of other assumptions), if two different wavelengths show different amounts of absorption (see Figure 2.17), the resulting brightness temperatures T_{b1} and T_{b2} can be related to the SST T_s by a simple linear formula, the most general version of which is

$$T_s = a_0 + a_1 T_{b1} + a_2 T_{b2},$$

where a_0, a_1 and a_2 are constant coefficients. In practice, the assumptions, particularly that of high transmission for all atmospheres, do not hold (see Figure 2.17). Nevertheless, this form for the algorithm has been widely used, but with the coefficients being obtained empirically or via radiative transfer modelling (see below), rather than analytically.

For AVHRR's early operational multichannel SST (MCSST) algorithms (McClain *et al.*, 1985), daytime observations used the so-called 'split window' channels at 10.8 μm and 12 μm for T_{b1} and T_{b2}, and for night-time observations the algorithm was extended by adding another term (equivalent to $a_3 T_{b3}$ in the above equation) for the 3.7 mm channel. For these algorithms, the coefficients were found by an empirical method. A large set of $(T_{\text{Strue}}, T_{\text{b}i})$ combinations was obtained for a wide range of different atmospheres (i.e. locations and seasons), where T_{Strue} are *in situ* measurements of bulk SST coincident in time and space with AVHRR's brightness temperature measurements $T_{\text{b}i}$ for the various channels, where the subscript i represents the channel number. The best coefficients were then obtained by fitting the algorithm to the data points. Excluding severe volcanic aerosol episodes, the residual rms uncertainty of ≈ 0.5 (night-time) and up to ≈ 0.8 K (daytime) in the resulting MCSST retrievals (Walton *et al.*, 1998) is due to a number of factors, the main ones being as follows:

Atmospheric effects

• Many atmospheres have a large water vapour content, and so are not optically thin as required by the theoretical algorithm.
• Some atmospheres have anomalous vertical profiles of water vapour that are not accounted for by the algorithm.
• Some atmospheres contain aerosol loadings that also absorb and emit radiation, but are not accounted for by the algorithm.
• Some pixels in the image may contain undetected cloud.

Instrumental effects

• The instrument radiances are not calibrated accurately enough.

Surface and in situ/*satellite match-up effects*

• There are variations in the skin–bulk SST difference.
• The $T_{\text{b}i}$ values used for match-ups are not calibrated accurately enough.
• The $T_{\text{b}i}$ values used for match-ups may include residual undetected cloud.
• The T_{Strue} values used for match-ups may have significant errors.
• Measurements of T_{Strue} and $T_{\text{b}i}$ are not completely coincident in time and space.
• It is difficult to find cloud-free match-ups for a sufficiently wide range of atmospheres.

There have been various attempts to improve the SST retrieval algorithm with AVHRR, as reviewed by Barton (1995). For example, Harris & Mason (1992) show that an algorithm making use of surface temperature variations within the scene can potentially reduce much of the residual atmospheric correction error caused by the effects of water vapour variations between atmospheres. However, Barton (1995) shows that such algorithms are difficult to implement in practice. For operational use with AVHRR, modified algorithms have been introduced, whose coefficients are dependent on an initial estimate of the SST. These have improved daytime accuracy (≈ 0.5 K) compared with the earlier MCSST versions (Li *et al.*, 2001). However, these *nonlinear SST* (NLSST) algorithms still use the empirical method for obtaining the best coefficients (Li *et al.*, 2001). Minnett (1990), however, uses a simulation method (as used by ATSR; see below) to produce the algorithm co-efficients for a regional study, thus eliminating all the errors associated with the surface and match-up effects of the empirical method. Barton (1998), in a study aimed at ATSR but also relevant to AVHRR, shows that improved accuracy can be obtained when a suite of algorithms is available, each with coefficients optimised according to parameters such as latitude, total water vapour and upper level water vapour.

ATSR uses essentially the same wavelength channels as AVHRR, but improves the atmospheric correction by imaging every piece of the ocean surface via two different atmospheric paths, with different optical thicknesses. The first observation is in the oblique along-track forward direction, and the second, about two minutes later, is in an approximately vertical direction, at nadir (Figure 2.16). This 'dual view' approach means that for each wavelength channel there are now two brightness temperatures, and this provides extra information on the atmosphere that can be used to reduce the residual atmospheric correction uncertainty. This improvement is particularly marked in the presence of aerosols.

The retrieval algorithm for ATSR (Zavody *et al.*, 1994) incorporates all the brightness temperatures in the same type of algorithm as the single-view algorithms:

$$T_s = a_0 + \Sigma a_{Ni} T_{bNi} + \Sigma a_{Fi} T_{bFi} \,,$$

where the extra subscripts N and F represent the nadir and forward views respectively. Again only the 10.8 μm and 12 μm channels are used in the day, and the 3.7 μm channel is added at night. However, an alternative simulation method of producing the coefficients is used, whereby the large set of $(T_{Strue}, T_{bNi}, T_{bFi})$ combinations is obtained by first procuring a large set of atmospheric vertical profiles (of temperature and water vapour) using radiosondes (for a wide range of seasons and locations). These atmospheres are then each fed into an atmospheric radiative transfer model (including surface reflection). For each atmosphere, and for a range of values of T_{Strue} (within a few degrees of the lowest radiosonde temperature value) and aerosol loadings, values of T_{bNi} and T_{bFi} are obtained from the model, using the known filter characteristics for each band. The best coefficients are then found by fitting the equation to the data as before. Compared with the purely empirical method outlined above, this simulation method of obtaining the coefficients eliminates all the problems associated with the surface and match-up effects. However, this alternative method does rely on the radiative transfer modelling being accurate enough. Also, the simulation method provides a measure of skin SST rather than bulk SST, so if bulk SST is the required parameter, a correction needs to be applied.

The initial modelling (Zavody *et al.*, 1994) was in fact inadequate in its incorporation of aerosol loadings (particularly volcanic stratospheric aerosols) and in its characterisation of the water vapour continuum absorption (Merchant *et al.*, 1999). However, new coefficients derived by Brown *et al.* (1997) made the algorithms more robust against aerosols, and those of Merchant *et al.* (1999) improved on this robustness, as well as dealing with the water vapour absorption problem and other minor modelling deficiencies. The resulting global algorithms were validated for severe conditions (i.e. high water vapour content and the presence of volcanic aerosol from Mount Pinatubo) using *in situ* data corrected from bulk SST to skin SST by a model. The SST retrievals were found to have an rms accuracy of ≈ 0.25 K and a bias of ≈ 0.2 K or less (Merchant and Harris, 1999), values which are low enough for climate research applications, though still worth improving. For example, Merchant *et al.* (1999) pointed out that global warming trends (currently ≈ 0.15 K per decade; IPCC, 2001) still cannot be monitored directly using ATSR data owing to small residual trends in the bias, though it seems likely that for this purpose researchers will anyway prefer to combine satellite SSTs with *in situ* SSTs, using remote sensing as a way of improving global coverage.

2.4.2.4 *SST monitoring with ATSR*

Various SST products are produced routinely from the ATSR instruments. The global Average SST (ASST) product contains ten arc-minute spatially averaged SSTs, grouped into half-degree cells, which can be used for a wide range of SST studies. Figure 2.18 (Plate 1) shows examples of monthly SST maps, illustrating SST variations in the equatorial Pacific due to ENSO (Section 2.1.2). In December 1997 and December 1998 there were strong El Niño and La Niña events respectively. During a La Niña event, strong easterly winds near the equator blow from a higher-pressure region over the eastern Pacific towards a lower pressure over the western Pacific. These winds cause cool surface water from near the South American coast to move out across the Pacific, with an associated sea level fall (rise) in the east (west). In an El Niño event, the surface pressure falls (rises) in the east (west), the easterlies are weakened or reversed, and warm water from the west moves eastwards, causing sea level to rise (fall) in the east (west). The variations in SST and pressure between the two extremes of ENSO affect the global circulation of the atmosphere, giving rise to climatic anomalies (e.g. drought or increased rainfall) in many areas.

2.5 Using Radar

Active radar instruments have antennas that send out pulses of microwave energy towards the Earth and detect the backscattered return pulses. Like passive microwave instruments they have the advantage of being able to observe the surface through cloud. They fall into three main types, namely scatterometers, synthetic aperture radars and radar altimeters.

Scatterometers are designed to measure the radar backscatter from the surface or the atmosphere, with differences in return power corresponding to variations in the scattering target characteristics. They have two main applications in climate research and meteorology. In the case of the ocean surface, the backscatter depends on the characteristics of the small capillary waves induced by surface winds, and so a set of directional antennas can provide information on surface wind speed and direction. In rain radars the return power depends on the concentration and size of the raindrops, and so can provide information on rainfall.

The synthetic aperture radar (SAR) has a large sideways-looking rectangular antenna. Though the diffraction-limited resolution of such an antenna is relatively poor, the SAR is able to produce high-resolution backscatter images of the surface as follows. In the across-track direction the narrow antenna dimension (typically ~ 1 m) means that the diffraction-limited beam is broad, and each radar pulse illuminates a wide swath on the surface. The high-resolution imaging scheme in this direction simply makes use of the increase in range (and hence in return pulse delay) with distance away from the satellite ground track. In the along-track direction the long dimension of the antenna (typically ~ 10 m) is effectively made into an even longer 'synthetic aperture' by the simultaneous analysis of successive return pulses, so that the effective beam is then narrow enough to have an improved diffraction limit to its resolution, and provide high-resolution imaging in that direction. SAR imagery has been used over land (e.g. for land use mapping), ocean (e.g. for internal wave monitoring) and ice (e.g. for sea ice mapping). Interferometry between different SAR images has also been used to study topographic variations such as glacier motion. However, like *Landsat* and *Spot*, current SAR instruments tend to be used for relatively small-scale studies, and their use in climate research has been somewhat limited.

The radar altimeter has a parabolic dish antenna pointed vertically towards the nadir point, and uses the delay time of the reflected pulse to measure the distance to the surface. It has important applications in climate research and is used as the main radar example in this section.

2.5.1 *Radar altimetry*

The basic concept of radar altimetry is very simple. To a first approximation the height h of the satellite above the surface is given by $h = ct/2$, where c is the speed of light and t is the round-trip time for a radar pulse to the surface at nadir and back. If the satellite orbit is accurately known (e.g. by using laser ranging from the Earth surface to the satellite), then the height of the reflecting surface can be found relative to some reference surface.

The main uses for radar altimetry in climate research involve measurements of the ocean and of the cryosphere (though there have also been altimetry studies of land and inland water, e.g. Rapley *et al.*, 1987). The topography of the ocean not only provides information about the geoid (the gravitational equipotential surface at mean sea level) but also measures dynamic topographic features on the ocean surface, including the height variations associated with ocean currents. Such measurements ideally need to be made to an accuracy of a few centimetres or better. Radar altimeters over the ocean can also measure the surface wind speed and wave height along the satellite ground track. In polar regions, altimeters can provide accurate data on the elevation of the ice sheets, which has important implications for how sea level is likely to vary with climatic change. They can also measure the thickness of sea ice, which is an important indicator of climatic change.

The choice of frequency for a radar altimeter depends on a number of factors, including atmospheric transmission, atmospheric refraction (see below), and the need for a high enough frequency to support a high bandwidth (i.e. a short pulse; see below). In addition, the frequency also determines (together with the antenna size) the diffraction-limited beam width. To date all satellite radar altimeters have used a frequency of ≈ 13.5 GHz (wavelength 22 mm). With a typical 1 m diameter antenna this gives a diffraction-limited full beam width of approximately $3°$. For a typical LEO altitude (e.g. 800 km) this would provide a *beam-limited footprint* on the ocean surface with a diameter D_{BL} of approximately 40 km (see Section 2.2.5). However, most applications require a beam size an order of magnitude smaller (e.g. 1–10 km for ocean applications; Chelton *et al.*), which for this wavelength suggests the need for a ~ 10 m diameter antenna. Such an antenna is not only impractical, but the resulting *beam-limited altimeter* would be highly susceptible to pointing errors; e.g. an uncorrected mispointing error of only $0.05°$ gives a range error of 0.3 m at an orbital height of 800 km. The solution to these difficulties is the *pulse-limited altimeter*. This design, used for all space-borne altimeters to date, retains a relatively small (e.g. 1 m diameter) antenna with its large diffraction-limited footprint, but uses a very short transmitted radar pulse. The observed area then becomes the smaller *pulse-limited footprint*, determined by the pulse duration, as described below. This design can tolerate quite large mis-pointing errors, as noted later.

Figure 2.19 shows the basic operation of such a pulse-limited radar altimeter. The short radar pulse expands as a spherical shell. As the shell intersects a flat surface, the area of the surface producing simultaneous backscattered power at the antenna has the form of a disc, which then expands as an annulus. For a diffusely reflecting surface such as the open ocean, the return power received by the antenna as a function of time thus rises rapidly to a fairly constant value that then falls to zero as the annulus reaches the edge of the antenna's diffraction-limited beam. The 'pulse-limited footprint' is therefore the disc on the surface for which the echo from the trailing edge of the pulse, at the centre of the disc, is received simultaneously with the echo from the leading edge of the pulse, at the edge of the disc. Its diameter D_{PL} (for a flat surface) is $2\sqrt{hc\tau}$, where τ is the pulse duration. (Note that this is $\sqrt{2}$, smaller than the diameter of the maximum simultaneously illuminated disc, because of the two-way trip of the pulse.) For typical values of $h = 800$ km and $\tau = 3$ ns, $D_{PL} = 1.6$ km for a flat surface.

The return waveform is sampled in a series of time 'bins'. The timing of the leading edge of the return waveform is used to determine the surface height, and the backscattered power level, converted to the normalised radar cross section (σ_0), gives information about the surface (e.g. wind speed in the case of the ocean).

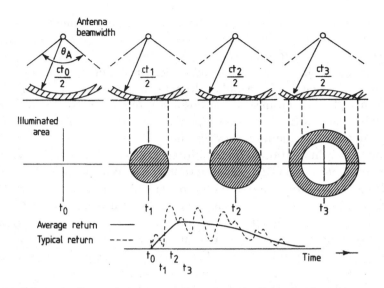

Figure 2.19 Schematic diagram of the operation of a pulse-limited radar altimeter of antenna beamwidth θ_A, showing the radar pulse intersecting a flat surface, the illuminated area, and the return pulse (assuming diffuse reflection), for various pulse round-trip times t_i. See text for discussion. From Rapley *et al.* (1983).

For a rough surface (e.g. a sea with large-scale gravity waves), the peaks reflect power sooner and the troughs reflect power later, having the effect of smearing out the leading edge of the waveform. Over the ocean, the slope of the leading edge therefore provides information about the height of the waves. The other effect of a rough surface is that peaks further away from the beam centre can now provide return power simultaneously with troughs at the centre. This increases D_{PL}, which for the example above would reach more than 10 km for a high sea state.

Despite the basic simplicity of the altimetry measurement concept, there are many factors that make accurate measurements difficult. There has therefore been much research devoted to tackling these effects in a series of altimetry missions over the last three decades. In particular, ocean height measurement precision and orbit accuracy have been substantially improved (Table 2.5). *TOPEX/POSEIDON* was optimised for altimetry over the ocean, whereas *ERS-1*, *ERS-2*, and particularly the RA-2 on *Envisat* (Table 2.1) have been designed to offer improved altimetry over more highly topographic surfaces such as ice sheets. However, these three missions still do not reach the latitudes above $\approx 82°$, so ESA's *Cryosat* (due for launch in 2004) will have an orbit that is very close to being polar, so as to carry out radar altimetry of ice sheets and sea ice to the highest latitudes. In the following sections we briefly discuss some of the factors affecting the use of radar altimeters for climate-related research. For more detail the reader is referred to the recent book, edited by Fu and Cazenave (2001), on the techniques and applications of radar altimetry.

Table 2.5 Summary of pre-*IEOS* satellite radar altimeters' ocean measurement precisions and orbit accuracies (from Chelton *et al.*, 2001).

Satellite	Mission period	Measurement precision (cm)	Best orbit accuracy (cm)
GEOS-3	Apr. 1975–Dec. 1978	25	∼ 500
Seasat	Jul. 1975–Oct. 1978	5	∼ 20
Geosat	Mar. 1985–Dec. 1989	4	10–20
ERS-1	Jul. 1991–May 1996	3	∼ 5
TOPEX/POSEIDON	Oct. 1992–present	2	2–3
ERS-2	Aug. 1995–present	3	< 5

2.5.1.1 Surface effects

Modelling the interaction with the surface

To extract useful information from the altimeter return pulses, it is necessary to model the interaction of the altimeter pulse with the surface. In the case of the ocean, it is usual to treat the backscattered radiation as being reflected from a large set of specularly reflecting surface facets within the antenna footprint that are oriented perpendicular to the beam direction, and distributed in height. Approximations to this height distribution enable the waveform to be modelled as a convolution of the pulse shape, the flat surface impulse response and the delay-time distribution of the surface facets. Averaging of many waveforms (typically ≈ 50) is necessary because individual waveforms display large fluctuations in intensity as a result of interference effects involving multiple unresolved surface facets. The leading edge of the waveform can then be analysed to obtain both the mean sea level and the *significant wave height* (usually defined as four times the standard deviation of the sea surface height, though definitions vary). The return power is inversely related to the wind speed, as decreased wind speed allows a larger fraction of the surface facets to be oriented perpendicular to the beam. (Note that this mechanism is different from the Bragg scattering mechanism associated with the wind scatterometer, which has 20–60° incidence angles.) Algorithms for determining wind speed with the altimeter are generally empirical, based on correlating *in situ* data with σ_0. Over ice sheets the radar pulse penetrates the ice to some extent, which alters the shape of the waveform leading edge. This effect therefore needs to be modelled in order to extract accurate surface heights from the return waveforms.

Surface corrections to mean ocean height

For climate research, the main interest in making ocean height measurements is to determine mean sea level changes and height gradients (i.e. pressure gradients) associated with ocean currents. However, there are a number of corrections that are needed. Tidal effects need to be corrected for, using tide models. Atmospheric effects on sea level must also be taken into account, e.g. by using meteorological data or by ignoring the affected data. An important example is the *inverse barometer effect*, whereby variations in surface atmospheric pressure will produce local changes in sea surface height (about 1 cm increase per millibar reduction in atmospheric pressure). An extreme but localised effect is the storm surge, where sea level increases of up to several metres are produced when the high winds from cyclones pile up water against the coast.

These surface height corrections are sufficient for measuring mean sea level changes and ocean current variability. To measure permanent ocean currents, however, it is also necessary to correct for the spatial variations in the geoid. Large-scale

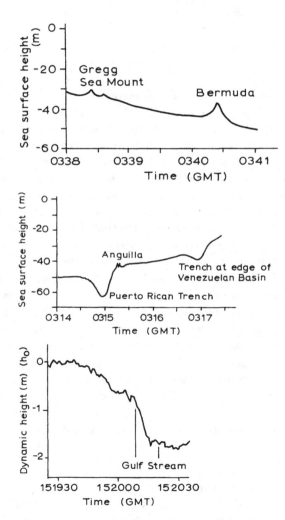

Figure 2.20   Ocean surface height from the *Seasat* altimeter, for tracks over sea-mount-type features, trench-type features, and an ocean current (the Gulf Stream). From Cracknell and Hayes (1991).

geoid undulations (with wavelengths of thousands of kilometres) have been measured independently using their effects on satellite orbits, but small-scale variations, often reflecting bathymetric features, are difficult to distinguish from height variations associated with the ocean currents. This can be seen, for example, in Figure 2.20, obtained from early altimetry data, where sea mounts (ocean trenches) can cause peaks (troughs) in the geoid as seawater is gravitationally attracted to the mount (sides of the trench). Currents such as the Gulf Stream, however, have height signatures of comparable size. (As can be seen, the signature consists of a step height change perpendicular to the current, because the pressure gradient force balances the Coriolis force to produce geostrophic flow.) Because of this difficulty, independent measurements of small-scale geoid undulations (with wavelengths of order 100 km) are now a high priority for climate research, using spacecraft to make high accuracy gravitational field measurements in orbit. The first is the *Gravity Recovery and Climate Experiment* (*GRACE*), launched in 2002, to be followed by the *Gravity Field and Steady-State Ocean Circulation Explorer* (*GOCE*) in 2006.

Finally, there are two sea state biases that need to be taken into account. The first is a bias caused by skewness in the surface height distribution of waves. The second is an electromagnetic bias, due to the fact that the height distribution of the specular facets, which determines the return pulse, is different from the height distribution of the whole sea surface.

Surface slope correction

For operation over ice sheets, the surface slope can become significant and give rise to an error in the surface height estimate. As shown in Figure 2.21(a), the altimeter

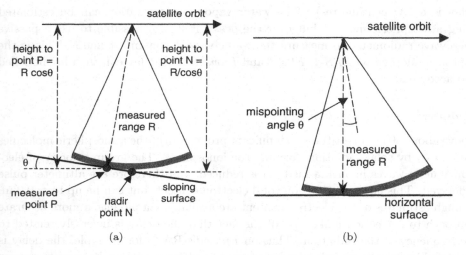

Figure 2.21 (a) Correction of radar altimeter heights for the effect of surface slope. (b) Tolerance of mis-pointing by a pulse-limited radar altimeter. Beam width highly exaggerated.

ranges to the nearest point on the surface rather than the point directly below the spacecraft. Provided that the across-track slope is known, this effect can be corrected for as shown. Assuming a linear slope, it is simple to calculate either the vertical distance to the measured point, or the vertical distance to the nadir point. In polar regions, ground track convergence means that the neighbouring track spacing is generally quite small, allowing reasonable across-track slope estimates to be made.

2.5.1.2   *Atmospheric effects*

The speed of light is reduced in any refractive medium such as the atmosphere. Thus the atmosphere introduces a delay into the altimeter pulses, as they travel to the surface and back, which needs to be corrected for. There are three main components to this correction, as follows.

Dry tropospheric height correction

The main constituent atmospheric gases (excluding water vapour) are well mixed, and so their effect on the pulse delay depends only on atmospheric density. This varies with time, geographical location and altitude. The resulting height error (typically ≈ 2 m) can be calculated accurately using measured and modelled along-track values of sea level pressure and the acceleration due to gravity.

Wet tropospheric height correction

Atmospheric water vapour results from surface evaporation, and is therefore highly variable with geographical location, being a maximum in the tropics (where it causes a height error of about 0.4 m) and a minimum in the polar regions (where the delay is a few centimetres). The water vapour concentration can be estimated using atmospheric models, but over the ocean it is also possible to use a passive microwave radiometer to measure the water vapour content at nadir. This is the approach adopted on *ERS-1*, *ERS-2* and *Envisat*. The delay can then be modelled and accounted for.

Ionospheric height correction

Above about 50 km in altitude, a significant proportion of the atmospheric molecules are ionised by solar radiation, forming the ionosphere. The group velocity of electromagnetic waves in such a medium is reduced, and therefore the altimeter pulse is delayed. This effect depends on total electron content, but can be up to ≈ 0.2 m. Though estimates of the electron content are available via models, a more accurate approach to correction makes use of the fact that the delay is inversely related to the frequency of the radiation. Thus on *Envisat*'s RA-2, for example, the delay is evaluated by using a second radar frequency (3.2 GHz) as well as the main frequency (13.575 GHz).

2.5.1.3   *Instrumental effects*

Orbit errors

A major difficulty with making absolute measurements of surface elevation is to determine the satellite's orbital elevation to sufficient accuracy. The basic method (Chelton *et al.*, 2001) is to model the satellite trajectory, taking into account such factors as gravitational perturbations, residual atmospheric drag, and radiation pressure, and to correct the model using independent observations of the trajectory. Observation methods include satellite laser ranging (SLR) from ground stations to passive on-board reflectors, the ground-to-satellite Doppler Orbitography and Radiopositioning Integrated by Satellite (DORIS) system developed for *TOPEX/POSEIDON*, the Precise Range and Range-rate Equipment (PRARE) on *ERS-2*, which involves two-way microwave transmissions, and the use of Global Positioning System (GPS) receivers on the satellite. Residual orbit errors are now a few centimetres in magnitude, comparable to the height measurement precision (Table 2.5). For monitoring variations in elevation with time using an Earth-synchronous orbit (see Section 2.2.1), the repeat period and ground track repeatability need to be considered. Repeat periods vary, depending on the application. The across-track variations in the positions of repeat tracks can usually be kept to ±1 km (Chelton *et al.*, 2001).

Height corrections

To ensure that absolute height measurements can be made, the geometry of the satellite and in particular the altimeter and the ranging equipment (e.g. targets for laser ranging) need to be accurately modelled. There are also various electronic and signal delays and biases within the instrument that need to be accounted for.

Pulse duration

To obtain a sufficiently small pulse-limited footprint (and for adequate height resolution), the pulse duration needs to be a few nanoseconds ($\approx$ 3 ns for all space-borne radar altimeters to date). To produce a pulse of this duration with an adequate signal-to-noise ratio, however, would require an unrealistically high transmit power. Therefore a 'chirped' system is used, whereby a longer pulse is produced, with continuously increasing frequency. The frequencies of the return pulse are then analysed with different delays, to re-create a short return pulse.

Tracking the surface

For efficiency the radar altimeter has a relatively small time window over which it samples the return pulse. Because the actual two-way travel time of the pulses varies widely around the orbit, due to orbital and geoid variations, it is necessary to incorporate an on-board system for tracking the surface, so as to maintain the leading edge of the return waveform within the sampling window. The tracking processor also provides estimates of two-way travel time, significant wave height and σ_0 (though all of these parameters can also be obtained by subsequent ground-based analysis, or *retracking*, of the waveforms). To acquire the surface initially, or after loss of tracking, the processing system uses a very broad time window to search for the return pulses, before initiating the standard tracking system.

Early tracking algorithms were optimised for the ocean, which provides relatively small rates of change of two-way travel time, and fairly uniform pulse shapes. More recent trackers such as that on *Envisat*'s RA-2 are also designed to track non-ocean surfaces (e.g. ice sheets and sea ice), where the two-way travel time can vary rapidly (e.g. with the topography of the ice sheet), and the return pulse shape can vary rapidly with the nature of the surface (e.g. as the altimeter passes over diffusely reflecting sea ice floes or calm, specularly reflecting water between them).

Mis-pointing

Unlike beam-limited radar altimeters (see earlier), pulse-limited altimeters can tolerate a certain amount of uncorrected mis-pointing (i.e. with the antenna not exactly pointing to nadir). (This feature was made use of, for example, by *Geosat*, whose gravity gradient attitude control led to residual mis-pointing of $> 0.5°$; Chelton *et al.*, 2001). This tolerance of mis-pointing is possible because the diffraction-limited beam width is larger than the pulse-limited footprint, and the pulse wavefronts are spherical. Thus, in Figure 2.21(b), provided that the off-pointing angle θ is less than half the beam angle, the measured range R is independent of θ, though the signal strength will be reduced compared with the pointing direction. Note, however, that the automatic algorithms for on-board processing of the data are generally affected by mis-pointing, though the correct values can be obtained by retracking (Chelton *et al.*, 1991).

2.5.1.4 *Examples of ocean and ice monitoring by radar altimetry*

As an example of radar altimetry over the ocean for use in climate studies, Figure 2.22 (Plate 2) shows sea level variations due to ENSO. The observed oscillation in sea level in the equatorial Pacific is related to changes in the pressure differences and winds across the basin, between the extreme phases of ENSO (see Section 2.4.2.4). Accurate radar altimeter data can be used to test theories of how ENSO occurs, as well as improving ENSO predictions (Picaut and Busalacchi, 2001). (The associated SST variations in the same two months are shown in Figure 2.18 (Plate 1).)

The world's main ice sheets on Greenland and Antarctica store vast quantities of water, and so have the potential to affect sea level as a result of global warming. Successive radar altimeter missions have improved our knowledge of the topography of these ice sheets. In particular, the *Geosat* and the *ERS* missions included

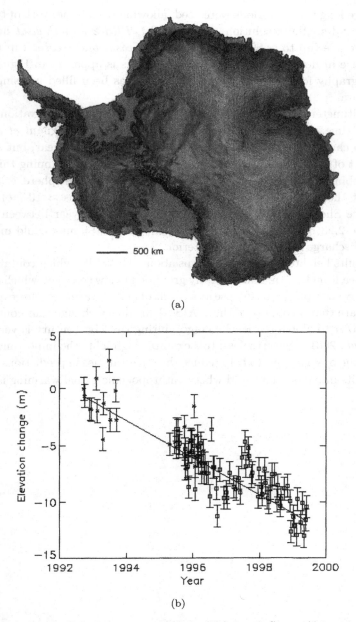

(a)

(b)

Figure 2.23 (a) Antarctic shaded topography from *ERS-1* and *Geosat* Altimetric Digital Elevation Model (with airborne data used south of 81.5°). The West Antarctic Ice Sheet (WAIS) is to the left. Image from Zwally and Brenner (2001). Copyright 2001, with permission from Elsevier. (b) Thinning of the Pine Island Glacier in the WAIS (13 km upstream of the grounding line), using altimeter data from *ERS-1* (stars) and *ERS-2* (squares). Reprinted with permission from Shepherd, A., *et al.*, 'Inland Thinning of Pine Island Glacier, West Antarctica', *Science* **291**: 862–864. Copyright 2001 American Association for the Advancement of Science.

phases where long repeat periods were used, allowing a dense network of tracks, and hence improved spatial resolution. Also, *ERS-1*, *ERS-2* and *Envisat* have higher latitude coverage (up to ≈ 82°) compared with *Seasat* and *Geosat* (up to ≈ 72°). The difference in detail using non-satellite methods is apparent in Figure 2.23(a), where topography for the region south of ≈ 82° has been filled in using airborne altimeter data.

Radar altimeter data can also be used to examine trends in elevation of the ice sheets with time. Between 1992 and 1996, for example, Wingham *et al.* (1998) observed no change in East Antarctica to within ±5 mm per year, but a negative trend in part of West Antarctica of −53 ± 9 mm per year. Continuing this research using a combination of *ERS-1* and *ERS-2* altimeter data, Shepherd *et al.* (2001) showed that the grounded Pine Island Glacier, which drains ≈ 10% of the West Antarctic Ice Sheet (WAIS), thinned by up to 1.6 m per year between 1992 and 1999 (Figure 2.23(b)). If sufficiently prolonged, such a thinning could increase the rate of ice discharge from the WAIS interior.

Such results heighten ongoing concerns about the WAIS, which contains enough ice to raise sea level by 6 m, and is mostly grounded below sea level, which may make it particularly susceptible to collapse as a result of global warming. However, current predictions are that a collapse of the WAIS (defined as a change that contributes at least 10 mm/yr to global sea level change) during the 21st century is very unlikely (Church *et al.*, 2001). Nevertheless, this example highlights the importance of using remote sensing observations to help reduce the uncertainties in predictions of climate change and its impacts, in a world where anthropogenic global warming is a reality.

Chapter 3

Planetary Science

Andrew Coates

3.1 Introduction

At the beginning of the last millennium, humankind knew of six planets, including our own. The invention of the telescope, and the beginnings of scientific thought on orbits and planetary motion, were in the 17th century. The next three centuries added Uranus, Neptune and Pluto to the list of known planets as well as many moons, asteroids and comets. Discoveries such as that the Earth was not the centre of the Universe and that planets orbit the Sun were key steps in increasing the understanding of Man's place in the Universe. But it was only in the latter part of the 20th century that we were privileged to carry out *in situ* exploration of the planets, comets and the solar wind's realm and to begin to understand the special conditions on the Earth that led to life prospering. This is leading to a detailed view of the processes that have shaped our solar system.

In this chapter we briefly review our current knowledge of the solar system we inhabit. We discuss the current picture of how the solar system began. Important processes at work, such as collisions and volcanism, and atmospheric evolution, are discussed. The planets, comets and asteroids are all discussed in general terms, together with the important discoveries from space missions, which have led to our current views. For each of the bodies we present the current understanding of its physical properties, and their interrelationships and present questions for further study.

What is in store for planetary exploration and discoveries in the future? Already a sequence of Mars exploration missions, a landing on a comet, further exploration of Saturn and the Jovian system, and the first flyby of Pluto are planned. We examine the major scientific questions to be answered. We also discuss the prospects for finding Earth-like planets and life beyond our own solar system.

3.2   The Solar System in the Last Four Millennia

Astronomy is one of the oldest observational sciences, having been in existence for about four millennia. This time is estimated on the basis that names were given to those northern constellations of stars that were visible to early civilisations. Two millennia ago, a difference between the more mobile planets and comets on the one hand and the fixed stars on the other was realised by Ptolemy but not understood. At the dawn of the present millennium six planets (Mercury, Venus, Earth, Mars, Jupiter and Saturn) were known. Aurorae were seen on the Earth and recorded but understanding still eluded us. In the first few decades of the last millennium comets were seen as portents of disaster (see Figure 3.1). Little further progress was made during the Dark Ages.

Science began to take giant leaps forward in the 16th century. Copernicus realised that the Earth was not at the centre of the solar system. Tycho Brahe's planetary observations enabled Kepler to formulate laws of planetary motion in the 17th century. In the same century, Galileo invented the telescope and Newton developed his theory of gravitation. The 18th and 19th centuries saw increasing use of this new technology, which deepened and increased understanding of the objects in the sky. Amongst many other discoveries were the periodicity of Halley's comet and the existence of the planets Uranus, Neptune and Pluto. The observations were all made in visible light to which our eyes are sensitive and which is transmitted through our atmosphere. The 19th century also gave us the basics of electromagnetism.

The 20th century witnessed many important scientific and technological advances. In terms of astronomy we have gained the second major tool for the exploration of the Universe in addition to the telescope — namely spacecraft. Space probes have not only opened up the narrow Earth-bound electromagnetic window, but it has also allowed *in situ* exploration and sampling of our neighbour bodies in the solar system. Using the techniques of remote sensing to look back at the Earth has added a new perspective. For the first time we can now begin to understand our place in the Universe and the detailed processes of the formation of the Universe,

Figure 3.1   The difference in perception of comets, at the beginning and end of the last millennium, is shown by a frame from the Bayeaux tapestry and an image from the Giotto Halley Multicolour Camera. The same comet is seen in each image. (Courtesy ESA/MPAe)

our solar system and ourselves. We are truly privileged to be able to use these techniques to further this scientific understanding.

Our sense of wonder in looking at the night sky has not changed. As we use better ground- and space-based telescopic techniques and more detailed *in situ* exploration, we can only feel a sense of excitement at the discoveries that the new millennium may bring. In this chapter we will review our solar system in general terms and in particular consider the interaction of the solar wind with the various planets and comets that inhabit the solar system. We consider scientific questions for the future and speculate a little on how they will be answered.

3.3   Origin of the Solar System

The present composition of the solar system reflects the history of its formation over 4.5 billion years ago. Because the planets are confined to a plane, the ecliptic, it is thought that the Sun and planets condensed from a spinning primordial nebula. The heavier elements in the nebula are thought to owe their existence to nearby supernova explosions. As the nebula collapsed and heated, the abundant hydrogen fuel ignited and fusion reactions started in the early Sun. Gas further from the centre of the nebula became progressively cooler and condensation occurred onto dust grains. This caused differences in composition due to the progressively cooler temperatures away from the Sun. Gravitational instabilities then caused the formation of small, solid planetesimals, the planetary building blocks (see Figure 3.2). Accretion of these bodies due to collisions then formed the objects familiar to us today.

This model predicts different compositions at different distances from the Sun, and this is seen in the different classes of solar system objects today. In the inner solar system, for example, the temperatures were of order 1000 K, near to the condensation temperature of silicates, while in the outer solar system temperatures

Figure 3.2   Artist's impression of the forming solar nebula. (NASA)

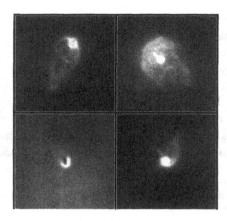

Figure 3.3   Star formation and the likely birth of planetary systems in the Orion nebula imaged using HST. (NASA/STSci)

were of order 100 K, nearer to the condensation temperature of water and methane ice. The forming planetesimals therefore had different compositions in the inner and outer parts of the solar system. The outer planets were derived from their much colder planetesimals, some of which remain as comets. The inner, rocky planets are associated with their own planetesimals, of which the asteroids are the partially processed survivors in the present solar system.

The formation process can be seen going on elsewhere, for example in the Orion nebula (see Figure 3.3). Protoplanetary discs have been imaged there using the Hubble Space Telescope, and structures that look like forming solar systems have been seen edge-on. Such patterns were also seen in the mid-1980s near the star Beta Pictoris. In that case the dust cloud was inferred to be due to pieces from forming or collided comets. Several other examples have been seen using infrared telescopes in space. It is thought that the basic formation of solar systems is widespread throughout the Universe.

3.3.1   *Processes: collisions, accretion and volcanism*

Clearly, collisions played an important role in the early solar system in the formation of planetesimals and larger bodies, and in collisions between these bodies. Accretion of material onto the forming bodies also played a key role. In the case of the outer planets, a rocky/icy core formed, and if the core was big enough, accretion of gas and dust from the solar nebula occurred. Since temperatures were cold in the outer solar system it was possible for the large cores to retain solar nebula material. This is why the current composition of the atmospheres of Jupiter, Saturn, Uranus and Neptune are close to solar proportions.

For the inner planets, on the other hand, temperatures were higher and it was possible for light atoms and molecules such as hydrogen and helium to escape.

The *Apollo* missions to the Moon provided information on the cratering rate with time since the solar system's formation. Cratering density was combined with dating of the returned samples from known positions on the surface, resulting in Figure 3.4. This shows the density of craters with time since 4.6 billion years ago. Clearly the cratering rate was high from 4.6 to 3.9 billion years ago, the 'late heavy bombardment' period, following which it slowed down between about 3.9 and 3.3 billion years ago, and it has been at an asymptotic 'steady' level since about 3 billion years ago. From this we can infer that there are two populations of Earth-crossing bodies. The first reduced with time as the population responsible was depleted by collisions, or the bodies were ejected from the solar system. The second group is the present population of Earth-crossing bodies. That population must be being replenished as well as lost.

The cratering process clearly depends on the size of the impacting object and its impact speed. Clearly the important parameter is the kinetic energy of the object. In general, impactors are travelling at tens of kilometres per second, so

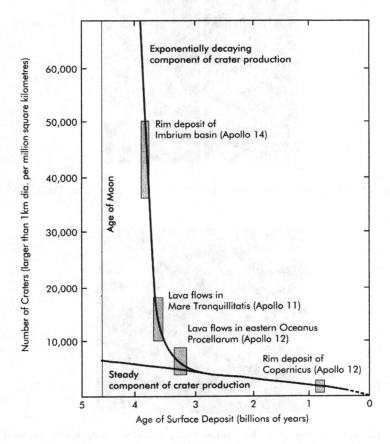

Figure 3.4   Dependence of lunar cratering rate on surface age.

impact kinetic energies are large. There are two physical shapes of crater seen —
simple craters (for small impactors) and complex craters (for large impact energy).
These types are illustrated in Figure 3.5. For larger energies, the complex craters
may include wall collapse and a central peak. Both crater types produce ejecta;
the distance these move from the crater depends on the gravitational force on the
target body.

The Moon contains a well-preserved cratering record, as there is no erosion by
atmosphere or oceans. An example from the lunar surface displaying both crater
types is shown in Figure 3.6.

In contrast, the Earth's cratering record is affected by erosion by the atmosphere,
by oceans and by volcanic and tectonic processes. Over 150 impact craters have

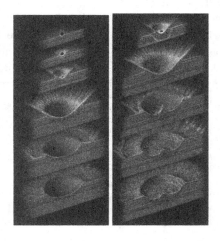

Figure 3.5   Simple (*left*) and complex (*right*) crater formation. (Don Davis, NASA)

Figure 3.6   Examples of complex (Euler, *top*) and simple (*bottom*) cratering on the lunar surface.
(NASA)

Figure 3.7   Mimas and its crater Herschel, some 130 km across. (NASA)

been discovered on the Earth. Of these about 20% are covered by sedimentation. Previously undiscovered craters are occasionally found, such as a new structure under the North Sea discovered by oil exploration geophysicists in 2002. However, in general only the remnants of relatively recent craters remain, though one of the large structures (Vredefort, South Africa) is 2.02 billion years old.

As well as the Moon and Earth, cratering is widespread in the solar system and is a feature of all of the solid bodies — planets, moons, asteroids and even comets — visited by spacecraft so far. Analysis of the cratering record on each body can be used in studies of a body's age and geological history.

Some extreme examples of cratering are worth mentioning. First, on Saturn's moon Mimas is the large (130 km) crater Herschel (see Figure 3.7). The impact that produced this crater may have almost torn Mimas apart, as there are fissures on the other side of the moon that may be related. Mimas is primarily icy according to albedo measurements, so this and the many other craters seen on Mimas show that craters are well preserved even on icy bodies.

A second extreme example is the Hellas impact basin in the southern hemisphere of Mars. This massive crater is some 2100 km across and 9 km deep. A ring of material ejected from the crater rises about 2 km above the surroundings and stretches to some 4000 km from the basin centre, according to recent data from the Mars Orbiter Laser Altimeter on the *Mars Global Surveyor* (see Figure 3.8 — Plate 3). As a rule of thumb, the diameter of any resulting crater is approximately 20 times the diameter of the impactor, so the original impactor in this case must have been ~ 100 km. The actual relationship between impactor size and crater size involves the kinetic energy of impact, the gravity of the impactor, the strength of the impacting material (relevant for small impacts), the angle of incidence and the composition of the target.

As a third example we can consider the formation of the Earth's Moon. Based on *Apollo* sample composition, and solar system dynamics calculations, it is now thought that the most likely explanation for the formation of the Moon was a large (Mars-sized) object hitting the Earth and the coalescence of material from that impact. Clearly collisions have a very important role in the solar system's history.

Collisions clearly still occur too: some 500 T of material from meteor trails and other cosmic junk hit the top of the Earth's atmosphere every year. In 1994, we witnessed the collision of Comet Shoemaker-Levy 9 into Jupiter — the first time such an event could be predicted and observed. Currently there are great efforts on the Earth to locate and track the orbits of near-Earth objects, as sooner or later a large impact will happen. The comfort we can take is that such collisions are very rare.

Another major, fundamental process in the evolution of solar system objects is volcanism. While only two bodies in the solar system have currently active volcanism (Earth and Io), there is evidence on many bodies for past volcanism, including on Mercury, Venus and Mars. We also see the effects of cryovolcanism on Europa and Triton and evidence of resurfacing at many other locations.

In general, an energy source is needed to drive volcanism. This may be from the heat of formation of the body, from radioactive decay of the constituents, from tidal forces, or from angular momentum. How much heat is left from formation depends on the size of the object. In particular, the area-to-volume ratio is larger for small bodies, so it is expected that these lose heat more quickly. Therefore, for example, Mars had active volcanism early in its history (e.g. Figure 3.9) but this has now died out, while the Earth still has active volcanism. On Io, the power for the volcanic activity comes from tidal forces. Venus on the other hand had rampant volcanism some 800 million years ago, when the planet was effectively resurfaced, but is now inactive. This is surprising as the size is similar to that of the Earth, but this could be associated with the slow current spin rate.

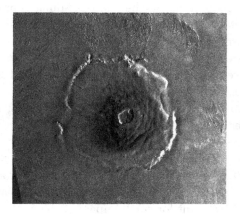

Figure 3.9   Olympus Mons on Mars, now extinct. At 27 km high and 600 km in diameter this is the largest volcano in the solar system. (NASA)

3.4 Evolution of Atmospheres

The main sources of atmospheric material are:

(1) Solar nebula
(2) Solar wind
(3) Small body impact
(4) Outgassing of accreted material

The clues to the relative importance of these mechanisms come from the relative abundances of the inert gases, of primordial (^{20}Ne, ^{36}Ar, ^{38}Ar, ^{84}Kr, ^{132}Xe) or radiogenic (^{40}Ar [from ^{40}K], ^{4}He [from ^{238}U, ^{235}U, ^{232}Th]) origin. The former must have been there since the beginning of the solar system as they are not products and do not decay, and the latter build up with time. These gases take no part in chemical processes, and most are too heavy to escape.

Minerals in the forming grains and planetesimals in the solar nebula contained trapped gases — volatile components. These included iron nickel alloys (which contained nitrogen), serpentine (which contained water, ammonia and methane) and several others. On accretion into a larger body, these cores would have attracted a 'primary atmosphere' of gases from the surrounding solar nebula (principally hydrogen, and otherwise reflecting the solar abundance).

As we saw earlier, in the first 0.5 billion years the impact rate by planetesimals was very high, bringing more material to the planetary surfaces (a 'veneer' of volatiles), the surfaces were heated by impacts, and the atmosphere was heated. There was also higher heating at that time, from gravitational accretion, and from radioactive decay of ^{40}K, ^{238}U, ^{235}U and ^{232}Th. There was also heating by differentiation as heavier substances dropped towards the centre of the recently formed bodies.

All this early heat would have increased tectonic activity at the surface, at least of the solid bodies. Larger planets have hotter cores, so gas should be released faster. The Earth's gas would have been released faster than on Venus (0.8 M_{Earth}) and much faster than on Mars (0.1 M_{Earth}). Isotope ratios at present show that both the Earth and Venus are effectively fully 'processed' (although there is still an active cycle involving volcanism and solution in the oceans on the Earth), while Mars is much less so and it is estimated that only about 20% of its volatiles have been released.

Several processes can release gases trapped in the interior. They are internal heating, volcanic activity, and impacts of large meteors — asteroids or comets.

Once released, gases would remain gravitationally bound assuming that (1) the body is big enough and (2) the gas is cool enough to remain gravitationally bound.

Other chemical reactions would start and the atmosphere would evolve over billions of years. Many changes have occurred since the formation of the solar system 4.6 billion years ago.

For the terrestrial planets, the isotope abundance ratios for the primordial isotopes are quite different to solar abundances, and more similar to those found in meteorites. This rules out 1 and 2 above as potential mechanisms. The abundance ratios for Venus, the Earth and Mars are about the same, but the absolute abundances decrease with distance from the Sun. This apparently rules out asteroid impacts as a dominant mechanism as the rates should be similar for these three planets, leaving outgassing as the dominant mechanism.

However, this picture is modified due to the mixing of the isotopes ^{36}Ar/^{132}Xe and ^{84}Kr/^{132}Xe. The values for Venus' atmosphere indicate a quite different origin, much more solar. For the Earth and Mars, however, the proportions show that the two planets are at either end of a mixing line: beyond the Earth end of the line, conditions are like those in comets, while beyond the Mars end, they are more like the primitive chondrite meteorites. This seems to indicate that a proportion of the Earth's current atmosphere is cometary in origin whereas in the Mars case the source was asteroids.

These are thus the expected proportions at the end of the bombardment about 4 billion years ago. Much of the water in the Earth's atmosphere and oceans may have come from comets.

The surface temperature of the planets at this stage then plays a role — controlled by the distance to the Sun. The temperature of a body (ignoring internal sources and atmospheric effects) can be estimated from the balance between solar input and radiation (assumed black body). Under that assumption, the temperature T can be written as $T = 279(1 - A)^{0.25}R^{-0.5}$ K, where A is the albedo and R in AU. If the kinetic energy of gas molecules is high enough it can escape the gravitational pull of the planet, if not it will be trapped to form an atmosphere. Inserting appropriate figures, it can be shown that Jupiter, Saturn, Neptune, Uranus, Venus, the Earth and Mars should easily have atmospheres, with Io, Titan and Triton on the borderline. Clearly the real picture is much more complex in terms of the energy balance, presence of an atmosphere, greenhouse effect, internal heating, etc. However, these simple considerations do remarkably well. The atmospheric evolution is then different for terrestrial planets and the outer planets; this discussion forms part of the next sections.

3.5   Terrestrial Planets

Due to the temperature variation in the collapsing primordial nebula, the inner regions of the solar system contain rocky planets (Mercury, Venus, the Earth and Mars). The condensation temperatures of the minerals forming these planets were higher than the icy material in the outer solar system. While it is fair to treat the inner planets as a group, the diversity of the planets and of their atmospheres is remarkable (see Figure 3.10). The three processes of impacts, volcanism and tectonics are vital ingredients in the evolution of the planets. Our understanding also depends critically on the sources and dissipation of heat.

Figure 3.10   Size comparison between Mercury, Venus, the Earth and Mars. (NASA)

The origin of the atmospheres of the inner planets is an important topic in itself. It is now thought that this is due to outgassing of primitive material from which the planets were made. As was pointed out previously, however, the cometary volatile content of the Earth's atmosphere seems to be higher than on Mars. The subsequent evolution of the atmospheres depends on two major factors: the distance from the Sun (which controls radiation input) and the mass of the body (controlling firstly the heat loss rate from the initial increase due to accretion, secondly the heating rate due to radioactivity, and thirdly the atmospheric escape speed). The presence of life on the Earth has also played an important role in determining atmospheric composition here.

Mercury is a hot, heavily cratered planet, which is difficult to observe because of its proximity to the Sun. Only one spacecraft, *Mariner 10*, has visited it, performing three fast flybys of the planet. Mercury is remarkable because of its high density, second only to the Earth. The other terrestrial planetary bodies — the Earth, Venus, Mars and the Moon — all fit on a straight line which relates density and radius, but Mercury is way off that line (Figure 3.11). It is unclear why this is — but may be associated with the origin of the planet or the stripping of its outer layers by the early T-Tauri phase of the solar wind.

Another remarkable and unexpected discovery from *Mariner 10* was the presence of a strong magnetic field and a magnetosphere around Mercury. It is likely that the planet has a larger iron-rich core in relation to its radius than the other inner planets. It is strange that there is so much iron, at this distance in the forming solar nebula. FeS (one of the major expected condensing iron-rich minerals) should not have even condensed. Mercury has an 'exosphere' rather than an atmosphere, since the pressure is so low that escape is as likely as a gas collision. An atmosphere was not retained because of the low mass (size) and high temperature (proximity to the Sun). The planet has an eccentric, inclined orbit. The rotation period is in a 2:3 ratio with its rotation. This indicates that the slightly non-spherical shape of the planet was important during its formation.

The surface of Mercury contains a well-preserved, little-disturbed cratering history. During the early bombardment in the first 0.8 billion years since formation, ending with a large impact that produced the 1300 km Caloris Basin feature, there

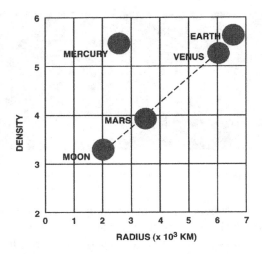

Figure 3.11   Density–radius plot for the inner planets showing the anomaly of Mercury. (ESA)

was important tectonic activity. There is also evidence of significant shrinkage of the planet due to cooling, causing 'lobate scarp' structures; this shrinkage was also early in the planet's history.

It appears likely from radar measurements that some of Mercury's polar craters, which are never heated by the Sun, may contain condensed ice from cometary impacts, similar to the deposits found using *Lunar Prospector* at the Moon. It would be interesting to estimate the cometary impact rate here in comparison with that at lunar orbit, to understand more about the collision history. But one of the fundamental unknowns about Mercury is what lies on its unseen side, not imaged by *Mariner 10*. Radar measurements indicate some structure, which could be a large volcano, but the *Messenger* and *BepiColombo* missions should explore this in detail.

The Venus–Earth–Mars trio are particularly important for us to understand because we know that life evolved at least on the Earth. In the case of Venus, the planet, although similar in size to the Earth, was closer to the Sun; water evaporated early because of the higher temperature and caused a greenhouse effect, resulting in further evaporation of water and eventually a runaway greenhouse effect. Lack of surface water meant that fixing of evolved carbon dioxide in the rocks as on the Earth was impossible. Carbon dioxide evolved into the atmosphere over the 4 billion years since formation, and continued the greenhouse effect after the hydrogen (from water dissociation) had escaped to space and the oxygen had oxidised rocks. The surface temperature is now 750 K, the atmosphere is thick and supports sulphuric acid clouds, and our sister planet is completely inhospitable to life. There is a total of only 10 cm equivalent depth of water in the Venus atmosphere, compared to 3 km in the atmosphere and oceans of the Earth. The clouds are observed to rotate much faster ($\sim$ 4 days) in the equatorial regions than the planetary rotation rate (once

Figure 3.12   Images from the bodies (Mars, Venus, the Moon and Eros) on which spacecraft have soft-landed so far. *Upper left*, from *Mars Pathfinder*; *lower left*, from *Venera 13*; *upper right*, from *NEAR-Shoemaker*; and *lower right*, from *Apollo 17*, showing Harrison Schmidt on the surface. (NASA, Russian Academy of Sciences)

per year); this 'super-rotation' is one of the aspects of the Venusian atmosphere that is not yet well understood.

In terms of the space programme there have been missions to study the atmosphere (*Mariner, Pioneer, Venera*), map below the clouds using radar (*Magellan*) and landers (*Venera*) have transmitted pictures from the surface (see Figure 3.12).

Mars, on the other hand, is smaller than the Earth and further from the Sun. Isotopic ratios between radiogenic ^{40}Ar and primordial ^{36}Ar indicate that only about 20% of the gas in the rocks has been evolved into the atmosphere, because of reduced tectonic activity (small size). Also, due to the size much of the early atmosphere was lost and the present atmospheric pressure is less than 1% that of the Earth. Substantial carbon dioxide ice deposits are present at the poles but there is not enough in the atmosphere to cause a greenhouse effect.

Martian oxygen isotopic ratios show that there must be a source of oxygen, perhaps frozen sub-surface water, which increases the proportion of the lighter isotope that would otherwise preferentially be lost to space, to the observed level. This has been partially confirmed recently by *Mars Odyssey*, which sees hydrogen, probable evidence of water, near the poles. However, the neutron analysis technique employed only works for the top 1 m of the Martian soil due to absorption, and it is hoped that the *Mars Express* ground-penetrating radar will survey down to 5 km. On the other hand, images from *Viking* and *Mars Global Surveyor* contain evidence that liquid water flowed on the Martian surface about 3.8 billion years ago (see Figures 3.13(a) and 3.13(b)).

The present Mars is much colder, dryer and much less hospitable to life than it once was. The equivalent depth of water in the atmosphere is only 15 microns. There is no possibility of finding liquid water on the surface due to the low pressure and average temperature. Any life on Mars was probably primitive and lived over 3.8 billion years ago. At that time, in addition to being warmer and wetter, Mars had a magnetic field, as shown by evidence of remnant magnetism in ancient rocks,

(a)

Figure 3.13   (a) Evidence from *Viking* for flowing water on Mars over 3.8 billion years ago. (NASA)

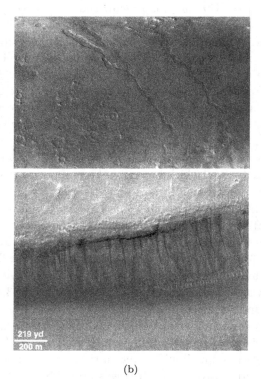

(b)

Figure 3.13   (b) Evidence from *Mars Global Surveyor* for flowing water on Mars over 3.8 billion years ago. Bottom image is evidence for possible water seepage to the surface within the last million years. (NASA)

by *Mars Global Surveyor*. This would have played a role in protecting any life at that time from galactic and solar cosmic radiation. That protection is no longer there. The possibility could remain, however, of there still being life on Mars in the form of sub-surface organisms near the water deposits, possibly similar to primitive organisms found in the Antarctic ice.

There is some recent evidence from *Mars Global Surveyor*, of water seepage from underground deposits to the surface, where it would immediately sublime. This discovery is significant — from the age of the surface this may have happened within the last million or so years (Figure 3.13). ESA's *Mars Express* and *Beagle 2*, as well as NASA's *Mars Exploration Rovers*, have a search for signs of water as a high priority when they arrive in 2003–4.

Some serious scientists propose that greenhouse gases could be introduced into the Martian atmosphere to warm the planet and release some of the trapped water and carbon dioxide, ultimately giving a hospitable environment for humans. This is an interesting idea in theory and a good target for computer simulations. In the opinion of this author terraforming would be the ultimate in cosmic vandalism if implemented.

The Earth was at the right place in the solar system and was the right size for life to evolve. The presence of liquid water on the surface meant that carbon dioxide could be dissolved in rocks as carbonates, some of this being recycled due to volcanism. As life developed, starting at the end of the bombardment phase, photosynthesis became important, leading to the production of oxygen and the fixing of some of the carbon in the biomass. Enough oxygen in the atmosphere led to production of stratospheric ozone, which allowed the protection of land-based life forms from harmful UV radiation.

The Earth is also the planet we know the most about. Looking at the Earth from space gave us a new perspective: an enhanced feeling that the Earth is special and indeed fragile. The average temperature of the Earth's surface is close to the triple point of water where solid, liquid and vapour may all exist. That is part of how we came to be here.

Our Moon is the first planetary satellite in terms of proximity to the Sun. Its density is much lower than the Earth's and there is effectively no atmosphere. The cratering record is therefore well preserved but the maria show that volcanic activity was important after the early bombardment and up to about 3.2 billion years ago. Despite intensive study by spacecraft (*Luna, Ranger, Surveyor, Apollo, Clementine, Lunar Prospector*), the origin of the Moon has not yet been determined completely from the competing theories (simultaneous formation, catastrophic impact on the early Earth, capture). However, the current favourite, which fits most of the evidence, is that a Mars-sized body hit and coalesced with the Earth, and the Moon formed from the impact ejecta.

As was already mentioned, the recent *Clementine* and especially *Lunar Prospector* missions returned evidence for hydrogen, perhaps in the form of water mixed with regolith, in craters towards the pole of the Moon. If this is water it would be from cometary impacts, and one of the exciting aspects of this is that there could be cometary material available for scientific study at a relatively close distance. Though studying this would be a large technological challenge, the scientific benefits of examining this pristine material from the early solar system are significant.

The other missions going to the Moon soon, including *SMART-1* and *Selene*, offer the possibility of better mineral maps and composition information over the whole lunar surface, adding significantly to the information gained from the *Apollo* and Russian lunar samples.

The satellites of Mars, namely Phobos and Deimos, may be captured asteroids based on their physical characteristics. However, the understanding of the dynamics of their capture is by no means solved.

Outstanding questions on our planetary neighbours

Despite the proximity of our planetary neighbours and the many space missions that have explored them, many important questions remain. Why is Mercury's core so large? Might a catastrophic collision early in its life explain this and its orbital eccentricity and inclination? How is seismic activity affected? Is there ice at Mercury's poles? Why is there super-rotation in the Venus atmosphere? What is the surface composition of Venus and Mercury? What is the geological history? How oxidised is the Venusian surface and what is the history of water in the Venusian atmosphere? What changes is humankind making to the Earth's climate and do these need to be ameliorated? What is the origin of the Moon? Do the Martian atmospheric loss rates to space support the models? Where is the water on Mars now? What is the history of other volatiles? Was there life on Mars? Could and should we terraform Mars?

3.6   Outer Planets

The outer planets group contains the gas giants (Jupiter, Saturn, Uranus and Neptune — Figure 3.14) and the icy object Pluto. The gas giants are heavy enough and were cold enough when they were formed to retain the light gases hydrogen and helium from the solar nebula, and these constituents form most of the mass of the planets, reflecting the early composition. The visible disc for telescopes and space probes is ammonia and water-based clouds in the atmosphere. At Jupiter, the largest planet and closest gas giant to the Sun, the cloud structure shows a banded and colourful structure caused by atmospheric circulation. The detailed cloud colours are not fully understood. There is no solid surface as such, but models of the internal structure of the gas giants show increasing pressure below the cloud tops; ultimately the pressure becomes so high that a metallic hydrogen layer forms between about 80% and 50% of the radius respectively. Dynamo motions in this

Figure 3.14   Size comparison for the gas giants Jupiter, Saturn, Uranus and Neptune, using *Voyager* images. (NASA)

layer, assuming it must be liquid, power the powerful planetary magnetic fields. A rocky/icy core is thought to be present at about 25% of the planetary radius. Jupiter rotates rapidly, providing some energy via the Coriolis force for atmospheric circulation. However, both Jupiter and Saturn have internal heat sources, which mean that they emit 67% and 78% more energy than they receive from the Sun respectively. This gives most of the energy for the atmosphere but the origin of the internal heat source is not fully understood. Models indicate that helium precipitation within the metallic hydrogen core, in which helium is insoluble, may be responsible. There are also strong zonal (east–west) winds near the equator on Jupiter and Saturn, stronger on Saturn, where they reach two thirds of the speed of sound. Their origin is not fully understood. The planets also have important long- and short-lived atmospheric features, of which the most prominent is the great red spot on Jupiter. This long-lived feature, seen for at least 300 years, is surprisingly stable and so far there is no adequate model to describe it. A similar spot feature appears on Neptune.

In situ results at Jupiter have recently been enhanced by data from the *Galileo* orbiter and probe. While the orbiter has discovered unexpected dipole magnetic fields in some of the Galilean satellites, the probe has sampled the atmospheric composition, winds, lightning and cloud structures at only one point, which turned out to be a non-typical location in the Jovian atmosphere. One of the discoveries was a lower than solar helium abundance, which provides some support for the idea of helium precipitation in the metallic hydrogen layer; a similar conclusion was arrived at based on *Voyager* data at Saturn. Also, there is less water than expected.

The gas giants each have important and fascinating moons. At Jupiter (see Figure 3.15 — Plate 4), Io has the only known active volcanoes other than the Earth, providing sulphur-based gases for the Jovian magnetosphere; Europa may have a liquid water ocean under its icy crust; Ganymede has its own 'magnetosphere within a magnetosphere'; and Callisto has a very old, cratered surface. Our knowledge of these has been revolutionised by *in situ* observation; before this only the albedos and orbital periods were known.

At Saturn, Titan is a tantalising object, planet-like in structure and the only moon with a significant atmosphere — 1.5 times the Earth's pressure at the surface. However, its face was shrouded from *Voyager*'s view by organic haze in its thick nitrogen–methane atmosphere. The atmosphere may hold clues about the Earth's early atmosphere; there may be methane- or ethane-based precipitation systems; and the ionosphere forms a significant source of atoms for Saturn's magnetosphere. *Cassini–Huygens* will study Titan and some of Saturn's 20 or so icy satellites in detail, starting in 2004. At Uranus, the moon Miranda graphically indicates the accretion theories, as it appears to be made up of several different types of structure seen elsewhere. Also, the moon system is out of the ecliptic because the spin axis of Uranus at 98° inclination is almost in the ecliptic itself. At Neptune, Triton is an icy satellite with a very thin atmosphere but it is in a retrograde orbit and is spiralling closer to Neptune; in tens of millions of years it may break up to produce spectacular rings. It may be similar in characteristics to Pluto and Charon.

Ring systems are present at all the gas giants but spectacularly so at Saturn. Saturn's rings were discovered by Galileo, found to be separate from the planet by Huygens, and observed to have gaps by Cassini, who also suggested that they were composed of separate particles. James Clerk Maxwell proved this idea mathematically two centuries later. Detailed exploration was begun by the flyby missions *Pioneer* and *Voyager*, which found remarkable structures in the rings, including warps, grooves, braids, clumps, spokes, kinks, splits, resonances and gaps. Whole new areas of solar system dynamics were opened up — including the study of electromagnetic forces, which may be important for spoke formation. The rings are less than a kilometre thick, as low as tens of metres in places, and are composed of billions of chunks of ice and dust ranging from microns to tens of metres in size. But the main question has not yet been satisfactorily answered: How did the rings form? Was it break-up of a smaller satellite or cometary capture?

And then there is Pluto, with its companion Charon. Following an elliptical, inclined orbit with a period of 248 years, and currently the furthest planet from the Sun, Pluto is an icy body rather than a gas giant. It may be closely related to, but larger than, the icy Kuiper belt objects, the outer solar system planetesimals, and it may also be related to Triton. Much will be learned by the first spacecraft reconnaisance of the Pluto–Charon system. But as Pluto moves towards aphelion its tenuous and interesting methane-based atmosphere will condense and become much less dense. In 2010–20 it is expected that a rapid atmospheric collapse will occur. There is a good case to get to Pluto as soon as possible and another excellent case for a visit near the next perihelion in 2237.

Questions for the new millennium on outer planets and their satellites

What causes the cloud colours in the gas giants? Are the internal structure models correct? What causes the internal heat source? Why are the zonal winds so high on Jupiter and, particularly, Saturn? Why are atmospheric features, such as the Jovian great red spot, so stable? Does Europa have water oceans and is life a possibility there? Will the Titan atmosphere evolve futher when the Sun becomes a red giant? Why is the Uranian spin axis so tilted and what are the effects? Can the study of Saturn's rings give us more information about the radiation–plasma–dust mixture in the early solar system? What are the basic characteristics of Pluto–Charon?

3.7   Comets

Comets are the building blocks of the outer solar system. They formed at the low temperatures prevalent at these distances in the primordial nebula and they retain volatile material from the early solar system. They are relatively pristine bodies, making their study important. Some of them collided and coalesced to form the cores of the gas giants. From the orbits of comets we find that many ($\sim 10^{12}$) were formed in the Uranus–Neptune region but were expelled to form the spherical (and theoretical) Oort cloud (Figure 3.16) with a radius of approximately 50,000 AU. Passing stars may disturb the orbits of these distant members of the solar system.

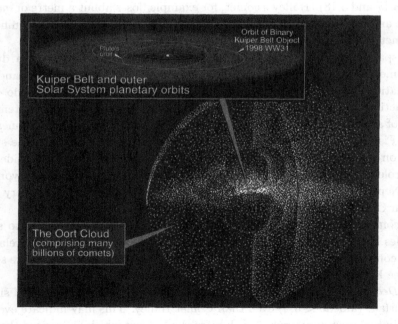

Figure 3.16   Illustration of the inferred location and shape of the Oort cloud and Kuiper belt. (NASA)

Figure 3.17   Comet Hale-Bopp in 1997, showing the dust tail (*straight*) and plasma tail (*to the left*). (NASA)

They may then plunge into the inner solar system, where their orbits have random inclinations. Others ($\sim 10^8$) form the Kuiper belt just beyond Pluto's orbit; their inclinations are close to the ecliptic plane. Whatever their origin, comet nuclei are dirty snowballs which, when they near the sun, emit gas and dust that form the plasma and dust tails seen in all comets (e.g. Comet Hale-Bopp in 1997; see Figures 3.17 and 3.18). Halley's comet, for example, loses about a metre of material per orbit and has orbited the Sun about 1000 times, but activity varies significantly from comet to comet.

The space missions in the mid-1980s confirmed that comets have a distinct nucleus, measured gas, plasma and dust composition and led to an understanding of tail formation. The main surprises were the darkness of the nucleus (Albedo $\sim 4\%$) and jet activity rather than uniform gas and dust emission. This is consistent with the idea of a cometary crust developing after multiple passes through the inner solar system. Fissures form in this as a result of phase changes underneath the surface brought on by the temperature variations over the comet's orbit. This idea also plays a role in the sudden flaring of comets, as seen at Halley when beyond the orbit of Neptune in 1990, and may be important in the break-up of cometary nuclei as seen in Comet Schwaschmann-Wachmann 3 in 1998.

The composition measurements showed that the abundance relative to silicon of volatiles like C, N and O is more similar to solar abundances, and so relatively pristine, compared to more processed bodies like meteorites and (even more so) the Earth (see Figure 3.19).

The *Deep Space 1* encounter with Comet Borrelly in 2001 showed a slightly darker surface (albedo $< 3\%$) even than Comet Halley. This may indicate even less volatiles in the crust as the comet, a Jupiter class object which has visited the inner solar system many times, is thought to be 'older' than Halley.

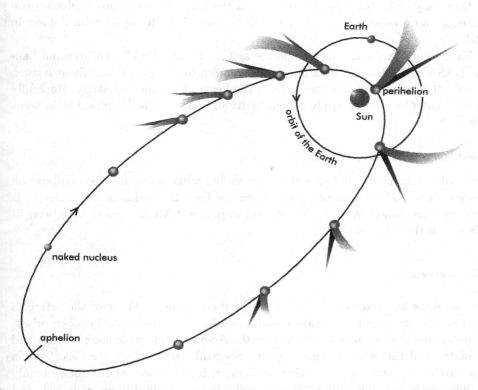

Figure 3.18   Illustration showing cometary tails — the plasma tail points antisunwards due to the solar wind, while the dust tail is curved due to radiation pressure pushing the grains to a larger heliocentric distance and therefore reducing the orbital velocity of dust grains.

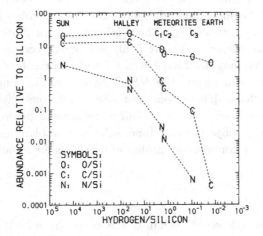

Figure 3.19   Comparison of abundance ratios of volatile elements for the Sun, comet Halley, meteorites and the Earth (see text). (ESA)

Cometary collisions played a key role in the inner solar system. Collisions can still occur now, as shown by the Comet Shoemaker-Levy 9 impact with Jupiter in 1994.

Future cometary missions include the ambitious ESA *Rosetta* Orbiter and Lander, NASA's *Stardust* to bring back dust samples to the Earth, and *Deep Impact*, which will release pristine material from the heart of a comet for study. Hopefully the ill-fated *Contour*, to study the diversity of comets, will be repeated in some form.

Questions for cometary missions

Given their importance in the early solar system, what is the detailed composition of several comets? Can we bring an icy sample back to the Earth for analysis? Is there an Oort cloud? What is the relation to planets? Might comets have brought volatiles to the inner planets?

3.8   Asteroids

In some sense the asteroids belong with the inner planets. Many of the asteroids occur in the main belt in between Mars and Jupiter. Some are in other orbits, including orbits that cross the Earth's path. A wide variation of eccentricities and inclinations of the orbits is also present. Spectral studies allow the classification of asteroids into several types: C-type, dark, rich in silicates and carbon, mainly an outer main belt population; S-type, rocky bodies, mainly inner main belt and Earth-crossing; M-type, iron and nickel. A few other asteroids do not fit this scheme. It seems likely that many asteroids are the remains of inner solar system planetesimals rather than being due to the destruction of a larger body. However, there have been collisions between some bodies since the early bombardment, leading to fragmentation and other processing. There is evidence for craters in the images of Gaspra, Mathilde and Ida (see Figure 3.20), and most recently from Eros. The moon of Ida, i.e. Dactyl, may itself have been formed as a result of a collision. Dust and boulders on Eros, seen during the *NEAR-Shoemaker* landing sequence, were a surprise but are perhaps debris from collisions. Collisions with the Earth and other planets may have been important, and as mentioned before, some asteroids are the nearest remaining objects to the inner solar system planetesimals. In future, some workers think that commercial mining of asteroids for minerals may become economically feasible.

Questions for asteroid missions

Why are asteroid types diverse? What is the composition? Which, if any, are planetesimals? How pristine? Do they contain interstellar grains from before the

Figure 3.20   Ida and its moon Dactyl, seen by the *Galileo* spacecraft in 1991. (NASA)

solar system? Is there any water? What is their origin? Which asteroids do meteorites come from? Might they be a future source of raw materials?

3.9   Magnetospheres

The aurora and cometary tails, the only visible clues to the solar wind, have been observed for centuries. As with many other scientific phenomena, the beginning of understanding of the aurora had to wait until the present era. Gilbert's ideas (in 1600) of the Earth as a magnet, Halley's (1698 and 1700) magnetic maps of the Earth from ships and his idea of the aurora being associated with the magnetic field, and George Graham's 1722 observation of the motion of compass needles were important early contributions. More recently, in Birkeland's terrella experiments in the early 1900s he fired electrons ('cathode rays') at a magnetised sphere in a vacuum, Appleton and others studied the ionosphere in the 1920s and Chapman laid the foundations of modern solar–terrestrial theory starting in 1930; these all contributed to our current understanding of the solar–terrestrial relationship. But at the dawn of the space age, the first and completely unexpected scientific discovery from the space programme was of the radiation belts of energetic charged particles trapped in the Earth's magnetic field (Chapter 1).

Observations of comets in the 1950s had led to the idea of a constantly blowing but gusty solar wind. The solar wind was confirmed by early space probes but it was not until the mid-1980s that the comet–solar-wind interaction was understood and backed up by *in situ* data. Solar–terrestrial research and solar wind interaction with other bodies remain active areas of research in the new millennium. One of the drivers for understanding is the effect that violent solar activity can have on the electromagnetic environment of the Earth and on humankind's technological systems. I have been privileged to be part of some of the exciting exploratory space missions in these fields.

In its present state, the Sun emits about a million tonnes of material in the form of plasma per second. At this rate it would take 10^{14} years to disintegrate; well before then, in about 5 billion years, the hydrogen fusion fuel will be exhausted and the Sun will become a red giant. In the meantime the Earth, other planets and comets are all bathed in the solar wind.

The solar wind is highly conducting such that to a good approximation the magnetic field is frozen into the flow. Plasma in the solar corona is hot enough to escape the Sun's gravity along the magnetic field where motions are unrestrained, and it can be shown that the solar wind becomes supersonic within a few solar radii. Beyond this the speed is almost constant for a particular element of solar wind but the value can vary between about 300 and 800 kms^{-1}. As plasma expands radially through the solar system it drags the solar magnetic field along, but this forms a spiral pattern in space due to the solar rotation. By the Earth's orbit the density is about 5 cm^{-3} but variable and disturbed by coronal mass ejections, shocks and discontinuities. As the solar cycle waxes and wanes, the source region of the solar wind changes, in particular, the coronal holes get smaller (increased magnetic complexity) and larger (decreased complexity) respectively. The electrodynamic environment of solar system bodies is extremely variable on timescales of seconds (ion rotation around the magnetic field) to 11–22 years (solar cycle).

The solar wind extends well beyond the planets. Ultimately a heliopause is required where the solar wind pressure balances that of the local interstellar medium, at about 150 AU on the upstream side. Before that, at about 100 AU, a terminal shock slows the solar wind from supersonic inside to subsonic outside, and upstream of the heliopause a bow shock may form if the local interstellar medium (LISM) motion is supersonic. The terminal shock, heliopause and bow shock are all hypothetical as the *Voyager* spacecraft, the most distant man-made object, has not yet crossed these boundaries. However, the inner heliosphere is becoming better understood from several spacecraft, notably *Ulysses*, which is measuring the structure out of the ecliptic plane in the Sun–Jupiter region for the first time.

The interaction of the solar wind with an obstacle depends critically on the obstacle itself: its state of magnetization, its conductivity and whether it has an atmosphere. We will consider two main types of object: a magnetised planet such as the Earth and an unmagnetised object such as a comet (see Figure 4.23 for a comparison of sizes of the interaction regions).

Other objects include interesting features of both extremes. Mars and Venus are unmagnetised but have some cometary features. Io is conducting, is within the subsonic Jovian magnetosphere, produces a plasma torus, drives huge field-aligned currents to Jovian auroral regions causing light emission there, and supports Alfven wings. Titan has a dense ionosphere. It is usually in the subsonic magnetosphere of Saturn but is sometimes in the solar wind, and field line draping seems to occur in Titan in the same way as it does at comets.

3.9.1 *Magnetised planet interaction*

The discussion concentrates on the Earth but is also relevant to Mercury, Jupiter, Saturn, Uranus and Neptune. For example, aurorae have been imaged on Jupiter and Saturn (see Figure 3.21 — Plate 5). Different magnetic dipole strengths and orientations, spin rates, and particle sources such as moons and ionospheres, and sinks such as rings, cause differences in the interactions. In the case of Mercury there is no ionosphere, so that at present we do not understand how the electrical currents close.

Magnetised plasmas do not mix. As the solar wind approaches the Earth a current sheet, the magnetopause, is set up to separate the regions of the solar and planetary magnetic field. The solar wind particle pressure outside balances the magnetic pressure inside (these are the dominant pressure components), so that the magnetopause is compressed on the dayside and extended like a comet tail on the nightside. This simple model was formulated by Chapman and Ferraro earlier this century and gives a good prediction for the magnetopause location. Outside the magnetopause a bow shock stands in the supersonic solar wind flow. The nature of this collisionless shock changes with magnetic field orientation as, to some extent, the magnetic field plays the role that collisions play in a fluid shock.

However, the real situation is not this simple. Shocked solar wind particles can penetrate the magnetopause directly via the funnel-shaped cusp regions on the dayside, to cause dayside aurorae. Also, if the solar wind magnetic field and the northward-directed terrestrial magnetic field are oppositely directed, then our nice fluid model is not enough, as first realised by Dungey. The explosive process of magnetic reconnection takes place at just the distance scale where the fluid approximation breaks down. This causes solar wind field lines to be connected to terrestrial ones and these are dragged over the polar cap like peeling a banana. Ultimately this leads to a build-up of magnetic energy in the tail, further reconnection in the deep tail and perhaps important trigger processes nearer to the Earth. The effect is that some plasma is shot down the tail and some towards the Earth, causing nightside aurorae. The reconnection process, and the electric field across the tail caused by the motion of the solar wind relative to the magnetised Earth, set up a convection system.

Another convection system is caused by plasma corotating with the planet. The atmosphere and ionosphere corotate with the Earth, as does the inner part of the magnetosphere — the plasmasphere. The overall circulation in the Earth's magnetosphere is given by the sum of the corotation electric field and the convection electric field. The convection part of this is extremely dynamic and is still the focus of intense research.

Measurements from ground and space are used together to diagnose the near-Earth environment. Despite the success of early satellites in mapping the various regions, one of the major limitations of these space measurements has been the use of single or dual satellites and the consequent space/time ambiguity. If one or two satellites see a signal due to a boundary, how can we know if the boundary has moved over the satellite or vice versa? The only way to resolve this is with more satellites. At present this is possible on the large scale with the ISTP (International Solar–Terrestrial Physics) fleet. Satellites upstream of the Earth monitor the Sun and solar wind (*SoHO, ACE, Wind*) while satellites in the magnetosphere (*Polar, Interball, FAST* and others) and in the tail (*Geotail*) monitor the overall effects.

This combination has proved extremely useful in following coronal mass ejections from the Sun, through the interplanetary medium and to the Earth. The increase of radiation belt particles (via an as-yet-unknown process) due to one of the events monitored with the ISTP fleet may have caused a commercial telecommunications satellite to fail. This was the first time that 'space weather' has been measured while causing catastrophic effects. As electronic integration and our dependence on technology increase, such events will become more important still. Space weather (Chapter 5) can affect satellites, electricity cables and oil pipelines, where currents are induced.

On a small scale the *Cluster* mission is the first coordinated multi-spacecraft fleet. This is a group of four spacecraft launched in 2000 following an abortive attempt on the first *Ariane 5* in 1996. They are flying in tight formation through the magnetosphere, typically a few hundred kilometres apart, in a polar orbit. The spacing has been as tight as 100 km and as wide as several thousand kilometres. At one part of the orbit the spacecraft are at the corners of a tetrahedron. With two spacecraft we can measure along a line in one dimension, with three we measure a two-dimensional plane, but with four spacecraft we can measure in three dimensions. For the first time the space–time ambiguity is being resolved and it is possible to measure vector gradients. This gives the exciting capability of being able to measure the parameters in Maxwell's equations of electromagnetism directly and to do real plasma physics with the results. The reconnection process, cusp entry processes and tail dynamics will be directly measured. The analysis from this important mission is just starting. Already, important data on reconnection, the magnetic cusp and the magnetotail have been gathered. Multi-spacecraft missions are certainly the way to go in this field, and NASA is planning the follow-on *Magnetospheric MultiScale* (*MMS*), which will cover separations down to electron gyroscales, 10 km, and visit different regions of the magnetosphere.

The recently launched *IMAGE* spacecraft is providing images of the inner magnetosphere and showing a two-dimensional collapse of part of the magnetosphere, called the plasmasphere, where plasma densities are high enough for such techniques to work. Ultimately, constellations or 'swarms' of identically instrumented

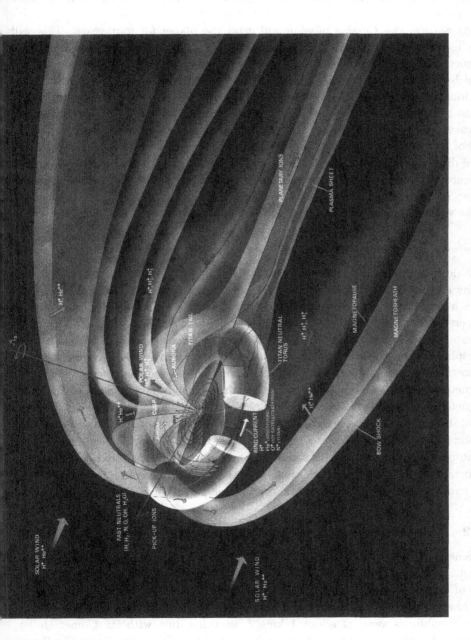

Figure 3.22 Schematic of Saturn's magnetosphere. (Courtesy John Tubb, LANL)

small spacecraft are required to get an instantaneous wide-scale view of the entire magnetosphere and understand the propagation of space weather fronts as they traverse it.

At Saturn, the *Cassini* mission will orbit the planet for four years. This will give us more information about the similarities and differences to the Earth's magnetosphere (see Figure 3.22). Saturn's magnetosphere is controlled by the solar wind but corotation is more important, the magnetosphere is bigger so timescales are longer, and the rings (see Figure 3.22) and moons impose important differences. This will be one mission that will contribute to the study of comparative magnetospheres, which will in turn help to tell us more about our own.

For the future, even the large-scale ISTP constellation and the small-scale three-dimensional *Cluster* measurements still lead to a significant under-sampling in near-Earth space. Plans are now being laid to address this limitation by means of two approaches: magnetospheric imaging and multi-spacecraft (30+) missions. These will be important techniques in the coming decades.

Questions on interplanetary medium and magnetised planet interaction

How do solar and interplanetary events relate? Will it be possible to forecast events at the Earth, particularly the sign of the interplanetary magnetic field? Is there a terminal shock, heliopause and bow shock far outside the planets? What triggers reconnection? What causes energy release in the tail? What do different timescales and other differences at the planets teach us about the Earth's magnetosphere? How are radiation belt particles accelerated? Can we develop better protection techniques for our satellites? How do field-aligned currents in Mercury's magnetosphere close? Is it important that the planet occupies much of the inner magnetosphere? How does this magnetosphere work? How do the ecliptic plane orientation of the Uranus geographic axis, the 60° tilt of the magnetic axis with respect to this, and the rapid rotation rate affect the Uranian magnetosphere? How is the planetary magnetic field produced at Mercury, Uranus and Neptune? How important to the survival of life on the Earth was the protection from solar wind and cosmic radiation afforded by the Earth's magnetic field?

3.9.2   *Comet–solar-wind interaction*

A comet interacts with the solar wind in quite a different way to a magnetised planet. The cometary nucleus is not magnetised and is a 'dirty snowball'; in the case of Comet Halley the nucleus is some 15 km by 8 km, as determined by the *Giotto* spacecraft. When the nucleus warms up near to the Sun, gas (mainly water vapour) sublimes away at about 1 kms^{-1}, carrying dust with it. Comet Halley, for example, produces about 20 tonnes per second while Comet Grigg-Skjellerup, *Giotto*'s second target, produced only 200 kg per second at the encounter time. The gas may ionise in sunlight or by charge exchange with the solar wind, on a timescale of a week or so, and the ions interacts with the electric and magnetic field in the solar wind.

The new ion is first accelerated along the electric field and then gyrates around the magnetic field. The motion of the particle is a cycloid, the same path taken by the valve on a bicycle wheel as it moves. The motion is well known to plasma physicists as 'E cross B drift' — and the speed of the centre of the motion is given by the ratio of the electric to magnetic fields. The motion in real space is equivalent to a ring in velocity space, centred on the drift speed in a direction along the magnetic field.

The ring causes plasma instabilities that excite Alfvén waves moving predominantly upstream along the magnetic field. We were able to show both experimentally and theoretically that, due to energy conservation, energy from the particles is given to the waves, causing the particle distributions to follow bispherical shells in velocity space, centred on the upstream and downstream waves. This can cause particle acceleration as well as deceleration and, contrary to expectation, comets are good particle accelerators.

The solar wind is slowed due to the added mass (it is 'mass-loaded'). This leads to draping of the magnetic field around the comet, as predicted by Alfvén, as the plasma is frozen to the flow. If the slowing proceeds rapidly enough a bow shock is formed in the flow. This was observed at the three comets visited by spacecraft so far, although in some cases the boundary was a 'wave' rather than a shock. Because the cometary ions have a gyration radius much larger than that of the solar wind particles, cometary bow shocks are the most complex in the solar system.

The regions inside the bow shock have provided several surprises and revealed boundaries not predicted by models. Nearest to the nucleus, however, a predicted boundary did appear where the magnetic field plummeted to zero: this is the cavity boundary. The number of cometary ions was so high here that the magnetic field was excluded completely.

Following the spacecraft encounters we have a detailed understanding of some of the physics at work in this interaction. The obstacle to the solar wind is very diffuse and dependent on the outgassing rate of the nucleus and its position in the solar system. At small comets we even discovered non-gyrotropic ring distributions.

After the spacecraft encounters, an exciting discovery was made that comets produce X-rays. The best explanation for this appears to be that the X-rays are due to the decay of excited states formed because of charge exchange between heavy solar wind ions and the cometary ions.

Questions on solar-wind–comet interaction

How permanent are the various plasma boundaries? Exactly how does the mass loading slow the solar wind? Is particle acceleration from the diffusing cometary ions enough or are other mechanisms needed? How does the cometary tail form, how does the magnetic cavity connect to it and what is the importance of tail rays seen in remote observations? Could we fly along a comet's tail to better understand it? What can comets tell us about instabilities in Earth-bound fusion machines or about astrophysical phenomena such as supernova explosions?

3.9.3 *Effect of charged particles on surfaces*

The effect of the solar wind in the region surrounding planetary bodies is an interesting topic, as we have seen. However, the effects are also of interest in the study of planetary surfaces. Prolonged exposure to the solar wind, or magnetospheric particles, and cosmic radiation, can have important effects. There are many processes, such as sputtering, secondary electron production, and secondary ion production at the microscopic scale, which can have macroscopic consequences. Examples of this are charging, surface conductivity changes, surface colour change, production of exospheres, and build-up of solar wind gases in significant quantities on surfaces (e.g. the Moon).

Such processes can happen at bodies with no atmosphere and no, or weak, magnetic fields, immersed in solar wind or magnetospheric plasma. Over long times this can lead to significant changes in the surface properties.

3.10 Missions

During our review we identified some important questions to be answered for each class of body or region in the solar system. To answer these questions, and others that will arise as some are answered, further space missions will be essential. Remote sensing techniques from the ground or from Earth orbit are unlikely to have sufficient resolution, the ability to penetrate clouds at the target or be able to see the far side of objects. In addition, the *in situ* measurements of plasma, dust, composition and direct sampling cannot be done remotely at all. Most of the questions and studies highlighted here play an important part in answering why humankind has evolved on the Earth. Some are directly related to the possible existence of life elsewhere. Answering these questions is thus of important cultural value as well as purely scientific curiosity.

Table 3.1 shows a list of past missions and those approved for the next decade or so. The natural sequence of solar system missions involves four stages: (1) initial reconnaisance by flyby, (2) detailed study by orbiters, (3) direct measurement of atmosphere or surface via entry probe and (4) sample return. The stage we have reached for each body is shown in Table 3.1.

The approved programme includes a mission to pursue the exploration of Mercury; an important series of missions to Mars culminating in *in situ* searches for life and sample returns to the Earth; the exploration of two possible future sites for natural life, namely Europa and Titan; an in-depth exploration of Saturn's system;

Table 3.1

Object	Past missions	Stage	Future missions (approved)
Mercury	*Mariner 10*	1	*ESA BepiColombo, NASA Messenger*
Venus	*Mariner, Pioneer Venus, Venera, Vega, Magellan*	3	*Japan ESA, Venus Express*
Earth	Many	n/a	Many
Moon	*Luna, Ranger, Surveyor, Zond, Apollo, Clementine, Lunar Prospector*	4	*Japan Lunar-A, Selene, ESA SMART-1*
Mars	*Mars, Mariner, Viking, Phobos, Pathfinder, Global Surveyor, Odyssey*	3	*NASA Mars Exploration Rovers, Japan Nozomi, ESA Mars Express, Beagle 2, Mars Reconnaisance Orbiter*
Jupiter	*Pioneer, Voyager, Galileo, Ulysses*	3	–
Saturn	*Pioneer, Voyager*	1	*NASA-ESA, Cassini–Huygens*
Uranus	*Voyager*	1	–
Neptune	*Voyager*	1	–
Pluto	–	0	*New Horizons Pluto*
Asteroids	*Galileo, NEAR, DS1*	1, 2, (3)	*Muses-C, Dawn*
Comets	*ICE, Sakigake, Suisei, VEGA, Giotto, DS1*	1	*Stardust, Rosetta, Deep Impact*
Sun + i/p medium	*WIND, ACE, ISEE, AMPTE SMM, Yohkoh, SoHO, TRACE, IMAGE, Cluster*	n/a	*Genesis, Solar-B, STEREO, MMS, Solar Orbiter, SDO, LWS, Double Star*

the first reconnaisance of the Pluto system; asteroid and comet landers and sample return missions; solar wind sample return.

This is an exciting and vibrant programme. In planning an exploration strategy, it is important to consider the balance between small spacecraft that can be constructed quickly and cheaply, but have limited capability, and larger, more expensive initiatives, which may involve multiple spacecraft operating in a coordinated fashion. One example of the latter approach is the *Cassini–Huygens* mission. Its strength is in its multidisciplinary approach: we are only likely to understand the complexities of Titan's atmosphere by using several techniques, the *in situ Huygens* probe and several different types of measurement from the *Cassini* orbiter.

One factor in solar system exploration is the time taken to get to the target. *Cassini–Huygens* took eight years to build and the flight time to Saturn is another seven years. While some opportunistic science was possible *en route*, in this case flybys of Venus, the Earth and Jupiter and measurements of the distant solar wind, a significant proportion of the careers of the scientists involved will have elapsed before measurements at Saturn are made. To get a reasonable payload to Pluto would take much longer, although a much smaller payload with limited capability could be propelled towards Pluto at a much faster rate, as in the *New Horizons Pluto* mission concept.

It is clear, however, that for detailed studies with large payloads, more rapid trajectories and advanced propulsion systems are needed, at least to explore the outer solar system. Ion propulsion or, when near the Sun, light-driven designs may be appropriate. Another problem for missions to the outer solar system and beyond is the provision of electrical power so far from the Sun. The only solution appears to be nuclear power.

If spacecraft within the solar system take a long time to reach the destination, is it realistic to consider missions beyond? Using current spacecraft technology, assuming the same speed as *Voyager* (3.3 AU per year), missions to the heliopause at 150 AU would take over 50 years, the Oort cloud at 50,000 AU would take at least 15,000 years, and Proxima Centauri, our nearest star at 4.2 light years away, would take about 80,000 years. For the time being, remote sensing seems to be the only tractable technique for the stars!

As we find out more about each object or region, further questions will be raised. It will also be necessary to explore further afield. Even if firm evidence is found for past life on Mars, we will need to understand how common the occurrence of life is in the Universe, hence the search for extra-solar planets. We need to understand our own solar system properly before extrapolation is possible.

We are making good use of space for peaceful meteorological, communications and positioning reasons on the Earth. Uses of space may also reach further into the solar system in the new millennium: for example, there may be economic sense in asteroid mining. Another example is mining the Moon for helium. Because the Moon is unmagnetised, the solar wind can impact it directly. The magnetic field diffuses relatively rapidly but the particles are buried in the regolith. Over billions of years, there may be enough ^{3}He buried to make mining of this isotope worthwhile for use in future fusion reactors on the Earth.

We may speculate on possible targets for robotic missions for the new millennium as follows: constant monitoring of the Sun and solar wind as part of an integrated space weather forecasting system; constant monitoring of weather on Mars, Venus and Jupiter to improve models; detailed explorations of Mercury and Pluto; return to sites of earlier exploration with better instruments, new ideas and atmospheric probes in multiple locations; sample return from nearby solar system bodies following detailed mapping and *in situ* composition measurements; exploration of Uranus and its extraordinary magnetosphere; explorations of outer planetary moons; investigation of the feasibility of asteroid mining; terminal shock, heliopause, heliospheric bow shock exploration; Oort cloud exploration; investigation of the feasibility of sending spacecraft to nearby stars and planetary systems.

3.10.1 *Planetary mission example* — Cassini–Huygens

As an example of a planetary mission we consider the *Cassini–Huygens* mission (Figure 3.23 — Plate 6) to Saturn (Figure 3.24). The *Cassini* orbiter represents the last of NASA's big planetary missions at the present time, while *Huygens* fitted nicely as a medium-sized ESA mission.

The scientific objectives, once *Cassini–Huygens* arrives at Saturn after its long journey, following gravity assists at Venus (twice), the Earth and Jupiter (see Figure 3.25), can be split into five categories:

(1) Saturn — atmosphere
(2) Saturn — rings
(3) Titan
(4) Icy satellites
(5) Magnetosphere

Clearly *Huygens* will contribute substantially to No. 3, but the *Cassini* orbiter during its four-year tour will contribute to all five. There are many exciting questions in each area, such as:

(1) What is the structure of the planet? Why does Saturn emit 80% more heat than incident on it from the Sun? Why do winds blow at 2/3 the speed of sound at the equator? What colours the clouds and why is this less marked than in Jupiter?

(2) What is the composition of the ring particles? How do the dynamics work, and what is the role of gravity, shepherding, electrostatics, collisions, etc.? How did the rings form?

(3) Does the cold temperature and atmospheric constitution mean that there is methane–ethane precipitation on Titan? What is the state of the surface? What atmospheric layers does this moon have and why is the surface pressure so high at 1.5 bar? How does the photochemistry work in the hydrocarbon haze? What is the role of magnetospheric electrons in the temperature balance in the ionosphere and atmosphere? Does Titan have a magnetic field?

(4) Why are Saturn's satellites so diverse? Why are some of the icy surfaces craters, and others reformed? Why are some surfaces, such as half of Iapetus, dark? Do any have a magnetic field?

(5) Is Saturn's magnetosphere intermediate between Jupiter (rotation-driven) and the Earth (solar-wind-driven? What difference does size make in the convection system? What is the interaction with the rings, and may it explain the ring spokes? What is the interaction with Titan, in the magnetosphere and when in the solar wind?

Figure 3.24   Spokes in Saturn's B ring as seen in *Voyager* images of scattered light. The size of the individual scattering particles is close to the wavelength of light ($<$ 1 micron). (NASA)

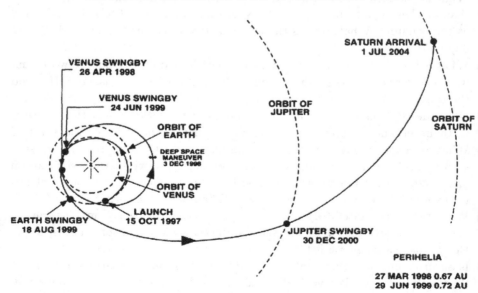

Figure 3.25   Interplanetary trajectory of *Cassini–Huygens*. (NASA)

To answer these questions and others, the orbiter carries 12 instruments: 6 remote sensing and 6 *in situ*. The *Huygens* probe also carries 6 instruments. There is UK participation in 6 of the orbiter instruments and 2 on the probe.

Arrival at Saturn in 2004 will provide a wealth of data to the worldwide planetary community. The results will revolutionise our knowledge of Saturn and its system, and will tell us more about the origin of the solar system.

3.11   Other Solar Systems

One of the most fundamental questions facing mankind at the dawn of the new millennium is: Has there been or could there be life elsewhere? The question has been haunting us for at least two millennia, but the answer is now closer given recent developments in technology, which are overcoming the vast difficulties for our observations. The answer will have profound and exciting implications for scientific, cultural, philosophical and religious thinking. Most scientists in the field now believe there was insufficient evidence for NASA's 1996 announcement of early life on Mars. Nevertheless, the announcement of the result was introduced by the President of the United States, generated huge interest and drew comment from leaders in many fields in addition to science. This illustrates the importance of the answer to humankind. Without life elsewhere, we feel alone and isolated in the Universe. Whether life has or has not evolved elsewhere yet, we would like to know why.

As we look into the night sky, at the 3,000 or so visible stars, it is natural for us to speculate on whether they harbour life. The knowledge that our galaxy alone is some 100,000 light years across, contains some 200 billion stars, and that the Universe contains some ten times as many galaxies as the number of stars in our Milky Way, makes the presence of life elsewhere seem possible and indeed likely. Attempts have been made to quantify this, and the Drake equation is still the best way of writing down the number of contemporaneous civilizations in our galaxy as the product of seven terms: the formation rate of stars, the fraction of those with planets, the number per solar system that are habitable, the fraction where life has started, the fraction with intelligence, the fraction with the technology and will to communicate, and the lifetime of that civilisation. Reasonable estimates for all of these parameters from astronomy, biology and sociology lead to values between zero and millions of civilizations within our own galaxy. Although we only know the first term with any accuracy as yet, my guess would be towards the higher end. But without firm evidence it is impossible that we will ever be able to determine most of these terms accurately enough, and the final answer will remain indeterminate by this approach. Unfortunately it seems that we are in the realm of speculation rather than science, creating a ripe area for science fiction.

Attempts to detect life elsewhere have included passive searches for organised signals and active approaches including the transmission of powerful microwave bursts and attaching plaques to the *Pioneer* and *Voyager* spacecraft leaving the solar system. Despite recently enlisting the assistance of much computing power around the world, the passive search approach feels forlorn, but is still worth a try. As for our own electromagnetic transmissions, they have reached only a tiny part of our galaxy in the 100 years since radio was invented and in the 26 years since the Arecibo telescope was used to shout to the cosmos 'We are here!' But we cannot get around the inverse square law for intensity of electromagnetic waves, making detection of our signals highly challenging — and in any case it would take a significant time for us to get an answer. Discounting an enormous stroke of luck, a more scientific and systematic approach is likely to be via the detection and analysis of extrasolar planets by remote sensing. The technology for this is just becoming available.

In the last few years we have seen the first tantalising evidence for solar systems other than our own, in different stages of formation. In 1984, observations of the star Beta Pictoris showed it to be surrounded by a cloud of gas and dust reminiscent of early solar system models. Other examples of dust–gas discs around stars have been detected recently using the Hubble Space Telescope. Have planetesimals, or comets, formed near these stars? Are observed structures in the cloud evidence for forming planets? We should begin to be able to answer these questions soon, and we should find many other examples.

So far, planet-hunters have found firm evidence for over 100 planets around Sun-like stars, and several additional candidates. There are even a handful of multiple planet systems.

The main technique used for this is to detect the anomalous motion, or 'wobbling', of the companion star, using large ground-based telescopes, although in one reported case to date an occultation, or reduction of the companion star's light due to the object crossing its disc, was detected. The planets are all inferred to be massive gas giants; most are several times heavier than Jupiter and the smallest so far is the size of Saturn. All orbit closer to the star than the giant planets in our own solar system, the main hypothesis for this being that they form further out and lose orbital energy via friction with the remaining gas and dust. Multiple large gas planets around another star have also been inferred from observations. One problem in the identification of any extrasolar planets is to distinguish them from 'brown dwarf' companions, failed stars, which were too light for fusion to start. Telescope technology is advancing, and dedicated observational space missions using interferometry have been proposed and will be flown within the next decade. We must choose the most promising stars to observe, starting with stars like our own Sun. We can expect that the present catalogue of 100 will increase soon and, significantly, the observable planetary size will decrease, becoming closer to the Earth's size, accurate distinction techniques between planets and brown dwarves will be developed, and statistics will be built up on size distributions and orbits. We will ultimately know whether our own solar system is unique.

Given the likely existence of small rocky planets elsewhere, and the fact that we are most unlikely to have the technology to make *in situ* measurements any time soon, we will need to establish methods for detecting the presence of life on these objects using remote sensing. Using data from the *Galileo* spacecraft during its Earth swingbys, Carl Sagan and colleagues detected the presence of life on the Earth using atmospheric analysis and radio signal detection. The challenge is enormously greater for the planets of remote stars. However, we may speculate that spectroscopic observations of ozone, oxygen and methane may hold the key, at least to finding clues for life as we know it. It will be difficult to prove conclusively that life exists there. However, surely amongst the many planets elsewhere there must be at least some which have similar conditions to the Earth? The accepted critical planetary properties for life to emerge are size, composition, stellar heat input and age. To this list we should add the presence of a magnetic field to protect against stellar winds and radiation.

Within our own solar system, where *in situ* exploration is possible, it is certainly worth searching for evidence of life on Mars during its early, warmer, wetter history. Several space missions are planned for this. In addition, two other possible sites for life within our own solar system sites should be reiterated here and explored. First, Europa may be a possible present site for life, in its liquid water ocean underneath an icy crust, deep enough to be shielded from Jupiter's powerful radiation belts. Second, Titan is a potential future site for life when the Sun exhausts its hydrogen fuel, becomes a red giant and warms the outer solar system. But, as well as looking for tangible clues within our own solar system, it is clear that we must make comparison with other solar systems and broaden the search.

3.12   Conclusion

In summary, there are many exciting and challenging ways we can explore solar systems in the new millennium: our own with *in situ* studies and beyond with remote sensing. The answers to be gained are fundamental to a better understanding of our place in the Universe.

Further Reading

Lewis, J.S., 1997, *Physics and Chemistry of the Solar System*. London: Academic Press.
Beatty, J.K., Petersen, C.C., and Chaikin, A., 1999, *The New Solar System*. Cambridge: Cambridge University Press.
Stern, S.A., 1992, 'The Pluto–Charon system', *Ann. Rev. Astron. Astrophys.* **30**, 185–233.

Chapter 4

Space Plasma Physics — A Primer

Christopher J. Owen

4.1 What Is Space Plasma Physics?

Space plasma physics is the discipline that seeks to understand our local space environment. This has particular application to understanding how mass, momentum and energy from the solar corona are transferred through the solar wind and coupled into the environments surrounding the Earth, the planets and other solar system bodies. It is, however, a basic science discipline. Most space plasma physical regimes cannot be mimicked in a laboratory set-up, yet these regimes have fundamental relevance to many wider astrophysical phenomena. The key to this discipline is that many of the theories and models underpinning our understanding of these environments have been directly testable or constrained by use of *in situ* observations by many satellites since the dawn of the space age some 45 years ago. This sets the discipline apart from many other space and astrophysical sciences, in which only remotely sensed observations are currently possible.

In this chapter we briefly explore some of the fundamental properties of space plasmas, and discuss some of the applications to the near-space environment of the Earth and other solar system bodies. However, the reader should be aware that many more comprehensive texts exist for this field, as well as an extensive research literature. Although knowledge of physics to undergraduate level is assumed, this chapter is intended only as an introduction for the novice or near-novice to the field of space plasmas.

4.2 So What Is a Plasma?

A plasma is a *quasi-neutral* gas consisting of positively and negatively charged particles (usually ions and electrons) which are subject to electric, magnetic and other forces, and which exhibit *collective behaviour*, such as bulk motions, oscillations and instabilities.

In general, ions and electrons may interact via short range atomic forces (during collisions) and via long range electromagnetic forces due to currents and charges in the plasma. (Gravitational forces may also be important in some applications.) In most space plasma applications, the plasma density is sufficiently low that collisions between individual particles are rare (the mean free path is extremely long compared to the size of the plasma system). It is thus the long range nature of the electromagnetic forces acting on and within it that allows the plasma to show collective behaviour. The simplest plasma is formed by ionisation of atomic hydrogen, forming a plasma of equal numbers of (low mass) electrons and (heavier) protons. However, plasmas may also contain some heavier ions and neutral particles (which interact with charged particles via collisions or ionisation). This is true, for example, of the terrestrial ionosphere which contains a significant fraction of both ionised oxygen and non-ionised particles from the upper atmosphere. Mathematically, an ionised gas of number density n_0 and temperature T can usually be considered a plasma if the following conditions hold:

$$\lambda_{\mathrm{D}} = \left[\frac{\varepsilon_0 kT}{e^2 n_0}\right]^{\frac{1}{2}} \ll L, \quad N_{\mathrm{D}} = \frac{4}{3}\pi\lambda_{\mathrm{D}}^3 n_0 \gg 1, \quad f_{\mathrm{CN}} \ll f_{\mathrm{Pe}} = \frac{1}{2\pi}\left[\frac{e^2 n_0}{m_e \varepsilon_0}\right]^{\frac{1}{2}},$$

where e and m_e are the charge and mass of the electron, k is Boltzmann's constant, ε_0 is the permittivity of free space and L is the scale length of the plasma system. The parameter λ_{D}, the *Debye length*, is the distance over which a plasma is able to shield out a charge imbalance within it, provided the number of particles N_{D} within a sphere of this radius is high. The ability to react to charge imbalances by providing an electromagnetic restoring force means that the electrons (which are lighter and more mobile than the ions) have a natural oscillation frequency, the *electron plasma frequency*, f_{Pe}. In order that the plasma continues to exhibit collective behaviour, the frequency of collisions of plasma particles with neutrals, f_{CN}, must be small compared to the electron plasma frequency.

4.3   The Realm of Plasma Physics

Space plasma physics can, of course, be thought of as a sub-discipline of plasma physics itself. However, the realm of plasma physics covers a huge range of parameter space, such that individual sub-disciplines can be rather disparate in their controlling physics. Figure 4.1 shows various plasma populations as a function of electron density and temperature. Note that the density scale covers more than 30 orders of magnitude, and the temperature scale about 10 orders of magnitude. Lines of constant λ_{D}, N_{D} and f_{Pe} are superposed on the plot to indicate regions in which the plasma conditions above are valid. Limits on laboratory vacuums mean that the laboratory plasmas appear at the high-density region of the figure. Most of the

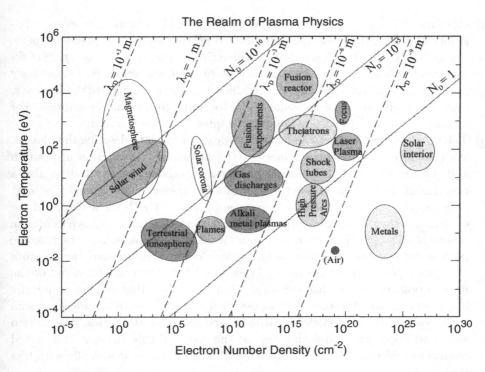

Figure 4.1 The realm of plasma physics. Note that space plasmas (solar wind, magnetosphere, ionosphere) fall in the low number density region of the plot, beyond the limits of laboratory vacuums.

(visible) universe is in the form of plasma: plasmas form wherever temperatures are high enough, or radiation is strong enough to ionise atoms. Examples of astrophysical plasmas are stellar interiors, atmospheres and winds, cosmic rays, interstellar medium, jets in active galaxies, pulsars and their magnetospheres and stellar accretion disks. Plasma regimes relevant to the study of space plasma physics, and thus to this chapter, include the solar atmosphere and solar wind, the Earth's (and other planets') magnetosphere and ionosphere (above 60 km for the Earth) and cometary comas. For the most part these plasmas, which we will consider more closely below, are located in the low-density region of Figure 4.1.

4.4 Ways to Understand Plasmas

Before we look in detail at the structures and dynamics of the space plasma regimes relevant to this chapter, it is useful to introduce a number of concepts pertaining to the underlying physics controlling such dynamics. In particular we need to understand how plasmas, and the individual particles within them, behave under the influence of the magnetic and electric fields (and other forces) that act upon them:

(a) Of course, the most 'exact' way to specify the state of a plasma is to give the positions and velocities of *all* the particles and *all* the fields at *all* points in space. For a system of N particles, the particle description would be a *phase space* of $6N$ dimension (phase space is the combination of *configuration space*, i.e. ordinary position, and *velocity space*, each of which has three components). However, even in low-density space plasmas, N is too large to use this description. For example, $N \sim 10^{15}$ in a 1 km cube in the interplanetary medium.

(b) The next-best method is to adopt a statistical approach to describe the plasma. We can define a *distribution function*, $f(x, v)$, which is a function of position and velocity in a six-dimensional phase space (although generally it is also a function of time). The number of particles in an elemental volume $d^3x d^3v$ at position (x, v) in phase space is then given by $f(x, v)d^3x d^3v$. The theory of the evolution of the distribution function through space and time is known as *plasma kinetic theory* (e.g. the Vlasov equation for collisionless plasmas). Macroscopic parameters such as plasma density, bulk velocity, pressure and temperature at a given point can subsequently be deduced by taking the moments of the distribution function. Although it is the best method that can be practically implemented, and the only one to use to understand more advanced plasma topics such as wave–particle interactions and plasma instabilities, it is still an advanced approach to use and beyond the scope of this chapter. Interested readers are referred to more advanced textbooks that deal specifically with this topic.

(c) A third (simpler) approach is to treat the plasma as a conducting fluid, and adapt the equations of fluid dynamics to include the effects of electric and magnetic forces. The resulting theory is called *magneto-hydrodynamics* (MHD). MHD can be derived from the kinetic theory by applying the appropriate (low-frequency) limits, but this involves averaging out almost all the kinetic effects. MHD is applicable to most space plasma regimes, but many interesting effects, including some which have critical importance in controlling the dynamics of the plasma, cannot be explained using this approach. The physics of the regions in which the MHD assumptions break down are thus at least as interesting as those in which it holds. The basis of MHD and its applicability are discussed in full in Chapter 9 of this book, and thus will not be covered in detail here.

Finally, a basic understanding of some of the behaviour of space plasmas can be gleaned from a knowledge of the behaviour of the individual charged particles making up the plasma, and how they respond to the large-scale electromagnetic fields in space. We will thus now briefly consider the motions of individual charged particles in these fields.

4.4.1   *Single particle dynamics — basic principles*

A single particle of charge q moving with a velocity v within an electric field $\mathbf{E}$ and magnetic field $\mathbf{B}$ is subject to the Lorentz force, $\mathbf{F}_{\mathrm{L}} = q(\mathbf{E} + v \times \mathbf{B})$. If we neglect to begin with the electric field ($\mathbf{E} = 0$), and assume that B is uniform, then it can be seen from this equation that the particle experiences a force which is perpendicular to both its direction of motion and the magnetic field direction. The component of the particle velocity which is parallel to the magnetic field, $v_\parallel$, is thus unaffected by the force, and the particle has a uniform motion along this direction. However, in the plane lying perpendicular to the field, the force acts to move the particle in an orbital motion about the field line. The speed of the particle around the orbit remains constant at $v_\perp$, the component of the particle velocity perpendicular to the field. This orbital motion is known as *gyromotion*. The radius, or *gyroradius*, or *Larmor radius* r_L of the motion about the field and the *gyro-* or *cyclotron frequency*, Ω_L, of this motion are given by the expressions

$$r_L = \frac{mv_\perp}{qB}, \quad \Omega_L = \frac{qB}{m}.$$

Since these expressions and the force $\mathbf{F}_L$ depend on the mass m and on the charge of the individual particles q, ions and electrons gyrate around the field in opposite senses and the gyroradius of the ions is much larger than for the electrons (and correspondingly the gyrofrequency much higher for electrons). Figure 4.2 (not to scale) summarises the basic motion of ions and electrons which have free motion at speed $v_\parallel$ along the magnetic field direction, and orbital motion at speed $v_\perp$ perpendicular to the field. Since the motion decouples in the parallel and perpendicular direction, it is useful to define the particle pitch angle, $\alpha = \tan^{-1}(v_\perp/v_\parallel)$, as a measure of whether the particle moves predominantly along the field or perpendicular to it. (Particles with $\alpha = 0°$ move exactly parallel to the field with no gyration, $\alpha = 90°$ particles gyrate around the field without moving along it, and $\alpha = 180°$ particles move exactly antiparallel to the field with no gyration. Intermediate pitch angles show a combination of the two types of motion.)

The basic motion of particles in a uniform magnetic field described above is disturbed if there are gradients in the field seen by the particle during its gyro-orbit,

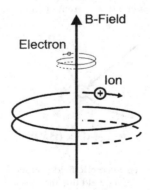

Figure 4.2 Basic particle motion in a uniform magnetic field with no electric field. Ions and electrons can move freely along the field, but gyrate around the field direction in opposite senses. The radius of gyration is much larger for ions. (Figure not to scale).

or if a force with a component perpendicular to the field direction acts upon it. (Forces acting parallel to the field will simply accelerate the particles in this direction.) Forces acting perpendicular to the field accelerate and decelerate the particle as it moves around its gyro-orbit. The associated change in speed of the particle causes a change in the gyroradius around the orbit, the net effect of which is that the particle no longer moves on a closed path in the plane perpendicular to the field. The presence of the force or gradient introduces a drift in the direction

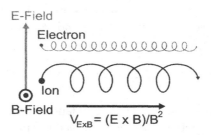

Figure 4.3   (a) Motion of ions and electrons in crossed electric and magnetic field — the *E*-cross-*B* drift. Ions and electrons drift with the same speed in the same direction, perpendicular to both the *E*-and *B*-fields. (Figure not to scale.)

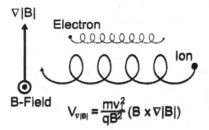

Figure 4.3   (b) Drifts of ions and electrons in a perpendicular magnetic field gradient. The gyroradius of both particles reduces in the higher field region, resulting in a drift in the direction perpendicular to both **B** and $\nabla|\mathbf{B}|$. This drift depends on particle charge, so ions and electrons drift in opposite directions, and thus create a current. (Figure not to scale.)

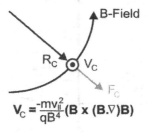

Figure 4.3   (c) Particle drifts in curved magnetic field geometry. The centrifugal force $\mathbf{F}_C$ associated with the particle motion along a curved field line induces a drift in the direction perpendicular to both the radius vector $\mathbf{R}_C$ and the local magnetic field direction. The drift is charge-dependent, so ions and electrons drift in opposite directions, giving rise to a current.

perpendicular to both the magnetic field and the force/gradient. Mathematically, the drift velocity $\mathbf{v_F}$ due to a general force $\mathbf{F}$ is given by $\mathbf{v_F} = (\mathbf{F} \times \mathbf{B})/qB^2$. In general, this is charge-dependent, so that ions and electrons drift in opposite directions. One exception to this is the electric, or E-cross-B, drift, which is associated with the electric force $\mathbf{F} = q\mathbf{E}$. In this case the charge terms cancel, such that both ions and electrons drift with velocity $\mathbf{v}_{E \times B} = (\mathbf{E} \times \mathbf{B})/B^2$, perpendicular to both the magnetic and electric fields. Figure 4.3(a) illustrates the E-cross-B drift motion of ions and electrons in the plane perpendicular to the magnetic field under the influence of an electric field pointing vertically up the page. Recall that in general the ions and electrons also have a free motion along the field direction (i.e. out of the plane shown).

Although the charged particles will drift under the influence of any number of possible forces (e.g. gravity in some applications) and gradients, there are two other drift motions that have particular application in space plasmas. These are the *magnetic gradient* and *magnetic curvature drifts*:

(a) The magnetic gradient drift arises when the particles gyrate in a magnetic field that has a significant gradient perpendicular to the field direction. In this case, the particles are influenced by a field strength that changes as they move around their gyro-orbit. Since the gyroradius is inversely proportional to the field strength, the local radius of curvature for the particle is higher in the low-field portion than in the high-field portion of its orbit. Again, this results in the particle gyration around a non-closing orbit. The resulting drift velocity $\mathbf{v}_{\nabla B}$ is given by

$$\mathbf{v}_{\nabla B} = \frac{mv_\perp^2}{qB^3}(\mathbf{B} \times \nabla B).$$

This drift is charge-dependent, so that ions and electrons drift in opposite directions when there is a magnetic field gradient perpendicular to the field. Consequently, there is a current associated with this drift, as would be expected in the vicinity of a gradient in the field (Ampere's law suggests $\nabla \times \mathbf{B} = \mu_0 \mathbf{j}$). The motion of ions and electrons in such a magnetic field gradient is illustrated in Figure 4.3(b).

(b) The magnetic curvature drift arises when particles have a parallel component of velocity that takes them along a magnetic field line that has a local curvature. Under these circumstances, the particles experience a centrifugal force that acts along the local radius of the curvature vector, and therefore perpendicular to the field direction (see Figure 4.3(c)). On the basis of the above discussion, we thus expect the particles to experience a drift in the direction perpendicular to the magnetic field and local radius of curvature (i.e. normal to the plane of Figure 4.3(c)). Analysis of the geometry of the field and forces leads to an expression for the drift velocity, $\mathbf{v_C}$, associated with the magnetic curvature given by

$$\mathbf{v}_C = \frac{-mv_\parallel^2}{qB^4}(\mathbf{B} \times (\mathbf{B} \cdot \nabla)\mathbf{B}).$$

Again, this drift is charge-dependent, such that the separation of ions and electrons constitutes a current. In practice, it is generally not possible to have large-scale curvature within the magnetic field without also having a gradient in that field. For this reason, the gradient and curvature drifts often operate simultaneously, such that a combined 'magnetic' drift velocity $\mathbf{v}_M = \mathbf{v}_{\nabla B} + \mathbf{v}_C$ should be considered.

Finally, it is useful to examine how the motion of the individual particles is affected by gradients in the magnetic field, which lie along the field direction. Such gradients arise naturally, for example in a planetary dipole field, where the field strength on the portion of a given field line crossing the planetary surface near the magnetic pole is much higher than the portion of the same field line crossing the magnetic equator at rather higher altitudes. A useful concept to introduce in order to understand the behaviour of such particles in such a field is the *magnetic moment*, μ, associated with the gyromotion of the individual particle. This is given by the expression

$$\mu = \frac{1}{2}mv_\perp^2/B,$$

i.e. the perpendicular kinetic energy divided by the local field strength. This is also known as the *first adiabatic invariant*, since it can be considered a constant of the particle motion provided that variations in the field that affect the particle are slow compared to the particle gyroperiod. In essence, the invariance of this parameter indicates that the particle motion responds to an increasing magnetic field strength in such a way as to keep the total magnetic flux enclosed within the gyro-orbit at a constant value. In the absence of other forces accelerating the particle parallel to the field, the total kinetic energy is also conserved. Hence a particle that moves along the field direction into a region of higher field strength must increase its perpendicular kinetic energy at the expense of the parallel kinetic energy. The perpendicular velocity, $v_\perp$, increases while the parallel velocity $v_\parallel$ decreases. In other words, the particle pitch angle $\alpha \to 90°$. Eventually, all of the particle energy is shifted to the perpendicular direction, such that motion along the field line ceases, and then reverses, such that the particle returns to the regions of lower field strength. The parallel energy then increases at the expense of the perpendicular. This process, illustrated in Figure 4.4, is known as *magnetic mirroring*, since the net effect is for the particle to be reflected at a given point where the magnetic field reaches a particular value. In effect, the particle experiences a force $F_s = -\mu dB/ds$, along the field. This force acts in the direction opposite to the field gradient dB/ds (s is defined as the coordinate along the field direction). For a particle observed with pitch angle α_0 where the field strength has value B_0, the point of particle mirroring

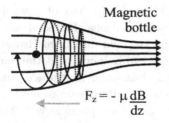

Figure 4.4   Magnetic mirroring — particles moving along magnetic field lines into regions of higher field strength experience a repulsive force which causes them to 'mirror' back towards regions of lower field strength. The magnetic moment, μ, is conserved if the field gradient is not steep.

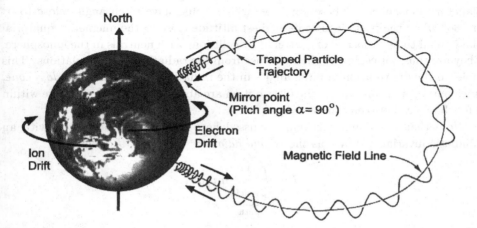

Figure 4.5   The motion of charged particles on dipole-like field lines in near-Earth space. Particles gyrate around the field, bounce between low-altitude mirror points in both the north and south hemispheres and drift around the Earth. Ions and electrons drift in opposite directions, thus contributing to a 'ring current' around the Earth. Particles are trapped on such field lines, so these are the main processes leading to the formation of the Van Allen radiation belts. (After Spjeldvik and Rothwell, 1985.)

will occur when the magnetic field reaches strength $B_M = B_0/\sin^2\alpha_0$ (at which point the particle pitch angle $\alpha = 90°$).

4.4.2   *Single particle dynamics — application*

The principles of single particle motion discussed above put us in a position to understand much of the behaviour of particles in the inner regions of planetary magnetospheres (the region of space occupied by the magnetic field of planetary origin). Particles in this region are energetic and trapped in torus-shaped regions about the planet. In the Earth's case, these *radiation belts*, first observed by James Van Allen in 1958, were among the earliest discoveries of the space age. Figure 4.5 indicates the motion of trapped particles in these regions, in which the magnetic field is relatively close to the form expected of a vacuum magnetic dipole field. The particles perform the basic motions discussed above. Firstly, each particle performs a gyromotion around the field direction. However, since the field strength seen by the particles tends to steadily increase as the particles move along the field lines to higher latitudes (and thus lower altitudes), conservation of the magnetic moment implies that the particles will migrate to pitch angles closer to 90° as they move into these regions. In fact, each particle will 'bounce' up and down along the field line between the two mirror points, one in each hemisphere, at which its pitch angle $\alpha = 90°$. The dipole-like field lines thus create a 'magnetic bottle' in which the particles are confined on the field line between the two mirror points. The exception to this is for particles whose mirror points are below the top of the planetary ionosphere. These particles, which would have pitch angles close to 0° or 180° when the particle is at its highest altitude close to the magnetic equatorial plane, tend to be absorbed or scattered by collisions with neutrals in the ionosphere. They are thus noticeable by their absence from the radiation belt populations. This 'hole' in the distribution of pitch angles in the radiation belt is called the *loss cone*, since particles in this part of the distribution are absorbed by the ionosphere within a few bounces between mirror points.

Note that the bounce motion discussed here can also be associated with an adiabatic invariant, known as the *second adiabatic invariant*, J, where

$$J = \oint_{\substack{\text{particle} \\ \text{bounce} \\ \text{path}}} mv_\parallel ds \,.$$

Provided that the magnetic field structure does not change significantly over the period that the particle takes to bounce at each mirror point and return to its starting point, this is also a constant of the particle motion.

Particles in the radiation belt region are also subject to drifts arising from magnetic gradients and curvature. Since both the main gradient and the local radius of curvature tend to be confined in the radial and/or latitudinal directions, these drifts are predominantly in the azimuthal direction around the Earth. These drifts thus create the torus-shaped regions of energetic particles that make up the radiation belts. Moreover, since these drifts are charge-dependent, the electrons drift from West to East, while the ions drift from East to West. The oppositely directed orbits of ions and electrons around the Earth also contribute to a loop of current, known as the *ring current* circling the Earth. The presence of this current does modify the shape of the magnetic field structure in the inner magnetosphere; in particular it acts to reduce the field strength in the equatorial plane in the region inside of the current loop. This can be detected by equatorial ground magnetometers, which can thus remotely sense the strength of the terrestrial ring current, thus providing a monitor of the fluxes of energetic particles in this region. Since the energy content of the ring current is strongly influenced by changes in the solar wind, particularly those driving magnetic storms (see below), monitoring of the ring current is used to produce an indication of the level of magnetic storm activity within the magnetosphere.

The orbital motion of the ions and electrons about the Earth can also be associated with an adiabatic invariant, the *third adiabatic invariant*, which is a constant of the motion provided changes in the field affecting the particle are slow compared to the orbital drift period. This implies that the particle orbit will alter to maintain the total magnetic flux enclosed within the orbit. Thus if strong ring currents act to reduce the field in the inner magnetosphere, the orbits of individual particles will tend to move to larger radial distances.

In practice, the real radiation belts are somewhat more complicated than the simple picture described here. There are inner and outer radiation belt regions (Figure 4.6), separated by a relative void in energetic particle fluxes. It seems that this void is most likely created by high levels of wave activity in this region, which scatters particle pitch angles into the loss cones, such that particles are rapidly lost to the ionosphere. Additional effects are also introduced by asymmetries in the magnetic field structure between the day- and nightside regions of the magnetosphere, and by solar-wind-driven time-dependent changes which can cause violation of one or more of the adiabatic invariants discussed above. Again, a complete discussion of these effects is beyond the scope of this brief introduction, so the interested reader is referred to more comprehensive texts on this subject.

Figure 4.6 The Van Allen radiation belts. Trapping of energetic particles on the dipole-like field lines near the Earth results in torus-like regions, i.e. the radiation belts. A number of distinct regions have been identified: an outer belt (shown as grey) where high electrons predominate, and an inner belt (stippled grey) composed predominantly of protons. Recent SAMPEX observations suggest that an additional belt exists (white), composed mostly of particles of interstellar origin. Gaps, or slots, in the belts are caused by wave–particle resonant interactions that act to scatter particles into their loss cones. (Mewaldt *et al.*, 1996. Reproduced by permission of American Geophysical Union.)

4.5   Space Plasma Applications

4.5.1   *The frozen-in flux approximation*

Before looking in detail at various space plasma regimes, it is useful to recap one particular principle arising from MHD (see Chapter 9). The *frozen-in flux approximation* is a central tool for understanding a wide variety of space plasma applications which will be discussed in subsequent sections. We start from the MHD induction equation describing the evolution of the magnetic field in a plasma with conductivity σ:

$$\frac{\partial \mathbf{B}}{\partial t} = \nabla \times (\mathbf{v} \times \mathbf{B}) + \frac{1}{\mu_0 \sigma} \nabla^2 \mathbf{B}.$$

(a) The first term on the right hand side describes the convection of the magnetic field with the plasma moving at velocity $\mathbf{v}$.

(b) The second term on the right hand side represents diffusion of the magnetic field through the plasma.

If the scale length of the plasma is L, such that the gradient term $\nabla \sim 1/L$, and the characteristic speed is V, then the ratio R_M of the two terms on the right hand side can be written as $R_M \sim \mu_0 \sigma V L$, which is known as the *magnetic Reynolds number*. If $R_M \ll 1$, then the diffusion term dominates, while for $R_M \gg 1$ convection dominates. In a typical space plasma, the conductivity is very high, and the scale lengths large. In the solar wind and the magnetosphere, $R_M \sim 10^{11}$. Hence the diffusion term is negligible in these contexts, and the magnetic field convects exactly with the plasma flow. This is often referred to as the 'ideal MHD limit' or the 'frozen-in flux approximation'. This is an extremely important concept in MHD, since it allows us to study the evolution of the field, particularly the topology of

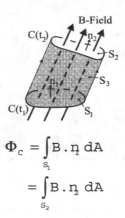

$$\Phi_c = \int_{S_1} \mathbf{B} \cdot \mathbf{n}_1 \, dA$$

$$= \int_{S_2} \mathbf{B} \cdot \mathbf{n}_2 \, dA$$

Figure 4.7   Frozen-in flux approximation 1. A closed contour C encloses a surface S_1 that is threaded by magnetic flux Φ_C. Under the frozen-in flux theorem, the surface may be moved and/or deformed, but will remain threaded by the flux Φ_C at all times.

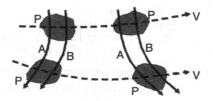

Figure 4.8 The concept of frozen-in magnetic flux 2. In a highly conducting plasma, in which the scale lengths are large, the magnetic field is frozen in. As a result, two volumes of plasma, P_1 and P_2, which are originally linked by fields lines A and B, will remain connected by field lines A and B, whatever the individual motions V_1 and V_2 of the individual volumes.

the field lines, by studying the plasma flow. Of course, the principle also works in the other direction: if we know how the magnetic field lines evolve, then we can deduce the plasma fluid flow. This concept is illustrated in Figure 4.7. A surface S_1 in the plasma bounded by a closed contour C encloses a specific amount of magnetic flux at a given time t_1. The surface may be subsequently deformed and/or relocated by motions of the plasma. However, under the frozen-in flux approximation, the surface will enclose the same magnetic flux Φ_C at a later time t_2 as it did at time t_1. (If the surface is reduced in area we can infer that the field strength will have increased during the convection.) We can also define a *magnetic flux tube* by taking the closed loop and moving it parallel to the field it encloses. The surface, or tube, S_3, thus created has zero flux through it, and consequently the fluid elements that form the flux tube at one moment, form the flux tube at all instants. Alternatively, if two fluid elements are linked by a *field line* (which can be defined as the intersection of two flux tubes) at one instant, then they will always be so linked (Figure 4.8). This property of space plasmas allows a basic understanding of many applications without the need to resort to explicit calculations of plasma motions and electromagnetic field parameters. For example, a strict application of frozen-in implies that there can be no mixing of fields and plasmas from different origins. The field and plasma of solar origin, which flow towards the Earth as the solar wind, are frozen out of the region occupied by field and plasma of terrestrial origin. The latter thus forms a bubble, the *magnetosphere*, which the solar wind is obliged to flow around. Note that the boundary between the two regimes carries a current to support the change in field and plasma properties across the boundary. Paradoxically, these boundaries tend to be very thin, such that R_M is small within them. Under certain circumstances, the magnetic flux may 'thaw' and slippage or diffusion of the field relative to the flow is possible. We will examine these interactions in more detail later in this chapter.

4.5.2   *MHD plasma waves*

The magnetic force $\mathbf{F}_M$ in a MHD plasma is represented in the momentum equation (Chapter 9) by the term $\mathbf{F}_M = \mathbf{j} \times \mathbf{B}$. Using the Maxwell equations in the MHD limit and some standard vector identities

$$\mathbf{F}_M = \frac{1}{\mu_0}[(\nabla \times \mathbf{B}) \times \mathbf{B}] = -\nabla \left[\frac{B^2}{2\mu_0}\right] + \frac{1}{\mu_0}(\mathbf{B} \cdot \nabla)\mathbf{B}.$$

The term on the right hand side can be further manipulated to cast the expression into terms acting perpendicular and parallel to the local magnetic field direction:

$$\mathbf{F}_M = -\nabla_{\perp} \left[\frac{B^2}{2\mu_0}\right] + \frac{B^2}{\mu_0} \frac{\mathbf{R}_C}{R_C^2},$$

where $\mathbf{R}_C$ is the local radius of the curvature vector, which points towards the centre of curvature of the field line. This equation shows that the magnetic force can be resolved into two conceptually simple components:

(a) A force perpendicular to the magnetic field which has the form of a pressure (i.e. it is the gradient of a scalar quantity, $B^2/2\mu_0$);
(b) A force towards the instantaneous centre of curvature that depends on the curvature R_C and the field magnitude B. This is the physical equivalent of a tension force B^2/μ_0 acting along the field lines.

Thus, forcing the field lines together results in an opposing perpendicular pressure force, while trying to bend the field lines results in an opposing tension force. This *magnetic pressure* and *magnetic tension* represent two kinds of restoring force which arise in an MHD plasma due to the presence of the magnetic field. These in turn support a number of wave modes in MHD, in addition to the sound wave that would arise in an ordinary gas. (They also form the basis on which the equilibria and instabilities of various MHD plasma systems may be investigated, although these topics are not covered here.)

The *Alfvén wave* arises in an MHD plasma entirely due to the tension force associated with the magnetic field line. It is essentially a magnetic wave as there is no associated compression of the plasma, as is the case in a sonic (pressure) wave. This wave propagates preferentially along the background field direction (and not at all across it) at the *Alfvén speed*, V_A, where

$$V_A = \frac{B^2}{\sqrt{\mu_0 \rho_0}},$$

B being the background field strength and ρ_0 the background mass density. This wave causes magnetic field and plasma velocity perturbations which are perpendicular to the background field (and the wave propagation vector), and thus is sometimes called the *transverse wave* or *shear wave*. It is analogous to waves on a string under tension.

In the MHD *magnetosonic wave* modes, both the magnetic field strength and the plasma pressure vary. The *fast mode magnetosonic wave* propagates preferentially in the direction perpendicular to the field, when it has a phase speed

$C_{\text{fast}} = \sqrt{C_S^2 + V_A^2}$ (i.e. faster than the Alfvén speed), where C_S is the sound speed in the plasma. In this wave the magnetic field strength and the plasma pressure vary in phase. The *slow mode magnetosonic wave* propagates preferentially along the field (and not at all across it), with a phase speed $C_{\text{slow}} \le V_A$. In this wave the magnetic field strength and the plasma pressure vary out of phase. Since $C_{\text{slow}} \le V_A \le C_{\text{fast}}$ the Alfvén wave is sometimes called the *intermediate wave*. As we will see in subsequent sections, when the plasma flow speeds exceeds the phase speeds associated with the magnetosonic waves, the latter may steepen to form *fast* and *slow mode shocks*.

4.5.3 *The solar wind and IMF*

The solar wind is a supersonic flow of ionised solar plasma and an associated remnant of the solar magnetic field that pervades interplanetary space. It is a result of the huge difference in gas pressure between the solar corona and the interstellar space. This pressure difference drives the plasma outwards, despite the restraining influence of solar gravity. The existence of a solar wind was surmised in the 1950's on the basis that small variations measured in the Earth's magnetic field (*geomagnetic activity*) were correlated with observable phenomena on the Sun (solar activity). The large-scale structure of the solar wind can be understood in terms of an MHD plasma. In 1958 Eugene Parker provided the first MHD solution for the continuous solar wind outflow from the corona — the solar wind can be considered the extension of solar corona into interplanetary space. It was first observed directly and definitively by space probes in the mid-1960's. Since that time we have been able to directly examine the solar wind plasma and the interplanetary magnetic field through *in situ* observations by many spacecraft in this region. Measurements taken by spacecraft-borne instruments have now yielded a detailed description of the solar wind across an area from inside the orbit of Mercury to well beyond the orbit of Neptune.

The solar wind, its origin, structures, dynamics and variations are discussed in some detail in Chapter 6. The solar wind itself is a fascinating plasma physics laboratory, with many outstanding research problems in its own right. The origin of the solar wind through interaction of the solar magnetic field with the expanding coronal plasma is a major topic in present-day research. However, in this chapter, we are primarily interested in the solar wind as the medium through which solar activity (or, in physical terms, changes in the solar magnetic field) is transmitted to planets, comets, dust particles, and cosmic rays that 'stand' in the wind. In particular, the solar wind is the input and controlling influence in the interdisciplinary subject known as *solar–terrestrial relations*.

Most observations of the solar wind have been made by spacecraft near the orbit of the Earth. Typical values for solar wind parameters at this distance (i.e. 1 AU) are:

Proton density	6.6 cm^{-3}
Electron density	7.1 cm^{-3}
He^{2+} density	0.25 cm^{-3}
Flow speed ($\sim$ radial)	450 km s^{-1}
Proton temperature	1.2×10^5 K
Electron temperature	1.4×10^5 K
Magnetic field strength	7 nT
Sonic Mach number	2–10
Alfvén Mach number	2–10
Mean free path	$\sim$ 1 AU

Since the magnetic Reynolds number in the solar wind is high ($R_M \gg 1$), the frozen-in flux approximation is valid. Hence the magnetic field at the surface of the Sun will be carried out into interplanetary space with the solar wind outflow while the footpoints of the field lines remain rooted in the solar corona. Separate plasma parcels emitted from the same location in the corona move radially outwards at high speed and are joined by a common field line; recall that plasma elements initially on the same field line remain on that field line at all times in the frozen flux regime. However, the Sun (and hence the footpoints of the magnetic field lines) rotates every 25.4 days. (The apparent period of rotation at the Earth is $\sim$ 27 days, equivalent to an angular frequency of 2.7×10^{-6} rad s^{-1}.) Thus, although the plasma parcels all move radially, the magnetic field line that threads them, with its footpoint fixed in the rotating corona, forms a spiral structure, as shown in Figure 4.9. This is known as the Parker spiral structure of the interplanetary magnetic field. The *interplanetary magnetic field* (IMF) is in reality rather variable in strength and direction, but on average conforms to the expectations of the Parker spiral model: at the orbit of the Earth, it lies in the ecliptic plane, and makes an angle of approximately 45° to the Earth–Sun line. In the ecliptic plane (i.e. perpendicular to the solar rotation axis), this spiral structure tends to wind tighter with increasing distance from the Sun. However, it is less tightly wound as one passes from the ecliptic plane towards higher heliographic latitudes. The latter point can be understood by considering plasma parcels emitted from a pole of the Sun (i.e. those moving along the solar rotation axis). In this case the coronal footpoint of the magnetic field line remains aligned with the plasma parcels at all times. The field lines emitted from the poles thus remain straight, and the solar wind expands freely outwards along that field direction. These regions in which the field structure allows unimpeded outflow of the solar coronal material are the source regions for the fast solar wind streams, which, in turn, are predominantly found at higher heliographic latitudes.

The above discussion summarises some of the basic structure of the solar wind and interplanetary magnetic field. However, there are many complicating plasma physical processes that occur on the Sun and within the solar wind. Many of the

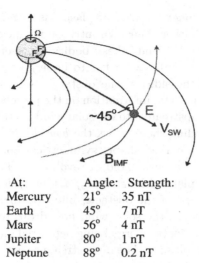

At:	Angle:	Strength:
Mercury	21°	35 nT
Earth	45°	7 nT
Mars	56°	4 nT
Jupiter	80°	1 nT
Neptune	88°	0.2 nT

Figure 4.9 The Parker spiral structure of the interplanetary magnetic field (IMF). The IMF is frozen into the radial outflow of the solar wind, and is therefore convected out into interplanetary space with the flow at speed V_{SW}. However, the feet of the field lines are also frozen into the solar surface, which rotates with an $\sim$ 27-day period. The foot of the field line frozen to a plasma parcel emitted from point F has rotated to point F' by the time the plasma parcel reaches e.g. the Earth (E). Thus the IMF is wound into an archimedian spiral, which makes a angle of 45° to the Earth–Sun line. Angles at other planets are given in the table — the spiral winds tighter with increasing distance from the Sun.

larger-scale processes (the interplanetary current sheet, the sector structure of the magnetic field, fast and slow flows, co-rotating interaction regions, coronal mass ejections, interplanetary shocks) are discussed in detail in Chapter 5. Smaller-scale processes, such as waves, turbulence and instabilities, are also the subjects of much current research, and the interested reader is again referred to the more comprehensive texts and research literature for details of such processes. For the remainder of this chapter, we will deal with the solar wind and IMF only as it pertains to the input of mass, momentum and energy to the immediate environs of various solar system bodies.

4.5.4   *Collisionless shocks and bow shocks*

We have seen above that the solar wind is a supersonic outflow of plasma from the Sun, with sonic Mach numbers typically in the range of 2–10. The solar wind flow speed also exceeds the *Alfvén speed* of the plasma, the characteristic wave speed arising from the presence of the magnetic field within the ionised gas. Alfvén Mach numbers have a similar range (2–10) to the sonic counterparts (although the two characteristic speeds are generally different). In addition, the frozen-in flow approximation implies that the IMF convects with the solar wind flow, and that by extension, any planetary magnetic field or plasma is frozen out of this flow. Many solar system bodies have either their own intrinsic magnetic field, or a population of plasma of internal origin. Thus these bodies represent real obstacles to the flow of the solar wind. The solar wind has to undergo a slowing and a deflection in order to pass around the sides of the obstacle. Given the highly supersonic, super-Alfvénic nature of the flow, information on the presence of the obstacle cannot generally propagate back upstream into the solar wind, so this slowing and deflection generally occurs at a standing shock wave, the *bow shock*, located just upstream of an obstacle. These have a curved shape, symmetrical about the Sun–planet line, which may approximate to a paraboloid of revolution. The exact position of the bow shock relative to the planet (the *stand-off distance*) depends on the solar wind ram pressure, and the shape of the obstacle; blunt obstacles have larger stand-off distances. Across such shocks the solar wind flow drops to subsonic, sub-Alfvénic speeds, with the upstream kinetic energy being deposited into thermal and magnetic energy in the downstream plasma. The downstream region, generally known as the *magnetosheath*, thus contains deflected, compressed and heated plasma, and tends to be a very turbulent region, both in the plasma and electromagnetic field parameters.

Although most of the discussion below will be focussed on bow shocks, there is also a counterpart, the interplanetary shock, which occurs within the solar wind itself. These shocks may occur, for example, when a fast solar wind stream runs into the back of a slower stream, and the plasma must be rapidly decelerated to obey the frozen-in flux condition (no mixing of the fast and slow plasma streams and magnetic fields). This results in a compressed and heated region within the solar wind that may persist for many solar rotations (i.e. they may be observed to pass over a spacecraft in the solar wind every ~ 27 days). For this reason they are termed *co-rotating interaction regions*. Secondly, interplanetary shocks may arise due to the supersonic expansion through the solar wind of ejecta material within a *coronal mass ejection* (CME). As this bubble expands, it may drive a shock both into the upstream solar wind and back towards the Sun. CMEs are thus sometimes observed to be associated with a 'forward-reverse' shock pair, which conditions the solar wind to the CME expansion into both the upstream and downstream regions.

Collisions in an ordinary gas serve to transfer momentum and energy among the molecules, and they provide the coupling, which allows the basic wave, i.e. the sound wave, to exist. However, in many space plasmas this collisional coupling is absent (the mean free path between collisions is typically greater than the size of the system). However, observations suggest that the thickness of the Earth's bow shock is only 100–1000 km. Whatever is happening at the shock, collisions cannot be important, and instead there are processes in operation which are unique to collisionless plasmas.

As discussed above, a plasma can support several different types of wave, involving fields and particles, rather than the single sound wave of a gas (e.g. the fast and slow magnetosonic waves and the Alfvén wave). Unlike the sound wave of a normal gas, these waves do not arise from collisions between particles, but rather the collective behaviour of these particles in electric and magnetic fields. The action of the magnetic field thus replaces that of collisions to 'bind' the particles of the plasma together. We can use the MHD approximation to describe the plasma in the regions far upstream and downstream of the shock. However, since MHD does not include effects due to individual particles (the 'kinetic' effects), it cannot tell us anything about how a shock provides dissipation, or what its small-scale structure will be. Again these are topics of current research interest, and more detailed information can be found in the scientific literature.

The character of a shock (or other discontinuity) in the plasma is controlled by the nature of the wave with which it is associated. Bow shocks and interplanetary shocks are generally associated with the fast magnetosonic wave, and so are often referred to as fast mode shocks. We will concentrate on the physics of this type of shock in this chapter. However, slow mode shocks, associated with the slow magnetosonic wave, are believed to occur in space plasmas, as are discontinuities associated with the Alfvén wave. In addition, as the IMF typically lies at an angle to the Sun–planet line, it intersects the fast mode bow shock at different angles around its curved surface. This results in major differences in the character of the shock at different points on its surface, depending on whether the field is largely normal or tangential to the local shock surface.

4.5.4.1 *MHD shock jump relations*

Although a complete understanding of the microphysics operating at collisionless shocks has yet to be achieved, we can use MHD to apply the requirements of conservation of mass, momentum and energy to the plasmas up- and downstream of the shock. The MHD equations described in Chapter 10 can be cast as difference conditions of the form

$$[\rho u_n] = 0 \,,$$

$$\left[\rho u_n^2 + P + \frac{B^2}{2\mu_0}\right] = 0 \,,$$

$$\left[\rho u_n^2 \mathbf{u_t} - \frac{B_n}{\mu_0}\mathbf{B_t}\right] = 0 \,,$$

$$\left[\rho u_n^2 \left(\frac{1}{2}u^2 + \frac{\gamma}{\gamma - 1}\frac{P}{\rho}\right) + \mu_n\frac{B^2}{\mu_0} - \mathbf{u}\cdot\mathbf{B}\frac{B_n}{\mu_0}\right] = 0 \,,$$

$$[B_n] = 0 \,,$$

$$[u_n\mathbf{B_t} - B_n\mathbf{u_t}] = 0 \,,$$

where we use the convention that $[X] = X_{\text{upstream}} - X_{\text{downstream}}$ is the difference between the quantity X in the upstream and downstream regions, and the subscripts n and t represent the components of vector quantities lying normal and tangential to the shock surface respectively. These equations are known as the shock jump relations, shock conservation relations or Rankin–Hugoniot equations, due to their relationship to the fluid dynamic conditions of the same name. The first equation expresses the MHD mass continuity equation, indicating that the flux of particles into the shock is the same as that leaving it in steady state. Consequently, if the flow speed u_n decreases at the shock, as expected, then the density must go up, explaining the observed compression of plasma in the magnetosheath. The second and third equations represent the conservation of momentum flux in the direction normal and tangential to the shock surface respectively. The fourth equation is an expression of conservation of energy, and how this is distributed between the plasma kinetic energy, thermal energy and magnetic field energy. The final two equations come from the electromagnetic field conditions; div $\mathbf{B} = 0$ implies that the normal component of the magnetic field is continuous, while the steady state Faraday law, curl $\mathbf{E} = 0$, and ideal MHD imply that the tangential component of the electric field $\mathbf{E} = -\mathbf{u} \times \mathbf{B}$, is also continuous. These equations can be simultaneously solved in single fluid MHD to uniquely determine the downstream conditions in terms of the observed upstream parameters. However, as indicated above, the nature of the shock is a strong function of the magnetic field direction relative to the shock surface.

4.5.4.2   *Shock structure*

In a *quasi-perpendicular shock*, the upstream magnetic field lies close to being *perpendicular* to the *normal* to the shock surface ($B_n \sim 0$). In this case, if the flow normal to the shock slows by a factor R, then by the first jump condition the density will increase by this factor. The final jump condition indicates that the strength of the magnetic field (which remains perpendicular to the shock normal direction) also increases by this factor. Further manipulation of the shock jump conditions can be used to find an expression for R. This is a function of both the sonic and Alfvénic Mach numbers, M_S and M_A, as well as the ratio of specific heats γ. Interestingly, for a monatomic gas ($\gamma = 5/3$), even for very high Mach numbers, $R \leq 4$. Thus the maximum compression ratio for these shocks turns out to be 4.

In a *quasi-parallel shock*, the upstream magnetic field lies close to being *parallel* to the *normal* to the shock surface ($\mathbf{B}_{Ut} \sim 0$). Manipulation of the above jump equations can be used to show that the downstream tangential field component, $\mathbf{B}_{Dt}$, is also ~ 0. Since the normal component of the field, $\mathbf{B}_n$, is a conserved quantity, the magnetic field magnitude remains approximately constant across the shock. In this case, there is a compression in the plasma density as the flow slows across the shock, but not in the magnetic field. Feeding this result back into the conservation relations results in all dependence on the magnetic field parameters dropping out of the equations. The equations thus assume their fluid dynamic rather than MHD form, such that it appears that the quasi-parallel shock is like an ordinary fluid shock, and the magnetic field appears not to play a role. In reality, the situation must be more subtle than this, since we have argued that in the context of a collisionless plasma the only way for dissipation to occur is via field-particle processes. The field must therefore play a crucial role that is not explicit in the MHD approach.

4.5.5   Shock acceleration

Shocks are also known to be a source of acceleration of particles to very high energies. Although the microphysics of collisionless shocks cannot be addressed with MHD, it is possible to reach some understanding of the small-scale structure and the acceleration process by examining the motion of individual particles in typical shock magnetic field geometries. Figure 4.10(a) indicates the field structure and a typical particle trajectory for a quasi-perpendicular shock surface. The field is near-perpendicular to the normal **n** to the shock surface. Consider a test particle in the upstream region. This particle gyrates around the field and convects

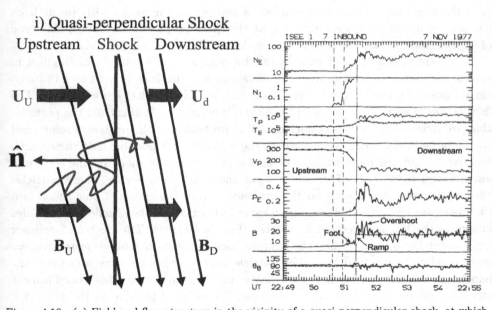

Figure 4.10   (a) Field and flow structure in the vicinity of a quasi-perpendicular shock, at which the upstream magnetic field, B_U, is close to being perpendicular to the shock normal n. The flow speed decreases through the shock such that the density and magnetic field strength increase in the downstream region. A fraction of particles (e.g. as indicated by the grey trajectory) may be reflected at the shock surface (ramp), but in this magnetic field configuration rapidly gyrate back into the shock and are transmitted to the downstream region. (b) Typical observation at a quasi-perpendicular shock (Sckopke *et al.*, 1983; reproduced by permission of American Geophysical Union), showing that the density N_E and field strength B increase suddenly at the shock, while the flow speed V_p falls. Quasi-perpendicular shocks exhibit readily recognisable structures such as the foot, the ramp and the overshoot. The shock foot, lying upstream of the main field increase, is attributed to the region containing the reflected particles.

towards the shock at the solar wind velocity. When it reaches the shock, it may pass straight through and continue gyrating downstream into the magnetosheath region. However, it appears that a fraction of particles may be reflected at the shock field gradient, known as the *shock ramp*. These particles now have a gyromotion, which initially takes them back into the upstream region. However, within half a gyroradius, the particle will reverse its motion and return to the shock ramp. Since the field lies close to being tangential to the shock surface, and is convecting towards it at the solar wind velocity, reflected particles are unable to perform more than one gyro-orbit before re-encountering the shock. Information on the presence of the shock surface cannot be carried more than two gyroradii upstream by these particles. This effect can be seen in typical observations of the quasi-perpendicular shock (Figure 4.10(b)). The sharp ramp is usually evident in the magnetic field as a spacecraft crosses such a shock, but a precursor signature, known as the *shock foot*, is evident for a short period immediately prior to the shock ramp. Scaling the signature by reference to the observed solar wind velocity reveals that it is indeed confined to within a few gyroradii of the shock ramp. Although other structures (e.g. the overshoot and downstream wave activity in Figure 4.10(b)) result from the microphysical processes occurring at the quasi-perpendicular shock, this type of shock remains relatively sharp, and is clearly distinguishable in data.

In contrast, examples of quasi-parallel shocks are quite hard to distinguish in typical data streams. Again, we can understand why this is so by reference to particle motion in a typical field geometry at a quasi-parallel shock (Figure 4.11(a)). The field is now close to being parallel to the shock normal **n**. In this case, the particles that are reflected at the shock surface will move back into the upstream solar wind with a large field-aligned component to the velocity. It is possible that these particles can escape upstream without being swept back to the shock by the continual antisunward convection of the solar wind and IMF. In this case, these particles effectively carry information on the presence of the quasi-parallel bow shock into the upstream medium. This results in extended regions of shock-reflected particles known as the ion and electron foreshocks (Figure 4.11(b)). The electron foreshock extends further upstream than the ion foreshock as the more mobile electrons move more quickly up the field line. The particle beams within these regions drive enhanced wave activity, such that the signature of the quasi-parallel shock itself is most often not distinguishable from the turbulent background plasmas in the foreshock regions.

The possibility that particles are reflected from the shock ramp gives a clue as to how the shock can generate some particles of very high energy that may appear both upstream and downstream of the shock. There are two generic mechanisms by which this may occur, both of which require that the particle interact with the shock many times. The first of these is illustrated in Figure 4.12(a). Multiple interactions with a quasi-perpendicular shock will lead to a net particle drift along the shock surface. For ions, this drift is in the direction of the convection electric field E_{SW},

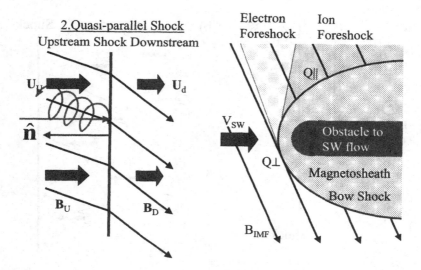

Figure 4.11 Field and flow structure in the vicinity of a quasi-parallel shock, at which the upstream magnetic field, $\mathbf{B}_U$, is close to being parallel to the shock normal $\mathbf{n}$. The flow speed decreases through the shock such that the density and magnetic field strength increase in the downstream region. A fraction of particles (e.g. as indicated by the grey trajectory) may be reflected at the shock and in this magnetic field configuration may rapidly escape along the magnetic field back into the upstream region. (b) Large-scale structure upstream of an obstacle to the solar wind flow. Shocked plasma between the bow shock and the obstacle itself forms the magnetosheath region. The bow shock itself generally has both quasi-perpendicular and quasi-parallel regions. Upstream of the quasi-parallel shock, particles escaping back upstream form the extended foreshock regions. Since electrons move much faster than ions, they are able to reach regions further upstream than the latter.

while for electrons it is oppositely directed to this electric field. Hence the particles gain energy as they move along the shock surface. If they gain sufficient energy, they will eventually escape from the region of the shock and will appear remotely from the shock as a high-energy population. The second mechanism may occur most readily at a quasi-parallel shock, and is a form of Fermi acceleration. This is illustrated in Figure 4.12(b). Particles reflected from the quasi-parallel shock may move readily back upstream into the foreshock regions. Turbulent magnetic field structures and/or wave–particle interactions may act to reflect the particle once more back towards the shock. If the reflecting structure is in motion towards the shock, the particle will pick up energy in the shock rest frame and thus return to the shock with a greater speed that at its initial encounter. Again, a small fraction of particles will undergo multiple bounces between the shock and the upstream 'moving mirrors', and thus may gain significant extra energy by the time they finally escape the vicinity of the shock. These and other processes may thus be responsible for the generation of a high-energy population in the plasma up- and downstream of bow shocks and interplanetary shocks. On a larger scale, they may also play important roles in particle acceleration in wider astronomical contexts, such as galactic jets.

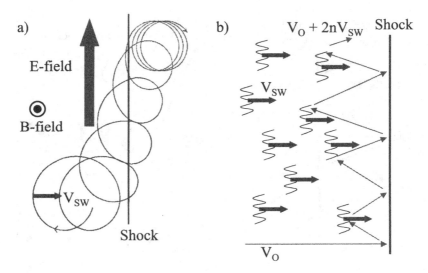

Figure 4.12 Acceleration mechanisms operating at shocks. (a) At a quasi-perpendicular shock, particles that interact with shock, either through reflection or by gyration motion in the different up- and downstream fields (as shown here), tend to drift along the shock. If (for ions) this drift is in the direction of the solar wind convection electric field, then the particles gain energy as they move along the shock surface. (b) At a quasi-perpendicular shock, particles reflected at the shock and moving back upstream in the foreshock region may be scattered back to the shock by irregularities in the solar wind. These particles may gain energy via the Fermi mechanism.

4.5.6 *Magnetic reconnection — a recap*

The concept of magnetic reconnection will be introduced in both Chapter 6 and Chapter 9. It is a key concept in understanding the coupling between the solar wind and planetary magnetospheres, as well as another potentially important mechanism for the acceleration of particles in space and astrophysical plasmas.

The frozen-in flux approximation (in the case of high magnetic Reynold's number) leads to the picture of space and astrophysical plasmas contained within separate regions. For example, field and plasma from the Sun (the IMF and solar wind) are frozen out of the region occupied by a planetary magnetic field. These separate plasma 'cells' are partitioned by thin current sheets, which support the change in magnetic fields across the boundary (recall $\nabla \times \mathbf{B} = \mu_0 \mathbf{j}$ in the MHD limit). However, exactly because of their thinness (small-scale lengths), the magnetic Reynold's number within the current sheet may be relatively small, such that diffusion of the magnetic field through the plasma may be important in these current sheets. If there is a strong magnetic field gradient and the fields on either side of the gradient are anti-parallel, then diffusion of the field at the gradient can lead to a loss of total magnetic flux, and this is termed *magnetic annihilation*. On the other hand, if there is an inflow into a spatially limited diffusion region, then field lines are continually being convected into the diffusion region and meeting field lines with an opposite orientation. In this case the field lines can 'slip' with respect to the plasma in the diffusion region, and *merge* with field lines from the opposite side of the gradient. The resulting 'reconnected' field lines are sharply bent through the current sheet (Figure 4.13). Magnetic tension forces associated with these hairpin-like field lines acts to accelerate the inflow plasma and the reconnected field lines along the current sheet and away from the diffusion region on each side. This situation is known as *magnetic reconnection*. This is an important dynamical process because it allows the two sides of the field gradient to be linked by the newly reconnected field lines threading through the current sheet, which allows plasmas from either side to flow along the field and mix with those from the other side. Also, magnetic energy is continually liberated in the process, causing accelerated and heated plasma flows.

Although the importance of reconnection is well recognised, it is fair to say that the microphysics of the diffusion process is *not* well understood. The causes of local, small-scale breakdown of the flux-freezing theorem which results in diffusion of field lines and the connection of plasma elements from different sources are currently largely unknown and certainly unobserved. However, the very small-scale

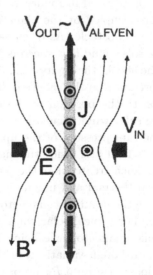

Figure 4.13 Magnetic reconnection geometry. In the simplest geometry, anti-parallel field lines are separated by a thin current sheet (shown as light grey) across which the field reverses. Due to the small scale lengths of such current sheets, the frozen-in flux approximation may break down. Magnetic flux is convected towards the current sheet from both sides, where it reconnects to form two hairpin-like field lines, which rapidly contract away from the neutral (or X-) point on both sides. Since magnetic energy is released to the plasma during this process, outflow jets of plasma are also formed moving away from the neutral line on both sides. More detailed information can be found in Chapter 9.

physics of magnetic diffusion (i.e. the physics of conductivity in a plasma with strong gradients) can affect the global topology of the field line configuration, and hence the global dynamics of large-scale plasma systems such as the Earth's magnetosphere.

4.5.7   The terrestrial magnetosphere

The magnetic reconnection process discussed above provides the primary coupling mechanism between the solar wind and the *terrestrial magnetosphere*, the region occupied by the magnetic field generated by the Earth's internal dynamo. In this section we describe how this process controls the structure and dynamics within the terrestrial magnetospheric system.

4.5.7.1   A 'closed' model magnetosphere

In the absence of the solar wind, the Earth's magnetic field would form an approximate dipole field, extending out into space. (Note, however, that the south magnetic pole is located near the northern geographic pole and vice versa, such that the Earth's magnetic field at the equator points closely northwards.) In the presence of the solar wind, we can invoke the frozen-in flux condition to understand the general form of the magnetosphere, as illustrated in Figure 4.14(a). Firstly, the solar wind flow is frozen out of the region occupied by the planetary field, such that it must slow down and deflect around the obstacle. As discussed above, this occurs at the upstream bow shock, since the solar wind flow speed is both super-sonic and super-Alfvénic. Downstream from the shock, the cavity occupied by the terrestrial field is confined by the high plasma pressures in the shocked plasma of the *magnetosheath*. In fact, a good approximation to the position of the dayside *magnetopause* (the sheet of current separating the solar wind plasma from the magnetospheric cavity) can be found by balancing the ram pressure of the upstream solar wind with the magnetic pressure of the Earth's field just inside the magnetopause. The ram pressure of the solar wind (which is much greater than the thermal or magnetic pressures in this region) is converted into thermal and magnetic pressure at the shock, and it is these pressures that balance the dominant magnetospheric magnetic field pressure across the magnetopause. For typical solar wind parameters, this balance in pressures results in the subsolar magnetopause being located on average around $10 \ R_E$ upstream of the Earth, although variations in the solar wind pressure can result in this boundary being located several R_E up or downstream of this position (extremely high solar wind pressures have occasionally driven the magnetopause in below geosynchronous orbit at $6.6 \ R_E$, such that spacecraft in this orbit have observed this boundary).

Around the flanks of the magnetosphere, the pressure in the magnetosheath tends to be lower. This is due in part to the actions of the bow shock being weaker further from the Earth–Sun line, where a smaller deflection of the solar wind flow is required for it to pass around the magnetosphere. Consequently, the magnetosphere

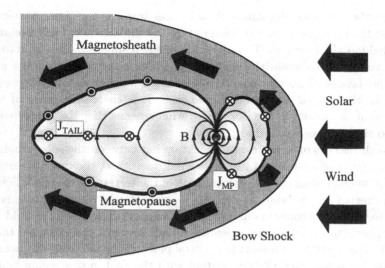

Figure 4.14   (a) The 'closed' magnetosphere. The terrestrial magetic field represents an obstacle to the supersonic, super-Alfvénic solar wind flow, which is frozen out of the cavity occupied by this field. The solar wind is thus slowed and deflected at a bow shock standing upstream of the cavity, and the shocked plasma downstream forms the magnetosheath. The pressure of the oncoming solar wind acts to compress the terrestrial magnetic field from its dipole configuration on the dayside, while the relative void of plasma in the downstream region results in a more extended nightside magnetosphere. A system of currents is set up on the magnetopause to support the magnetic field change at the boundary, and across the tail to support the extended magnetic field in this region.

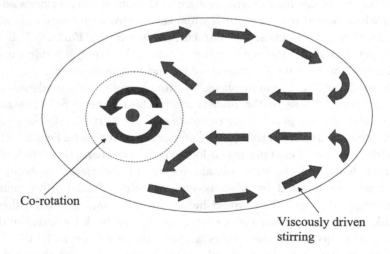

Figure 4.14   (b) Equatorial projection of the closed magnetosphere, indicating the form of magnetospheric plasma convection in this plane. Close to the Earth, the magnetic field strength is sufficiently high to impose co-rotation on the plasma (black arrows). At larger distances, viscous processes operating at the interface between the magnetospheric and magnetosheath plasma drive the latter anti-sunwards along the flanks of the magnetosphere (grey arrows). A return, sunward flow is set up within a more central channel on the nightside.

extends further towards the flanks than it does towards the Sun. Moreover, the pressure in the nightside region is much reduced as the solar wind has a relative void or wake behind the obstacle. The nightside magnetosphere is thus able to expand into this void, and so again the magnetosphere extends further on the nightside than on the dayside. This tear-shaped model of the magnetosphere (Figure 4.14(a)) is known as the *'closed' model magnetosphere*. It is the consequence of an exact application of the frozen-in flux condition, which means that the magnetosphere is 'closed' to the entry of solar wind plasma.

Although there is no significant direct coupling between the solar wind and the magnetosphere, the presence of the solar wind does affect the internal structure of the 'closed' magnetosphere. The deformation of the terrestrial magnetic field by the pressures exerted on the boundary has to be supported by a system of currents. In addition to the magnetopause current, which supports the change in field from its magnetospheric to its solar wind orientation and strength, the extended tail has to be supported by a current crossing the central plane of the tail from the dawn side to dusk. This *'cross-tail current'* is consistent with the field in this region being more sharply curved than would be the case in a pure dipole. Currents by definition must be continuous, i.e. have to flow around a closed circuit. In this case, the cross-tail current is closed by connecting to the magnetopause currents that flow back from dusk to dawn over the outer boundary of the nightside magnetosphere.

A second major current system within a closed model magnetosphere is the *ring current* system. The particle drifts described in Section 4.4 cause a separation of ions and electrons in the inner magnetosphere — the ions circulate clockwise around the Earth when viewed from the north while the electrons circulate anticlockwise. Hence there is a net clockwise circular current about the Earth. This current provides a net decrease of the equatorial magnetic field strength inside the current ring, and hence can be remotely sensed by ground magnetometers.

In this model of the magnetosphere, we would expect only moderate internal plasma convection. Close to the planet, where the magnetic field is strong and dominates the dynamics, plasma may *co-rotate* with the Earth due to the coupling of angular momentum from the atmosphere/ionosphere up magnetic field lines into the *plasmasphere*, a region of cold plasma co-located with the energetic particles forming the radiation belts. In the more distant regions of the magnetosphere, plasma convection may be induced by 'viscous'-type coupling of solar wind momentum across the magnetopause. This would induce an antisunward plasma flow down either flank, with a return flow up the centre of the tail back towards the dayside. The equatorial projection of these flows is illustrated in Figure 4.14(b).

The low levels of magnetospheric plasma convection that are expected in this model in fact tell us that these are not the only processes occurring in the real magnetosphere. In particular, very strong magnetospheric convection is observed when the IMF has a southward component. Since this field direction is anti-parallel to the direction of the terrestrial magnetic field just inside the magnetopause, this is a strong indication that magnetic reconnection plays a key role in the structure and dynamics of the real magnetosphere under these IMF conditions.

4.5.7.2 *The 'open' magnetosphere*

Under northward IMF the magnetosphere may well approximate to or display localised characteristics of the closed magnetosphere described above. However, the southward IMF direction provides a strong gradient in the magnetic field at the dayside magnetopause, where the field reverses in direction. As discussed above, these circumstances can lead to a localised breakdown of the frozen-in flux condition, leading to the occurrence of magnetic reconnection. Although this breakdown is very localised, perhaps occurring over a few tens of kilometres, it has a global effect on the structure of the magnetosphere. The occurrence of reconnection at the dayside magnetopause is the central process in the formation of an *'open' magnetosphere*, which is illustrated in Figure 4.15.

An 'open' magnetosphere still represents a significant obstacle to the solar wind flow. Thus, as discussed above, a bow shock again stands upstream of the magnetosphere in order to slow and deflect the flow around this obstacle. It also acts to

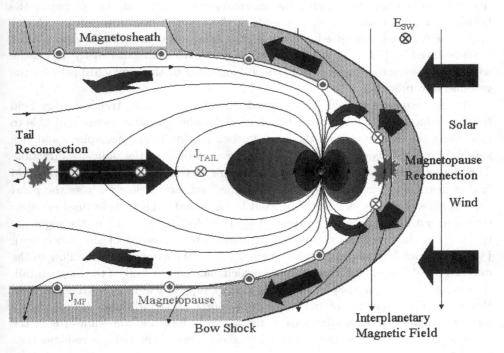

Figure 4.15 The 'open' magnetosphere. The breakdown of the frozen-in condition at the thin magnetopause current sheet allows magnetic reconnection to occur. The solar wind and magnetosphere thus become directly connected, and solar wind mass and energy are transferred across the magnetopause. For full details see text.

compress and thus strengthen the southward-directed magnetic field in the magnetosheath. The dayside magnetopause is again located approximately at the point where the upstream ram pressure balances the internal magnetic pressure (i.e. around 10 R_E upstream of the Earth). The occurrence of reconnection in the vicinity of the subsolar dayside magnetopause creates two 'open' magnetic flux tubes, which have one end located in the north and south polar ionospheres, and the other ends out in the solar wind. Just after they have undergone reconnection, these field lines have a sharp kink as they thread through the dayside magnetopause. The magnetic tension of these field lines at the magnetopause acts to try to straighten them. Hence these field lines peel back away from the reconnection site (the *neutral line* or *X-line*), heading north and south over the two polar caps respectively. At the same time, plasma from the external magnetosheath is able to pass along the field lines, through the magnetopause, and populate a boundary layer on the portions of the reconnected flux tubes lying just inside of the magnetopause. These populations extend down the field lines into the *polar cusps*, the funnel-shaped field regions formed above the poles. Similarly, previously trapped energetic magnetospheric particles are able to escape from the magnetosphere and into solar wind along the open field lines. The energy associated with the kinked magnetic field structure is liberated to the plasma crossing the magnetopause, such that the plasma in the boundary layers is accelerated to speeds of the order of the local Alfvén velocity above the external flow speed. Formally, this energy transfer from field to plasma is represented by $\mathbf{j} \cdot \mathbf{E} > 0$, where $\mathbf{j}$ is the magnetopause current density and $\mathbf{E}$ the solar wind convection electric field which is applied to the magnetosphere by the reconnection process.

As the reconnected field lines unwind over the poles and straighten, the field tension reduces and plasma acceleration ceases (although particles are still able to pass along the field lines from the magnetosheath into the magnetosphere and vice versa). Beyond this point, the field lines become bent in the opposite direction. This is due to the continuing anti-sunward flow of the solar wind end of the field line at near solar wind speeds, while the other end moves slowly across the polar cap ionosphere and so is effectively fixed to the Earth. These field lines are thus stretched out and form a very extended ($\geq 1000\ R_E$) magnetotail in the nightside region. On these regions of the tail magnetopause (tailward of the two cusps), $\mathbf{j} \cdot \mathbf{E} < 0$, which indicates a transfer of energy from the external plasma flow to the magnetic field inside. The two tail lobes, which make up the bulk of the magnetotail, thus represent a repository of magnetic energy tapped from the solar wind flow. However, the stretched open field lines in the north and south lobes are oppositely directed (pointing closely sunwards in the northern lobe and anti-sunwards in the south lobe). An extended cross-tail current sheet lying in the tail centre plane thus separates them.

As reconnection proceeds at the dayside magnetopause, magnetic flux is transported from the dayside magnetosphere and into the nightside. The dayside magnetosphere is thus eroded, so the magnetopause may move closer to the Earth. Clearly, this situation cannot continue indefinitely — the dayside magnetosphere never disappears completely. Hence there must exist a mechanism whereby magnetic flux is returned to the dayside from the nightside. This mechanism is again magnetic reconnection, this time occurring between the oppositely directed magnetic fields of the two tail lobes. The cross-tail current is again relatively thin (i.e. a small-scale structure), and so the frozen-in condition may break down at some point in the tail. Reconnection at this point causes the oppositely directed lobe field lines to change topology. On the Earthward side of the tail neutral line, a sharply bent field line is created with ends in the north and the south polar ionosphere. Tailward of the neutral line, a sharply bent field line has both ends extending into the solar wind far downstream from the Earth. The tension in both these field lines again causes them to recede away from the neutral line on both sides, liberating lobe magnetic energy back to the plasma on the field lines as it does so. Again $\mathbf{j} \cdot \mathbf{E} > 0$ within the cross-tail current layer, and the direction of E implies that the plasmas in the lobe $\mathbf{E} \times \mathbf{B}$ drift towards the centre plane on both sides in order to supply this process. Consequently two plasma jets are formed moving towards and away from the Earth on the Earthward and the tailward side of the neutral line respectively. These jets will be flowing at speeds of the order of the lobe Alfvén velocity, which, due to the relatively strong field and low plasma densities typically found in the lobes, may be of order 500–1000 km s^{-1}. The resulting hot, accelerated plasmas form a population known as the *plasma sheet*, which surrounds the immediate vicinity of the cross-tail current in the tail. From observations, it appears that the average flow in the tail reverses from Earthward to tailward at a downtail distance of 100–$140\ R_{\mathrm{E}}$, such that this distance represents the typical location of the tail reconnection process. The reconnected field line formed tailwards of the tail neutral line eventually straightens and rejoins the solar wind. The field line formed Earthwards of the neutral line contracts back towards the Earth, eventually slowing in the near-Earth magnetosphere as it meets a region of stronger, more dipolar field. These field lines then convect around the Earth and back to the dayside, where they may again reach the dayside magnetopause and once again undergo reconnection. In this manner a continual cycle of magnetospheric convection is set up, with open field lines being created by dayside reconnection. These initially unwind, accelerating dayside boundary layer plasmas as they do so, but then are dragged by the solar wind flow over the poles of the Earth and added to the magnetotail lobes. From here they $E \times B$-drift towards the centre plane of the tail, where they may again undergo reconnection. The closed field lines thus created convect back towards the Earth along the equatorial plane, accelerating plasma sheet plasma as they do so. Eventually the closed flux tubes migrate back to the dayside when the cycle begins again.

Recall that this convection cycle is dependent on the direction of the IMF. In effect, under southward IMF, reconnection enables a significant fraction of the solar wind convection electric field to be coupled along the open field lines directly into the magnetosphere, where it drives levels of convection far higher than can be achieved in a closed magnetosphere (or under northward IMF conditions). It is also interesting to note that since the open-field lines extend down into the polar ionosphere, the convection electric field also maps down into this region. Hence the polar ionosphere has a convection pattern which maps the elements of that of the magnetosphere (Figure 4.16). The region containing the footpoints of the open flux tubes that have undergone dayside reconnection is known as the *polar cap*. This is an approximately circular region, typically extending down from the magnetic poles to $\sim 75°$ geomagnetic latitude. The magnetospheric motion of open flux from the dayside reconnection site over the poles and into the magnetotail maps as an anti-sunward motion of the open field line footpoints and associated plasma across the polar cap from midday to midnight. The motion of closed field lines created at the nightside reconnection site maps to the ionosphere as a return flow to the dayside at latitudes equatorwards of the boundary of the polar cap. Note that forces necessary to drive these ionospheric flows are communicated from the magnetosphere by a relatively complex system of *field-aligned currents*. These currents flow down the magnetic field lines, across (at least a portion of) the polar ionosphere and back up the magnetic field to the magnetosphere, where they close as part of either the ring or cross-tail currents. The magnetic force, $\mathbf{j} \times \mathbf{B}$, associated with the ionospheric section of these current systems provides the impetus for the ionospheric convection pattern described above. Readers interested in the details of these current systems and their role in magnetosphere–ionosphere coupling are referred to the more advanced texts on this subject.

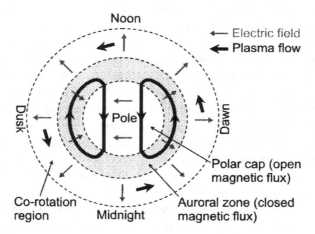

Figure 4.16   Schematic of the polar ionospheric convection pattern associated with an open magnetosphere. The footpoints of open magnetic flux created by dayside reconnection move anti-sunwards (from top of figure) in the polar cap under the influence of the solar wind flow. Magnetotail reconnection recloses the field lines, which convect back sunwards at low latitudes in the auroral zones. At the lowest latitudes, the flux tubes co-rotate with the Earth. The stresses required to drive these flows are transmitted from the magnetosphere by a system of field-aligned currents.

4.5.7.3 *Dynamics — flux transfer events (FTEs), storms and substorms*

The open magnetospheric system described in the previous section is essentially steady state in nature. Magnetic flux and magnetospheric plasma are driven around the system by reconnection at the dayside creating open flux, and reconnection on the nightside returning open flux tubes to a closed topology at essentially the same rate. Although this picture may be correct when averaged over time, it is clear that in the real magnetosphere, reconnection often occurs in a patchy manner in both time and space, and that the dayside and nightside reconnection rates are rarely balanced, even when the IMF remains southward. Convection within the magnetosphere may thus be highly time-dependent, and proceed in a sporadic rather than continuous manner.

Evidence that the reconnection process itself may occur in a more patchy manner than described above, even when the IMF is consistently southward, is regularly observed by spacecraft crossing the dayside magnetopause. An outbound crossing of this current layer is generally indicated by the observation of a sharp discontinuity in the field direction and a change in the plasma properties from the hot, rarefied magnetospheric plasma to the cooler, denser magnetosheath plasma. However, in the immediate period before and after the crossing, a repetitive characteristic bipolar signature is often observed in the magnetic field data, most obviously on the less variable magnetospheric side of the boundary. This bipolar signature consists of a brief (typically < 1 min) deflection of the field towards the magnetopause followed by deflection of similar duration away from the magnetopause (or vice versa). This is often associated with a concurrent enhancement of the total field strength. In a coordinate system with one of the axes aligned with the normal to the magnetopause, the magnetic field component in this normal direction will show a +/− (or −/+) deflection. Inside the magnetosphere, these signatures may also be associated with the brief appearance of plasma with characteristics of the external magnetosheath plasma population. As indicated above, these signatures are of short duration, but often repeat every 4–8 min in the interval around the spacecraft crossing of the magnetopause current layer. The occurrence of these events is also a strong function of the IMF direction, with events occurring predominantly under southward IMF conditions, indicating their relationship to the reconnection process. An example of these signatures observed by the *Cluster* quartet of spacecraft is shown in Figure 4.17(a).

The basic interpretation of these signatures is illustrated in Figure 4.17(b). Although the detailed interpretation varies with different researchers, the underlying explanation is common. It is clear that these signatures are the result of a short duration burst of reconnection, probably over a rather localised section of the dayside magnetopause. This creates two tubes of open flux of limited cross-section (~ 1 R_E across). These tubes peel back across the dayside magnetopause as they unwind and release the associated magnetic tension. A spacecraft near the

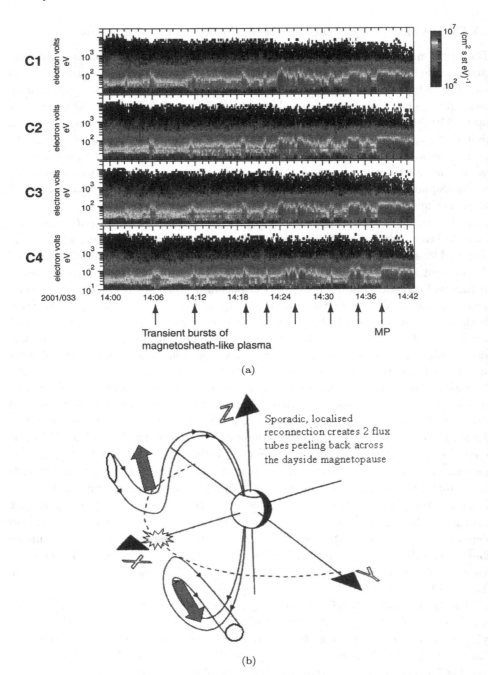

Figure 4.17   Flux transfer events (FTEs). (a) Transient bursts of magnetosheath-like plasma with the magnetosphere can be explained by (b) the occurrence of sporadic and lightly localised reconnection, which creates two tubes of open magentic flux. These peel back across by dayside magnetopause as the magnetospheric portions fill with magnetosheath plasma.

magnetopause may intersect one or more of these reconnected flux tubes, or *flux transfer events* (FTEs), as it unwinds across the dayside magnetopause. The bipolar field deflections are associated with the tube structure and the deflection in the overlying external field caused by its motion. The observations of magnetosheath-like plasma are consistent with the entry of this plasma through the magneto-pause along the newly opened field lines that make up the FTE. Although these signatures have a clear interpretation based on a time-dependent reconnection rate, it is not currently understood why or where the reconnection should occur, why it should occur only in brief bursts, even when the IMF is relatively steady, or why there is a consistent repetition rate for these events. However, these uncertainties mimic our current lack of detailed understanding of the reconnection process itself.

The effects of unbalanced dayside and nightside reconnection rates have more global implications for the magnetospheric structure and dynamics. In particular, this causes a cycle of energy storage within the magnetosphere followed by an explosive dissipation of this energy that is known as a *magnetospheric substorm*.

A southward turning of the IMF generally leads to an increase in the net reconnection rate at the dayside, whether through an increase in reconnection at a quasi-steady neutral line or by an increase in the size or number of FTEs. However, the effects of the southward IMF are not immediately communicated to the night-side, and so reconnection proceeds at a relatively slow rate at the neutral line in the distant tail. As a consequence, the dayside magnetopause is eroded, and a net open magnetic flux is added to the tail lobes. Thus the magnetotail (and the ionospheric polar cap) grows in diameter and the lobe field strength increases. At the outer boundary, this increase in open flux in the tail causes an outward flaring of the magnetopause surface (Figure 4.18). Near the equatorial plane, the plasma sheet thins and the cross-tail current intensifies to support the increased field strengths. In the inner magnetosphere, field lines that would usually have a near-dipole orientation may become stretched out into a more tail-like geometry. The net effect during this interval, which is known as the *substorm growth phase* and typically lasts for ~ 1 h, is to tap a vast amount ($\sim 10^{16}$ J) of solar wind kinetic energy and store it in the form of magnetic energy in the two magnetospheric tail lobes.

The eventual response of the magnetospheric system to this increased load of magnetic energy is the subject of much current research, and there are a number of competing theories about the details and temporal ordering of events during what is known as the *substorm expansion phase*. What is clear, however, is that the stored energy is suddenly and rapidly dissipated, and that reconnection again plays a major role in the dissipation process. It appears most likely that rather than support an increase in the reconnection rate at the distant tail neutral line (DNL), the cross-tail current sheet becomes unstable in the near-Earth tail where a new reconnection site forms, typically at distances of 20–30 R_E downtail. Reconnection proceeds rapidly at the *near-Earth neutral line* (NENL); Figure 4.19), which has a number of consequences for the structure of the magnetotail.

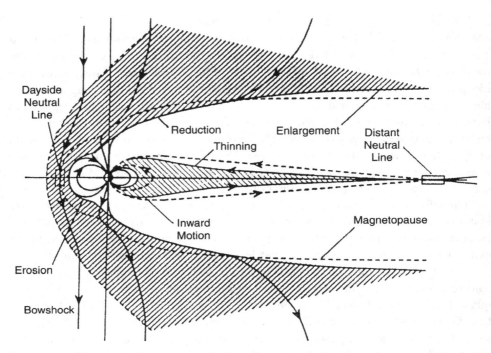

Figure 4.18   Changes in the magnetosphere configuration during the substorm growth phase. Enhanced reconnection rates at the dayside magnetopause (due to a southward turning of the IMF) result in the erosion of magnetic flux from the dayside magnetosphere and thus extra flux added to the magnetotail. The magnetopause thus becomes increasingly flared and the magnetotail increases in diameter. The plasma sheet thins as the tail magnetic field becomes increasingly stretched away from a dipole. (McPherron, 1995; courtesy Cambridge University Press.)

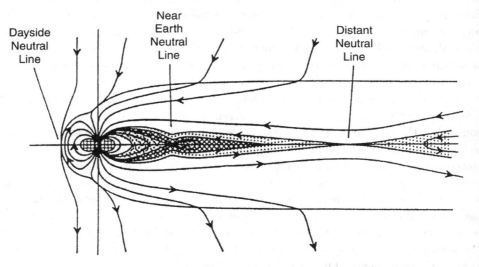

Figure 4.19   Formation of the near-Earth neutral line (NENL) at the end of the substorm growth phase. (McPherron, 1995; courtesy Cambridge University Press.)

Firstly, in the region tailwards of the NENL, a nested set of closed loop magnetic field lines is formed from field lines that have been reconnected first at the DNL and are subsequently pinched off closer to the Earth at the NENL. These field lines trap hot accelerated plasma sheet plasma, thus forming a plasma bubble, known as a *plasmoid*. Once the NENL overtakes the DNL, and begins to reconnect lobe field lines that have yet to be reconnected at the DNL, this plasmoid is overlaid by field lines which have both ends in the solar wind far downstream. A combination of the tension force associated with these field lines and the high field and plasma pressures in the near-Earth magnetosphere acts to expel the plasmoid (and the DNL if it is still active) tailwards and out into the downstream solar wind (Figure 4.20). The size of these plasmoid structures may be several tens of R_E in length (comparable to the distance between the NENL and the DNL) by 5–10 R_E in height and width. Consequently, these massive structures contain significant amounts of magnetic field and plasma thermal and kinetic energy, and thus act to dissipate a significant fraction of the energy stored in the magnetotail during the growth phase.

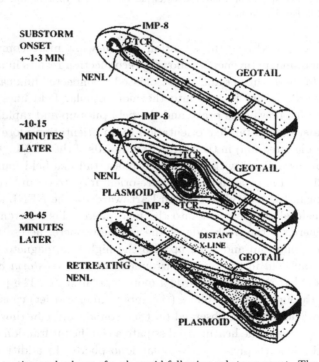

Figure 4.20   Formation and release of a plasmoid following substorm onset. The formation of the NENL results in the disconnection of a large section of the plasma sheet, which forms a magnetic bubble, or plasmoid, of closed loop magnetic field lines, between the near-Earth and distant neutral lines. As reconnection proceeds at the NENL, the plasmoid is expelled downtail due to the plasma pressure gradient in the near-tail and the magentic tension forces associated with the overlying field lines reconnected at the NENL after the plasmoid is disconnected. Tailward motion of the plasmoid causes compressions of the lobe magnetic field (TCRs — travelling compression regions). The plasmoid eventually joins the downstream solar wind. (Slavin, 1998; reproduced by permission of American Geophysical Union.)

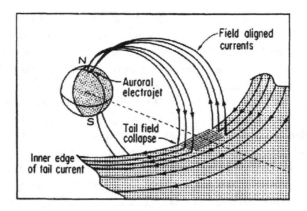

Figure 4.21 The substorm current wedge. At substorm onset the magnetic field in the near-Earth tail rapidly becomes more dipolar. The cross-tail current, which supported the stretched field configuration, is diverted down the magnetic field on the dawn side, across the auroral ionosphere, and back up into the magnetosphere on the dusk. (McPherron *et al.*, 1973; reproduced by permission of American Geophysical Union.)

Earthwards of the NENL, the situation is somewhat more complicated. As usual, liberation of magnetic energy following reconnection results in an Earthward jet of plasma. However, this jet and the associated reconnected flux cannot proceed very far Earthwards before it runs into the more dipolar field lines in the inner magnetosphere. The flows thus brake and the field piles up and rapidly dipolarises (changes from a stretched tail-like orientation to an orientation more aligned with the expected dipole direction) in the region Earthwards of the NENL. The enhanced cross-tail current associated with the growth phase tail-like field must be rapidly removed from the tail in order for this field re-orientation to occur. To do this, the cross-tail current in the central tail region Earthwards of the NENL appears to be diverted via a field-aligned current into the ionosphere. This current flows down into the ionosphere on the dusk side of the magnetosphere, across the midnight boundary of the polar cap and back up into the nightside magnetosphere on the dusk side. The tail field in the region surrounded by this *substorm current wedge* (Figure 4.21) is able to collapse towards a more dipolar state. The portion of this current flowing through the ionosphere (the substorm electrojet) causes disruption to the magnetic field strength observed on the ground, and can thus be remotely sensed. It also acts via Joule heating to dissipate a significant fraction of the energy stored during the growth phase in the polar ionosphere. In addition, significant fluxes of energetic particles are injected into the ring current and radiation belt regions during the substorm expansion phase, again contributing to the dissipation of the stored energy. Finally, particle precipitation into the nightside polar ionosphere also provides a dissipation mechanism. This is the most visible manifestation of the substorm process, as it results in a major increase in auroral activity. Figure 4.22 shows a series of photographs of the auroral region taken from the *Polar* spacecraft

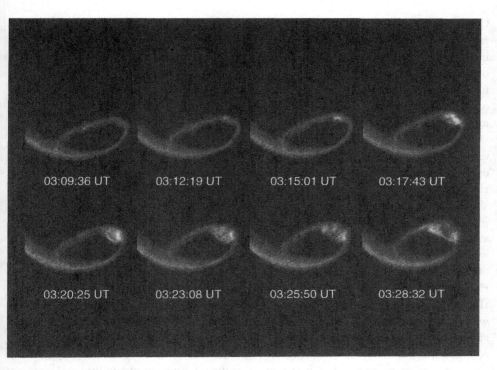

Figure 4.22   Auroral substorm activity viewed from space by the University of Iowa Visible Imaging System on the *NASA/POLAR* spacecraft on 23 September 1996. At substorm onset, the quiet time auroral oval develops a localised brightening, which expands polewards and westwards as the expansion phase proceeds. (Courtesy of L.A. Frank and J.B. Sigwarth, University of Iowa and NASA/Goddard Space Flight Center.)

during a substorm. Prior to expansion phase onset, the auroral oval forms a simple narrow ring. At expansion phase onset, a localised auroral brightening is evident in the midnight sector of the auroral oval, which then expands both polewards and westwards until it covers a large fraction of the polar cap region. (The expansion of the aurora is the original source of the name of this phase.) Viewed from the ground, this sequence would be seen as the expansion of a narrow curtain of auroral light until it fills most or all of the visible sky overhead.

The rapid dissipation of energy during the expansion phase leads to a reduction of the field strength in the magnetotail lobes and a net return of closed flux tubes back to the dayside magnetosphere. It appears that a reduction in the stress on the magnetotail allows the reconnection rate at the NENL to drop, and for its location to migrate tailwards to replace the pre-substorm DNL. This portion of the substorm cycle, known as the *substorm recovery phase*, thus sees the magnetosphere returning to its quiet time configuration.

As indicated above, there are a number of substorm model controversies that make this a currently active area of research. Again, our lack of understanding of the reconnection process itself contributes to this situation. We do not understand why the expansion phase begins when it does, or the exact role of reconnection in it. Does the commencement of reconnection coincide with the expansion phase onset, or does the current disruption occur first and subsequently drive reconnection further tailwards? Is there an external trigger for a substorm expansion phase onset? Interested readers are referred to the plethora of current literature on these and many other substorm questions.

A prolonged period of southward IMF can often lead to a magnetic storm, especially if it is associated with high-speed solar wind flows. Such conditions are common during the passage of a coronal mass ejection (CME) past the Earth, since these often contain a *magnetic cloud*, an ordered magnetic field structure, probably similar to a plasmoid, which applies first a northward-directed magnetic field for many hours, followed by a similar duration southward-directed magnetic field (or vice versa). The magnetic storm lasts for several days, in contrast to the substorm, which lasts several hours and may repeat several times a day. Although the arrival of the CME can cause an initial compression of the magnetosphere, and hence a rise in the magnetic field strength known as a *storm sudden commencement* (SSC), the main characteristic of the storm is a strong enhancement in the ring current, which manifests itself as a large reduction in the surface equatorial magnetic field strength. Repeated substorms during the *storm main phase* inject significant fluxes of energetic particles into the ring current. These decay over several days during the *storm recovery phase*, but it seems a significant fraction of the injected electron population is pumped up to relativistic energies in the radiation belts during this phase. These can pose a problem to orbital spacecraft, as they are a significant hazard for on-board electronic systems.

4.5.8   *Solar wind interaction with other bodies*

4.5.8.1   *Other planetary magnetospheres*

The basic solar-wind–magnetosphere interactions described above occur at every magnetised planet. Hence Mercury, Jupiter, Saturn, Neptune and Uranus each have a bow shock and a magnetospheric cavity. The main difference is in the size of each magnetosphere (illustrated in Figure 4.23), which is a function of the strength of the planetary dipole moment and the solar wind parameters at the planet's orbit. However, each planet has one or more unique and significant distinguishing features, which make it interesting to magnetospheric physicists in its own right. For example, Mercury has no ionosphere and an insulating regolith, yet still seems to show evidence of substorm type activity, even though there can be only limited flow of field-aligned currents to the planet. As we have seen above, FACs play a key role in substorm dynamics at the Earth. Jupiter's magnetosphere is the largest in the solar system, and strong centrifugal forces associated with the rapid rotation of the planet cause significant distortions in the inner magnetosphere. In addition,

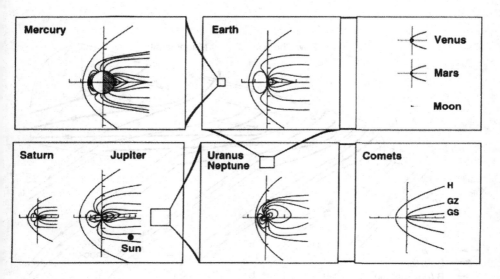

Figure 4.23 Comparison of the sizes of the magnetospheres of the magnetised planets (left four panels), the solar wind disturbances caused by some non-magnetised bodies (Venus, Mars and the Moon in the upper right panel) and comets (bottom left panel; H = Halley, GZ = Giacobini–Zinner, GS = Grieg–Skellerup). (From Coates, 1999.)

the moon Io is a significant internal source of magnetospheric plasma, which forms a *plasma torus* about the planet, adding an extra feature to the magnetospheric dynamics. Neptune and Uranus have dipole axes which have significant tilts away from the rotation axes. Also, at Uranus the rotation axis lies close to the ecliptic plane. Consequently these two gas giants have times when the dipole axis points directly into the solar wind, which may have direct access to the planet down the funnel of the magnetic cusps.

4.5.8.2 *Solar-wind–ionosphere interaction*

Venus has no significant magnetic field, but has a very thick atmosphere. The intensity of sunlight at Venus orbit leads to high levels of photo-ionisation, resulting in a dense ionospheric plasma. This plasma constitutes a conducting obstacle to the solar wind flow. The solar wind and IMF are frozen out of the cavity occupied by the ionospheric plasma. Thus there is a bow shock upstream of Venus that acts to slow and deflect the solar wind and a magnetosheath between the topside of the ionosphere and bow shock, where the flow is subsonic. The obstacle to the flow presented by Venus's ionosphere is of course much smaller than the obstacle presented by the Earth's magnetosphere (even though the two planets are similar-sized bodies). The boundary between solar wind and ionospheric plasma, the *ionopause*, is located at the point where the solar wind dynamic pressure is balanced by the thermal pressure of the Venus ionospheric plasma. Note, however, that a small

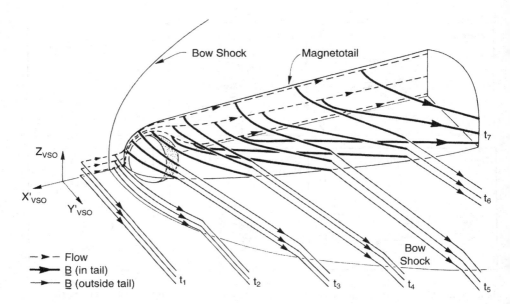

Figure 4.24   Formation of an induced magnetic tail at Venus, which has no intrinsic magnetic field but a dense ionosphere. IMF field lines passing close to the planet are mass-loaded, and thereby slowed, by newly ionised particles from the planetary atmosphere. Parts of the field lines further from the planet continue at the solar wind flow speed, such that the magnetic field forms a tail-like structure. (Saunders and Russell, 1986; reproduced by permission of American Geophysical Union.)

fraction of neutral particles from the atmosphere may reach altitudes above the ionopause, before they are photo-ionised. These newly created ions are '*picked up*' in the flowing magnetosheath plasma, and rapidly travel downstream from Venus. This process thus removes ions of planetary origin from above the ionopause, and is termed '*scavenging*'. However, the pick-up process causes a *mass loading* of the IMF field lines in the vicinity of the planet. Hence the solar wind is further slowed in this region since the solar wind momentum must be shared with the new ions. Since the more distant portions of the field line are unaffected by this process, these sections continue at the undisturbed solar wind speed. This has the effect of draping the IMF field lines over the ionospheric obstacle, thus creating an induced magnetotail structure in the region downstream from the planet (Figure 4.24).

4.5.9   *Insulator bodies (e.g. the Moon)*

The Moon is a non-conducting body, and with the exception of some isolated areas of weak remnant magnetism, it has no significant magnetic field. Thus there are no processes which act to deflect solar wind plasma from the Moon. The solar wind impinges directly on the lunar surface at supersonic speeds. There is no bow shock or magnetospheric or ionospheric cavity. The particles are absorbed at the surface such that lunar soil must contain a record of the ancient solar wind, although spluttering of lunar material following impact may also result in a scavenging process by the solar wind.

Conversely, the lunar surface and interior conductivity are thought to be very small, so the magnetic diffusion time for magnetic fields through the Moon is very small. Thus the IMF passes through the Moon virtually unaffected. This magnetic flux emerges from the nightside of the moon into a plasma-free lunar wake in the downstream region. Close to the moon this wake has the same cross section as the moon, but this fills in as solar wind plasma that missed the Moon flows along the field lines into the void. The length of the wake thus depends on the ratio of the solar wind flow speed to thermal speed, since the latter is the characteristic filling speed (a hotter plasma will migrate along the field more quickly to fill up the depleted wake). It also depends strongly on the magnetic field configuration. During periods when the field is closely aligned with the flow, the plasma needs

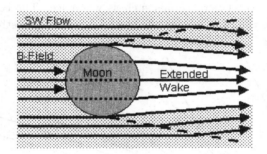

(a) Flow-aligned B-field

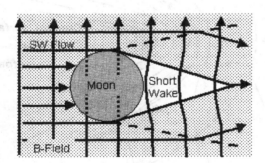

(b) Flow-perpendicular B-field

Figure 4.25 Solar wind interaction with the Moon, an insulating body with no intrinsic magnetic field and no significant ionosphere. The solar wind impinges directly on the sunlight surface, while the IMF passes through the Moon almost unaffected. The extent of the wake region downstream of the Moon depends on the relative orientation of the IMF to the flow: (a) for a flow-aligned field, the solar wind plasma cannot easily convect perpendicular to the field and so the wake is extended; (b) for an IMF perpendicular to the flow, the solar wind plasma quickly fills in the wake as it may move along the field direction.

to migrate across the field to fill in the wake, and we have seen this is not easy for a high-conductivity plasma. Thus an extended wake persists (Figure 4.25(a)). Conversely, when the field is near-perpendicular to the flow, the wake will fill in relatively quickly (Figure 4.25(b)).

Note that the Moon does have small regions of remnant surface magnetisation, perhaps formed during meteor impacts in the past. These have recently been shown to form mini-magnetospheres, which cause a small deflection in the solar wind near the surface. Note also that the Moon spends part of its orbit inside the terrestrial magnetosphere, such that its interaction at these times is with the various magnetospheric fields and plasma populations discussed above, rather than the solar wind.

4.5.10 Comets

A comet nucleus is a relatively small body ($\sim$ a few tens of kilometres in size) made up predominantly of volatile materials such as water ice. The comet spends most of the time far from the Sun, where this material remains mostly inert. However,

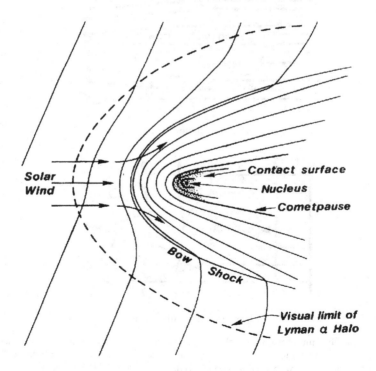

Figure 4.26   Solar wind interaction with a comet. Mass loading of IMF field lines by newly ionised outgassing cometary material acts to slow the solar wind flow in the vicinity of the comet. Since the IMF is frozen to the solar wind flow, this slowing induces a cometary magnetic tail downstream of the coma. (Luhmann, 1991; reproduced by permission of American Geophysical Union.)

near orbit perigee, when the comet passes rapidly close to the Sun, some of the volatiles are heated sufficiently to be released as neutral atoms, which form a vast ($\sim 10^6$ km) gas cloud, or coma, around the nucleus. The solar wind does not interact significantly with the neutral atoms, or with the nucleus. However, atoms in the gas cloud are photo-ionised to form cometary (water group) ions. These heavy ions are picked up in the solar wind. More ions are picked up where the gas cloud is denser (i.e. nearer the nucleus) and, as at Venus, the momentum associated with the solar wind must be shared with the additional mass of cometary plasma. Thus a solar wind flux tube that passes through the cometary plasma will be retarded in the vicinity of the comet by this mass-loading process. The IMF field lines are draped around the nucleus and thus create an induced magnetotail downstream from the comet (Figure 4.26). Note that two other plasma boundaries are often defined for the solar-wind–comet interaction. The contact surface defines the region from which the solar wind is effectively excluded by a balance of the solar wind dynamic pressure with that of the outflowing cometary plasma. The cometopause defines the region beyond the contact surface in which the plasma is predominantly of cometary origin.

4.6 Concluding Remarks

By necessity, this chapter has only grazed the surface of a number of research areas under the umbrella of space plasma physics. We have been able to do little more than introduce each application, although it is hoped with the basic concepts of fluid and particle motions set up in the early sections, the fundamental processes controlling the structure and dynamics of space plasma may be understood. Many other areas have received no treatment at all in this chapter, yet are important areas in their own right. These include plasma waves, plasma equilibria and instabilities, the details of the magnetospheric processes at other planets, the details of magnetic storms and space weather, and the details of the coupling processes (the field aligned currents) between the magnetosphere and the ionosphere. Any one of these areas could form a chapter or more in its own right, so again it is hoped that this chapter has provided a basic conceptual introduction on which to build with further reading.

Chapter 5

Space Weather

Sarah Matthews

5.1 What Is Space Weather?

The NASA National Space Weather Program defines space weather to be 'those conditions on the Sun and in the solar wind, magnetosphere, ionosphere and thermosphere that can influence the performance and reliability of space-borne and ground-based technological systems, and can affect human life or health. Adverse conditions in the space environment can cause disruption of satellite operations, communications, navigation, and electric power distribution grids, leading to a variety of socio-economic losses.'

Understanding the origins of space weather is a topic that has been receiving increasing attention over the last few years. Since the advent of space flight in 1957, mankind's exploitation of space has continually increased. Many decades of space missions have led us to realise that space is often far from a vacuum. The Sun emits a constant stream of plasma and magnetic field, known as the solar wind, which traverses interplanetary space, ultimately reaching and interacting with the edge of the Earth's magnetosphere. In addition to disturbances that can be created by the 'normal' solar wind, solar activity also produces events such as flares and coronal mass ejections (CMEs) that can produce spectacular effects on the near-Earth space environment. Some of these effects include: acceleration of charged particles in space, intensification of electric currents in space and on the ground, enhanced auroral displays and global magnetic disturbances on the Earth's surface. Despite the recent heightened interest in this topic, it has in fact existed for many years, and encompasses the fields of solar, solar–terrestrial and atmospheric physics. What is new about the umbrella of space weather is the coupled approach that is now being taken in order to understand the whole chain of processes that produce these effects.

Solar–terrestrial physics evolved from studies of geomagnetism in the 19th century, whose aim was to understand the origins of the electric currents that generate global geomagnetic disturbances, such as the geomagnetic storms and substorms that we refer to today. It has been clear since the end of the 19th century that these magnetic storms are generated by currents flowing in the magnetosphere and upper atmosphere. However, at that time the link with events on the Sun was less clear. In 1851 Carrington and Hodgson were independently performing routine observations of sunspots when they observed a sudden and rapid brightening in a particular sunspot group. This 'white light' flare was followed almost immediately by a compass deflection and a few days later by an intense geomagnetic storm. Universal acceptance of any causal link between these two events took rather longer to arrive.

The existence of space weather at all is due to the existence of both a solar and a terrestrial magnetic field. The Sun's magnetic field underpins all of solar activity without which there would be no solar wind, no solar flares and no coronal mass ejections. Without the terrestrial magnetic field we would have no protection from all that solar activity has to throw at us.

5.2   Solar Activity

For as long as we have been observing the Sun, the manifestations of solar activity have been apparent. The longest records document the cyclic variation of sunspot number, Hale's polarity law and the butterfly diagram, all of which are described in detail by Harra (Chapter 6). The solar magnetic field is generated below the photosphere by a dynamo process involving induction, produced by the motions of the conducting plasma below the Sun's surface. What we observe at the solar surface is one of the loss terms of this dynamo coming from magnetic buoyancy. The properties of this emerging flux can then be used to infer details of the dynamo process.

From the space weather point of view the importance of solar activity lies in the changes in the corona that are induced by changes in the magnetic field over the solar cycle and the propagation of these effects out into the interplanetary medium. The structure of the corona is determined by the extension of the emerged magnetic flux up into the solar atmosphere. Consequently the basic characteristics of the corona are controlled by the solar dynamo and show variations that reflect the changes in the magnetic field over the course of the solar cycle.

At solar minimum the X-ray corona is often completely devoid of active regions and is much fainter and weaker than the solar maximum corona. Polar coronal holes are large and prominent at this point of the cycle and white-light observations show that the streamer belts are concentrated in the equatorial plane. At solar maximum the X-ray corona is bright and highly structured with many active regions present that dominate the emission. The polar coronal holes are somewhat diminished and the white-light corona is also much more highly structured, with streamers present at all latitudes. Figure 5.1 illustrates the dramatic change in

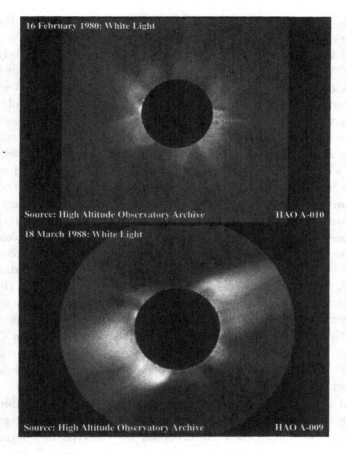

Figure 5.1 Images of the white-light corona during total solar eclipse at solar minimum in 1988 (*bottom*) and solar maximum (*top*) in 1980, illustrating the difference in coronal topology. Notice the prominent coronal holes and equatorial streamer belt in the solar minimum corona compared to the much greater latitudinal distribution of streamers in the solar maximum image. (All eclipse images digitised from photographs in the archives of the High Altitude Observatory, National Center for Atmospheric Research, Boulder, Co.)

the corona at solar minimum and maximum. At times in between maximum and minimum the most obvious difference in coronal structure is the appearance of equatorial coronal holes that can persist for several rotations, bringing regions of open magnetic field down close to the streamer belt.

5.3 The Solar Wind

In 1957 Chapman showed that the high thermal conductivity of the Sun's corona meant that it should extend out into interplanetary space rather than being confined near the surface like planetary atmospheres. In the following year Parker showed that in fact an inconsistency in Chapman's analysis meant that the corona actually expands outwards forming the solar wind. Thus the existence of the solar wind was predicted and Parker went on to show that the expansion was likely to have a velocity of 1000 km s^{-1} at 1 AU. Despite the fact that there were already indications that a solar wind should exist, including the 27-day recurrence period seen in early studies of strong aurorae, and Biermann's arguments in 1951 that solar radiation pressure alone was insufficient to explain the orientation of comet tails away from the solar direction, Parker's supersonic wind remained somewhat controversial until the first definitive *in situ* observations from the *Mariner 2* spacecraft launched in 1962.

The solar wind is a continuous outflow of ionised plasma pervaded by magnetic field from the Sun. The huge pressure difference between the corona and the interstellar medium drives the plasma outwards, despite the countering force of gravity. The wind expansion is modulated by the magnetic field. There are two types of solar wind: high speed ($\geq$ 600 km/s) and low speed. Observations from *Skylab* confirmed the association of the high speed streams with coronal holes,coronal holes where the field lines are to all intents and purposes open to the interstellar medium. As described above, the nature of the corona, and thus of the coronal holes, changes with the solar cycle. At solar minimum large polar coronal holes dominate and the fast speed solar wind originates at high latitudes, as can be seen in Figure 5.2 using results from the Solar Wind Observations Over the Poles of the Sun (SWOOPS) instrument on the *Ulysses* spacecraft. The presence of equatorial coronal holes at times of declining solar activity brings these fast speed streams down into the equatorial plane in the region of the streamer belt.

The origin of the slow speed solar wind streams is less well determined. These streams are seen to originate from lower latitudes and thus from regions where the field is predominantly closed. In order for the material that forms the slow

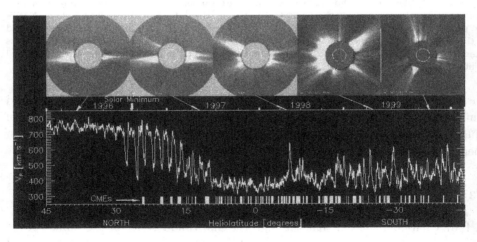

Figure 5.2   From McComas *et al.*, 2000, showing LASCO images of the white-light corona and the solar wind speed from the SWOOPS instrument plotted as a function of heliolatitude. (Reproduced by permission of American Geophysical Union.)

solar wind to escape, Fisk (1996) has proposed that the combination of the Sun's differential rotation and super-radial expansion brings high latitude field lines down to lower latitudes where reconnection occurs with closed field lines, allowing plasma within closed loops to escape into the slow solar wind.

The combination of radial expansion and solar rotation on the solar wind causes the solar magnetic field to become wound up into an Archmidean spiral similar to the effect that you would get by swinging a garden hose. The angle between a radial vector and the IMF is about 45° to the Earth's orbit. Despite its limitations, Parker's solar wind model is still a good description of what we actually observe. However, note that Fisk's magnetic field model is better able to account for e.g. the slow solar wind.

At times of solar minimum the solar magnetic field can be approximated to that of a simple dipole. Observations of the Sun's magnetic field at the surface clearly reveal that on small scales the Sun's field is much more complicated than this. However, this complicated field can be described by a spherical harmonic expansion, the higher order terms of which fall off more rapidly with distance than the first order dipolar term. Hence, a dipole is a good first order description of the solar field at minimum.

Figure 5.3 illustrates the closed field lines in the equatorial regions, while the field lines at higher latitudes are carried out into the heliosphere by the solar wind. Near the equator is a region where oppositely directed field lines meet. Maxwell's equations tell us that a current sheet must form in this location, and this is known as the heliospheric current sheet.

Even at solar minimum there is a small offset between the Sun's rotation and magnetic axes. As the magnetic axes precess about the rotation axis, the heliospheric current sheet also rotates with the Sun. The resulting structure of the current sheet is often referred to as the 'ballerina skirt', as shown in Figure 5.4. In principle then, a spacecraft in the equatorial plane of the heliosphere should observe

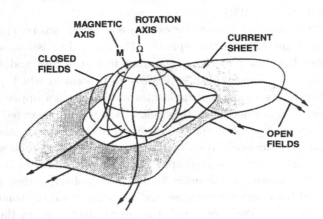

Figure 5.3 Formation of the heliospheric current sheet. (Smith, 1997; reproduced courtesy of University of Arizona Press.)

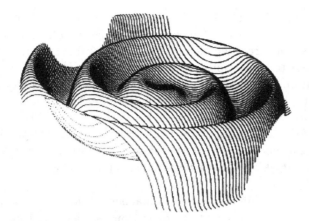

Figure 5.4   Ballerina skirt topology of the heliospheric current sheet. (Forsyth and Gosling, 2001.)

outwardly directed field lines for half the time and inwardly directed field lines for the other half. In practice the current sheet has localised warps that complicate the field polarity pattern, so that this is not the case.

5.4   Aurora

Amongst the oldest-known and certainly most spectacular displays of the effects of space weather are the aurorae. The *aurora borealis* and *aurora australis* are, respectively, mostly permanent features of the northern and southern polar night. Occasionally, during periods of enhanced solar activity aurorae are seen to extend to much lower latitudes, giving the more densely populated areas of the globe the opportunity to observe these spectacular light shows. Figure 5.5 (Plate 7) shows an example of the aurora australis (southern lights) taken by the crew of the space shuttle *Endeavour* in April 1994.

The origin of the aurora relies on the existence of three factors: the solar wind, magnetic fields and the Earth's atmosphere. The auroral emission occurs as the result of collisions between particles originating from the Sun and those in the Earth's atmosphere. The mechanism is a two-step process in which precipitating electrons and ions excite atoms and molecules in the Earth's upper atmosphere, leading to the emission of photons as they relax back to an un-excited state.

As described above, the solar wind carries with it the solar magnetic field. On reaching the region of influence of the Earth's magnetic field, the solar wind distorts the Earth's field, producing a comet-shaped magnetosphere as described by Owen (this volume). The terrestrial magnetic field acts as a barrier that protects the Earth from most of the energetic particles and radiation from the solar wind. The traditional explanation of the polar aurora suggests that some of these particles are able to penetrate into the magnetosphere via the polar cusps, where they are accelerated along the Earth's field lines and reach the neutral atmosphere in a rough circle, called the auroral oval. This annulus is centred over the magnetic pole of the Earth between the 60° and 70° north and south latitudes. However, this explanation is overly simplistic.

When the solar magnetic field encounters the terrestrial magnetic field, it is possible for reconnection to occur, producing reconnected field lines that interconnect across the magnetospheric boundary. The field lines from the Earth that are involved in this process originate from the area contained within the auroral oval, which is approximately centred at the pole. The solar wind thus flows both along the magnetospheric boundary and across these reconnected field lines in a generator-like process. This solar wind-magnetosphere generator can generate more than 10^6 MW of power, with an accompanying induced voltage of 20–150 kV. Since charged particles can only move easily along magnetic field lines, the magnetic field lines behave like current-carrying wires, and since the reconnected field lines originate from an oval area centred on the pole, the terminals of the generator are connected to the boundary of this oval by magnetic field lines. As a result the discharge circuit links the morning side of the magnetospheric boundary (positive), through the morning half of the oval boundary, most of the way around the auroral oval and out to the evening side of the magnetospheric boundary (negative). This system is called the region 1 current system. Induced by the region 1 current system is the region 2 current system, which is the part of the region 2 current that is discharged back to the magnetosphere from the equatorward boundary of the morning side oval and from the magnetosphere to equatorward boundary of the oval on the evening side. Currents are carried by electrons, so that there is a downward flow of electrons carrying an upward current from the magnetosphere to the ionosphere along the polar boundary of the auroral oval on the evening side, and along the equatorward boundary of the oval in the morning sector, as shown in Figure 5.6. It is these downward-flowing electrons that excite the atoms and ions in the upper atmosphere that produce the auroral emissions.

The field-aligned currents close by connecting to horizontal currents in the ionosphere that are concentrated at two main locations along the auroral oval: westwards in the morning side and eastwards in the evening side. These currents are called the *auroral electrojets* and are commonly on the order of 10^6 A. A large fraction of the power created by the solar-wind–magnetosphere generator is dissipated via Joule heating in the ionosphere, creating large-scale atmospheric circulation of the upper atmosphere.

There is a good analogy between the image produced in a television and the aurora where the aurora corresponds to an image on the screen of the cathode ray tube. Movement of the image is produced by movement of the impact point of the electron beam on the screen, which is produced by changes in the electric and magnetic fields in the modulation devices in the tube, which are in turn produced by input changes. As a result of this similarity many workers have been trying to

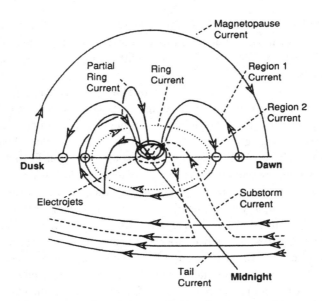

Figure 5.6   Schematic representation of currents linking the magnetosphere and ionosphere (from Clauer and McPherron, 1980; reproduced by permission of American Geophysical Union). The partial ring current flows near the equatorial plane near dusk and closes through the ionosphere. The substorm current wedge is a diversion of the tail current that also links into the ionosphere via field-aligned currents. In the ionosphere these current systems flow in channels of high conductivity at high latitudes and are called the auroral electrojets. The sheets of field-aligned currents near dawn and dusk are called the region 1 and region 2 current systems, with higher latitude region 1 currents flowing into the ionosphere from the dawn sector and out at dusk. The lower latitude region 2 system has opposite polarity.

infer changes in the electric and magnetic fields of the magnetosphere on the basis of auroral motion.

5.5   Auroral Substorms

In quiet geomagnetic conditions the auroral ovals are relatively fixed in position, with each oval extending furthest towards the equator on the night side. On the day side, the auroral ovals appear closer to the geomagnetic poles as a result of the distortion caused by the interaction of the solar wind with the magnetosphere. At the onset of an auroral substorm the auroral curtain in the midnight sector will suddenly increase in brightness by more than order of magnitude and start to move towards the pole at a few hundred ms^{-1}. As a result the thickness of the midnight part of the auroral oval will increase rapidly, producing a large bulge. The result of this can be observed as a travelling surge with a wavy structure. At the same time aurorae in the morning sector often divide into rays. This auroral activity can last for 2–3 h and may be repeated several times by further substorms.

The cause of the auroral substorm is an order of magnitude increase in the solar-wind–magnetosphere generator from $\sim 10^5$ MW during quiet times to 10^6 MW for a period of a few hours. The direction of the solar magnetic field is extremely important in determining this increase. If the solar field is directed southwards reconnection between the two fields is more effective, producing more reconnected field lines and thus a greater intensity of magnetic field across the magnetospheric boundary. This is equivalent to increasing the magnetic field in the generator.

Intensification of the solar wind magnetic field can occur as the result of coronal mass ejections (CMEs) and lead to very large increases in the generating power if the CME is directed towards the Earth and the magnetic field that it carries away is oriented southwards. This type of increase will cause larger discharge currents and brighter aurora. In this case, the discharge current produces intense, rapidly varying magnetic fields that constitute the geomagnetic storm field.

5.6   Co-rotating Interaction Regions (CIRs)

As discussed above, the dipole component of the Sun's magnetic field is drawn out radially and wound into a spiral at low latitudes by differential rotation. Between the opposite polarity hemispheres lies the heliospheric current sheet. At solar minimum the magnetic axis of the Sun is aligned with the rotation axis, but as the solar cycle progresses it tilts and eventually inverts. As a result of this tilt the high speed wind streams that are normally found at high latitudes can be brought down into the ecliptic where the slow solar wind streams originate. As the Sun rotates, high speed wind can then be emitted in the same direction as the slow streams. When the high

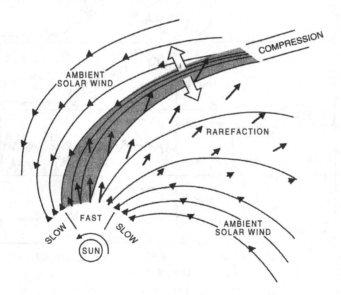

Figure 5.7   Schematic of a co-rotating interaction region in the equatorial plane. Solid lines show the magnetic field lines and the length of the arrows is a measure of the flow speed. (Forsyth and Gosling, 2001.)

speed streams overtake the slow streams an interaction occurs, plasma and magnetic field fluctuations are compressed and intense magnetic fields can be produced. Since the whole system rotates with the Sun, this is called a co-rotating interaction region, and such regions can persist for many solar rotations. At the edges of the CIR a pair of shocks form: a forward shock that propagates outwards and a reverse shock that propagates inwards into the high speed stream. A schematic of a CIR is shown in Figure 5.7. Observations have shown that particles can be accelerated at both shocks. During the declining phase of the solar cycle low latitude coronal holes

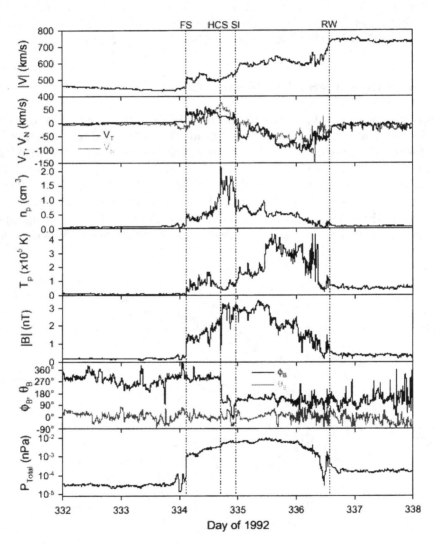

Figure 5.8   Solar wind parameters during a CIR observed by *Ulysses* in 1992.   (Forsyth and Gosling, 2001.)

are common and the interaction of high speed streams from these with slow solar wind streams is the cause of most of the 27-day recurrent geomagnetic activity that we observe, including small geomagnetic storms and high intensity long duration continuous auroral electrojet activity (HILDCAA). An example of a CIR is shown in Figure 5.8. This shows a CIR observed by *Ulysses* at 5 AU in 1992; the dashed lines show the forward and reverse shocks bounding the CIR, the heliospheric current sheet and the stream interface. Plotted from top to bottom are the solar wind parameters.

CIRs do not tend to have fast forward shocks at 1 AU, so there is no sudden storm commencement (SSC) phase associated with them as in magnetic storms. Also, although the magnitude of the B_z component of the magnetic field is often quite high, it is rapidly fluctuating, which means that the storm intensity is correspondingly low.

The recovery phase of these events can often last for weeks as a result of this fluctuating field. Every time the field turns south reconnection occurs, preventing the ring current from decaying and producing the continuous substorm activity known as a HILDCAA. As a result of this behaviour CIRs can actually be more geoeffective in transferring energy into the magnetosphere than interplanetary CMEs at solar maximum.

The origin of the fluctuations in B_z (i.e. the solar magnetic field in the North–South direction) has been revealed by studies using data from the *Ulysses* mission. Cross correlation analysis of fluctuations in B and the solar wind velocity showed high correlation at zero lag, demonstrating that they are Alfvén waves. Alfvén waves occur when the vibration of the magnetic field part of the wave is perpendicular to the direction of travel of the wave.

5.7 Solar Flares

Solar flares are the largest explosions in the solar system, producing enhancements across the electromagnetic spectrum and accelerating charged particles to high energies. For many years solar flares were regarded as the drivers behind all non-recurrent geomagnetic activity. Since their frequency is highest at solar maximum and most non-recurrent geomagnetic activity also occurs then, this was a natural connection to make prior to the discovery of CMEs. While it is now apparent that CMEs probably play the greatest role in producing the largest geomagnetic events, flares do produce space weather effects and may in fact be intimately related to the CME process.

Solar flares are described in detail in Chapter 6 but we will briefly review some of their characteristics here. They can be broadly classified into impulsive and long duration events. Long duration events can last for many hours and are generally associated with eruptive behaviour. The current 'standard' model for long duration eruptive flares has been developed over several decades by a number of workers

(Carmichael, Hirayama, Sturrock, Kopp, Pneumann, Cargill, Priest and Shibata). A cartoon of one of the most recent versions of this model is shown in Figure 9.6.

In this model a filament is activated, becomes unstable and starts to rise. The opened field lines following its eruption close back down by reconnection producing the flare, particles are accelerated and the reconnection jet collides with the top of the soft X-ray loop producing a fast MHD shock that leads to the production of a loop-top hard X-ray source. Accelerated electrons and protons spiral down the legs of the loop; the electrons produce radio emission as they propagate and hard X-ray emission via Bremsstrahlung action with ambient protons in the dense chromosphere. As a result of the energy deposition in the chromosphere, a radiative instability leads to the ablation of heated plasma, which produces enhanced soft X-ray emission in the flare loop, in a process known as chromospheric evaporation. Accelerated protons lead to the production of γ-ray emission via interactions with ambient heavy nuclei.

These large and long duration flares are most commonly associated with CMEs. However, all flares produce enhanced soft X-ray and EUV radiation as well as high energy particles, and it is these properties, even in the absence of any associated mass expulsion, that produce space weather effects.

5.8   The Ionosphere

The ionosphere is that part of the Earth's upper atmosphere that contains free electrons and protons caused by the Sun's ionising radiation. It lies at the base of the magnetosphere, and above 60 km the electron density is high enough to influence radio wave propagation, which is one of the primary reasons for its importance.

The existence of the ionosphere was most convincingly shown after Marconi demonstrated in 1901 that radio waves could propagate across the Atlantic. As a result of this demonstration Kennelly and Heaviside proposed that the waves must be undergoing reflection by a conducting layer high in the atmosphere. The final proof of this came in 1924, when Appleton and Barnett and Breit and Truve demonstrated direct reflection of radio waves from ~ 100 km altitude.

From a space weather point of view the importance of the ionosphere comes from its role in many telecommunications and surveillance transmissions. The maximum useable frequency of such transmissions depends on the electron density in the ionosphere, which is extremely variable. Low frequency waves will be absorbed in the lower ionosphere and the amount of absorption also depends on the electron density. This is strongest during the day as a result of the incident sunlight and sporadic high ionisation events associated with enhanced flare emission (in the EUV and X-ray range) and can lead to complete radio blackouts over short periods.

Although radio transmissions between the Earth and spacecraft are at frequencies not absorbed by the ionosphere, they can still be subject to scintillation effects that can degrade the signal. Phase path changes and propagation delays can also occur, making adjustments to precise measurements necessary, e.g. in radar altimetry and GPS measurements.

The strong ionospheric currents that are induced by large magnetic storms can in turn induce additional currents on the Earth that can flow through grounded electrical power grids and damage transformers or trip circuit breakers. Electrical heating above 120 km during storms raises temperatures, which in turn reduces the rate of decay of the electron density with height. Satellites in low Earth orbits (< 1000 km) can experience increased atmospheric drag as a result, which in turn can lead to them being temporarily lost to tracking services, or in severe cases to premature re-entry.

5.9 Solar Energetic Particle Events (SEPs)

High energy particles from the Sun were first observed as sudden increases in intensity in ground level ion chambers during the large solar events of February and March 1942 by Forbush. Balloon and rocket experiments in the 1950s and 1960s demonstrated that the cause of these events was mainly protons. In order to reach the Earth these protons must have energies in excess of 1 GeV. Protons with lower energies than this will be deflected by the terrestrial magnetic field and channelled into the polar regions. This was long before CMEs were discovered and so these events were naturally attributed to solar flares, and indeed observations from solar flares in the γ-ray region of the spectrum certainly indicate that protons are accelerated to at least MeV energies during these events. The proposition that the energetic proton events originated from solar flares was perpetuated for many years, until evidence started to amass that CMEs were important for these events. In the late 1970s and early 1980s Kahler and colleagues found that there was a high (96%) correlation between large SEP events and CMEs. Also, Pallavicini and co-workers were finding a distinction between impulsive and long duration X-ray events, the latter being associated with CMEs.

SEPs are now understood to be of two different varieties: impulsive and gradual events. The former are believed to be associated with solar flares while the latter occur as the result of a CME. However, given the uncertainties in our understanding of the flare–CME relationship this distinction may not be black-and-white. In terms of this distinction impulsive events produce 1000-fold enhancements in the $^3\mathrm{He}/^4\mathrm{He}$ abundance ratio relative to that measured in the corona, as well as enhancement of the heavy ion population. This is believed to be the result of resonant wave–particle interactions at the flare site. In contrast, gradual events produce the most intense high energy particles as the result of acceleration at CME-related collisionless shocks. These events usually directly reflect the abundances and temperature of the ambient, unheated corona.

The differences between the two types of events described above clearly indicate two different acceleration processes. The earliest evidence of this came from radio observations made by Wild and colleagues in the 1960s. The emission mechanism in radio bursts is plasma emission, so that the frequency of the radio burst, $\nu \propto \sqrt{n_e}$, where n_e is the electron number density. Type III radio bursts showed a drift downwards in frequency with time. They were consequently ascribed to 10–100 keV electrons streaming out from the corona with a velocity of ~ 0.1 c, the decrease in frequency being the result of decreasing electron density with height in the corona. In contrast, type II radio bursts showed a much slower drift rate that seemed to correspond to acceleration at a shock wave moving out through the corona at ~ 1000 km/s. The difference between gradual and impulsive events is illustrated well in Figure 5.9, from Reames (1999). The gradual event was associated with the eruption of a quiescent filament and the impulsive events are associated with impulsive flares that have no evidence of accompanying CMEs. The extended intensity–time profiles of gradual events are due to continuous acceleration, whereas in impulsive events this variation is determined by scattering of the particles as they traverse interplanetary space.

The high correlation between SEPs and CMEs by Kahler does not by itself tell us that SEP acceleration occurs at a CME shock. For example, the halo CME event of 6–10 January 1997 is an interesting case. In this event the CME occurred near central meridian on 6 January and a well-defined magnetic cloud reached the Earth on 10 January, causing a severe geomagnetic storm. However, no interplanetary protons with energies ≥ 1 MeV were observed. The shock transit speed was calculated to be ~ 385–490 km/s, which is not much greater than the ambient solar wind

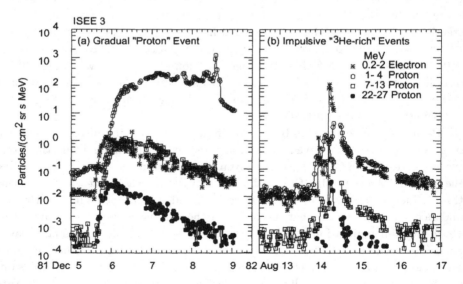

Figure 5.9   Intensity profiles of electrons and protons in (a) gradual and (b) impulsive SEP events. The gradual event was associated with a disappearing filament and CME but no flare, while the impulsive events came from a series of flares with no related CME. (From Reames, 1999; reproduced by kind permission of Kluwer Academic Publishers.)

speed. Only when shock transit speeds are in excess of 500 km/s do SEP events become likely, and it appears that speeds greater than 750 km/s always produce SEP events. Therefore only the fastest 1–2% of CMEs cause particle acceleration.

However, Kiplinger (1995) found a good correlation between interplanetary proton events and progressive hardening of the hard X-ray spectrum in solar flares with 82% of flares that displayed spectral hardening having associated 10 MeV proton events or enhancements, and 95.4% of events that did not display spectral hardening having no associated proton event. Also, recent work by Klein *et al.* (2000) suggests that even when a CME is involved acceleration in the low corona is still important, with temporal variations in radio emission showing good correlation with temporal variations in the proton flux at 1 AU.

The production and propagation of high-energy particles is clearly important because of the potential threat to both humans and spacecraft sub-systems. In terms of spacecraft effects SEPs are most commonly associated with causing severe radiation damage and single event upsets (SEUs). For example, the 'snow' that

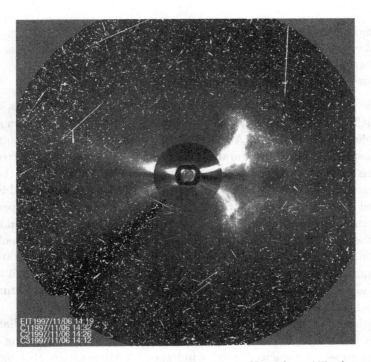

Figure 5.10 A coronal mass ejection that took place on 6 November 1997, observed with the LASCO and EIT instruments on *SOHO*. The highly energetic protons produced from the flare caused numerous bright points and streaks on the images, giving the snow-like appearance. A consortium of the Naval Research Laboratory (USA), the Max-Planck-Institut fuer Aeronomie (Germany), the Laboratoire d'Astronomie (France) and the University of Birmingham (UK) produce the SoHO/LASCO data used here. SoHO is a project of international co-operation between ESA and NASA.

is seen in images from the instruments on the *Solar and Heliospheric Observatory* (*SOHO*) following solar flares and CMEs is the result of proton bombardment of the detectors (see Figure 5.10 for an example). Less obviously visible but more damaging are the SEUs that occur in electronic sub-systems when a charged particle ionises a track along a fraction of the circuit and causes the circuit to change state. From a human point of view astronauts in low Earth orbit are particularly at risk from tissue damage caused by high energy particle fluxes. This is especially true during space walks or during events where a CME has caused compression of the magnetosphere. Also at risk, but to a lesser extent, are members of the air crew and passengers on high altitude aircraft, who can expect to receive a greater cosmic ray dose.

5.10   Other Sources of Energetic Particles

As well as the energetic particles produced in SEPs, space weather effects are also produced by:

- Galactic cosmic rays ($E > 100$ MeV)
- Trapped energetic ions in the inner radiation belts (10–100 s MeV)
- Relativistic electrons in the radiation belts (0.2–5 MeV)
- Substorm-produced electrons (10–100 keV)

Galactic cosmic rays, produced for example in supernova explosions, can penetrate the heliosphere, where they are decelerated to some extent and then guided by the heliospheric magnetic field. At the peak of the sunspot cycle, the magnetic field of the heliosphere is strongest and most effective at deflecting galactic cosmic rays; there is thus an inverse correlation between galactic cosmic ray flux and the sunspot cycle. These produce similar effects to SEPs, e.g. radiation damage and SEUs.

Trapped ions and protons in the inner radiation belts are also responsible for producing radiation damage and SEUs. The toroidal van Allen radiation belts that surround the Earth are aligned with the Earth's magnetic axis, which is offset from the rotation axis by 11°. On one side of the Earth the inner belt is 1200–1300 km from the surface while on the other side it is 200–800 km. Over Brazil the dip is greatest and spacecraft in low Earth orbit passing over this region actually enter the radiation belt. This is called the South Atlantic anomaly.

The high energy (MeV) electrons trapped in the Earth's outer radiation belts are subject to highly variable changes in relativistic electron flux during magnetic storms and times of enhanced solar wind speed with negative B_z, and present a severe threat to astronauts and electronic components. In particular they can penetrate spacecraft walls and embed themselves in dielectric materials, leading to a process called deep dielectric charging, where induced potential differences can lead to surges inside circuits. The processes that lead to these relativistic electron enhancements are not currently well understood but could be due to acceleration of lower energy electrons already present in the radiation belt or particles injected during substorms. The acceleration mechanism is currently unknown. Candidate acceleration mechanisms are waves caused by (1) externally driven waves on the magnetopause as a result of high solar wind speeds, and (2) internally generated waves due to unstable particle distributions.

Substorm-produced electrons are responsible for surface charging effects in spacecraft as a result of interactions between the ambient plasma medium and the spacecraft surface. In a cool, dense plasma there is a balance between incident particles and emitted particles, producing a low, and usually slightly positive, spacecraft potential, whereas this current balance is much harder to attain in a hot plasma and large potentials can occur.

5.11 Coronal Mass Ejections and Geomagnetic Storms

In 1991 Jack Gosling showed that there was an excellent correlation between major geomagnetic storms ($K_p \geq 7$) and the interplanetary manifestations of CMEs. The recognition of the role of CMEs in creating geomagnetic disturbances led Gosling to coin the term 'solar flare myth', in which he proposed that contrary to the previous wisdom flares had nothing whatever to do with geomagnetic disturbances. This proposal led to some years of fierce rivalry between the flare and CME factions of the solar physics community, with calls for resources to be diverted away from solar flare research and into CMEs.

Some workers argued that CMEs were merely a large-scale coronal response to solar flares. However, results from the *Solar Maximum Mission* (*SMM*) suggested that flares can occur both before and after CME onset (e.g. Harrison, 1990). More recent results from the *SOHO* and *Yohkoh* missions suggest that in fact flares and CMEs are probably intimately related, with both likely to be the result of large-scale destabilisation in the coronal magnetic field. Currently, the trigger for neither event is well understood.

Hundhausen (1993) proposed that a CME be defined as 'an observable change in coronal structure that (1) occurs on a timescale between a few minutes and several hours and (2) involves the appearance of a new discrete, bright white-light feature in the coronagraph field of view'. They involve the expulsion of plasma and magnetic field from the solar corona into interplanetary space and the conversion of magnetic energy to kinetic. Typical energies of CMEs are similar to those seen in large flares, approximately 10^{25} J, although it should be noted that in flares this energy is mainly distributed in heating and acceleration of particles, while in CMEs it is mostly kinetic.

CMEs are still best observed with coronagraphs and while there are still a number of ground-based facilities (e.g. Mauna Loa Solar Observatory (MLSO) in Hawaii and Mirror Coronagraph for Argentina (MICA) in Argentina), the biggest advances in our understanding of CMEs over the last 30 years have come from space-based instruments. The first of the space-based coronagraphs was on *Skylab*, which operated between 1973 and 1974 with a field of view between 1.5 and 5 solar radii. Since then new instruments have provided improvements in sensitivity and field of view. After *Skylab* the Solwind instrument on *P78-1* operated from 1978 to 1985, observing over 2.5–10 solar radii. It was followed by the *SMM* coronagraph, which was operational in the years 1980, 1984–89 over 1.6–5 solar radii. Mostly recently the LASCO (Large Angle and Spectrometric Coronagraph) instrument on *SOHO* has been observing with three nested coronagraphs (C1: 1.1–3; C2: 1.5–6; C3: 4–30 solar radii) at the L1 point since 1996, with only a short break in 1998, when the satellite was temporarily lost.

In terms of morphology CMEs are much like flares in that each one is different. However, for both flares and CMEs it is possible to make general classifications, and one of the most important of these is the three-part CME structure that is characteristic of some events (e.g. Illing and Hundhausen, 1985). The earliest observations that displayed these characteristics were made by *SMM*. For these CMEs the pre-event configuration had a prominence (see the solar physics chapter for a definition of a prominence) above which was a dark void (the prominence cavity) and both were located at the base of a helmet streamer. As the CME occurred a three-part structure (bright front, dark cavity and bright prominence) corresponding to each of these regions was seen to move out into space. A good example of a three-part CME observed by LASCO is seen in Figure 5.11.

The importance of this three-part structure was identified by Low (1996) when he suggested that the prominence cavity was in fact a flux rope. A flux rope is an organised, twisted region of magnetic field that is distinct from its surroundings. The prominence is supported inside this flux rope, which otherwise contains only low density plasma. The emission from the prominence thus dominates and the region above appears dark.

The occurrence rate of CMEs depends on the phase of the solar cycle, as is the case with flares. At solar minimum the average is a rate of ~ 0.5 per day. At maximum this rises to between 2 and 3 per day (e.g. St Cyr *et al.*, 2000). The phase of the solar cycle also has a bearing on the latitudinal distribution of CMEs. Near solar minimum CMEs are most concentrated at the equator, while they originate over a wide range of latitudes at maximum. This suggests a strong connection with streamers, since these show a similar cycle dependency. CMEs are huge structures with a distribution of angular widths that peaks at $\sim 40°$, but has a tail extending up to 360°, with these events being called 'halo' CMEs. The mass expelled during a CME is typically a few times 10^{12} kg. Given that the solar wind mass flux is 10^{14} kg per day, this makes CMEs a small contribution to the Sun's total mass loss. Velocities range from 100 to 2000 kms^{-1}, with an average value of ~ 350 kms^{-1}, measured in the plane of the sky.

It is widely believed that CMEs are accelerated out into the corona by magnetic forces. As a result of the high temperatures associated with the corona, it is

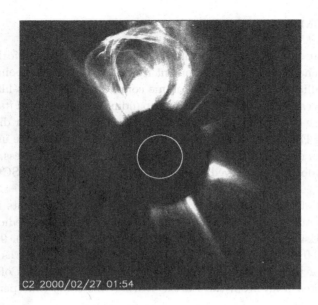

Figure 5.11   The 'light bulb' CME observed by LASCO, showing the common three-part structure of bright leading front followed by dark cavity and bright prominence. A consortium of the Naval Research Laboratory (USA), the Max-Planck-Institut fuer Aeronomie (Germany), the Laboratoire d'Astronomie (France) and the University of Birmingham (UK) produce the SoHO/LASCO data used here. SoHO is a project of international co-operation between ESA and NASA.

not possible to measure the magnetic field there directly using the Zeeman effect. However, measurements of the CME acceleration profile can be used to investigate the hypothesis that the CME is driven by magnetic forces. St Cyr *et al.* (2000) have determined the acceleration profile in 17% of the CMEs observed by LASCO over the period 1996–99. Averaged over the C2 and C3 fields of view, they found accelerations in the range of 1.4–4 ms^{-2}. They found no evidence of deceleration. However, some cases of deceleration have been reported (Sheeley *et al.*, 1999) for fast CMEs moving perpendicular to the plane of the sky.

Cargill (2001) has pointed out that these results are unexpected. If you consider a CME moving through the corona, then it is subject to an outward Lorentz force, an inward gravitational force and a net drag, which can be in either direction, depending on the relative motion of the CME with respect to the solar wind. One would therefore expect a rapid deceleration unless the density in the CME is very high or the magnetic forces are very strong.

The discovery of the three-part CME structure prompted the search for evidence of magnetic flux ropes in coronagraph data that should appear as circular intensity patterns that move out through the corona. Intensity patterns of this kind have been seen by Dere *et al.* (1999), suggesting that flux ropes do indeed exist as part of CMEs in the inner corona. In the first two years of LASCO operations, about 25–50% of CMEs showed this helical structure.

5.12   Halo CMEs

The events that fall in the tail of the distribution of angular widths, the 'halo' CMEs, are potentially the most important from a space weather point of view but are the most difficult to observe. The name comes from their halo-like appearance around the Sun caused by the CME front expanding outwards from its origin on the solar disc. These events can be produced on either side of the solar disc, but if this origin is on the Earth-facing side of the disc, then these events are usually headed our way. This is not a sufficient condition to ensure geo-effectiveness, however. An example of a halo CME observed with the C2 coronagraph on LASCO is shown in Figure 5.12.

As a result of their potential to be Earth-directed, halo events have received a great deal of attention in recent years as potential tools for predicting geomagnetic storms. Using criteria proposed by Thompson *et al.* (1999), 92 halo CMEs were identified from the LASCO data by St Cyr *et al.* (2000) as fulfilling the requirements for potentially hazardous space weather. About half of these turned out to be directed away from the Earth. During this same time period about 21

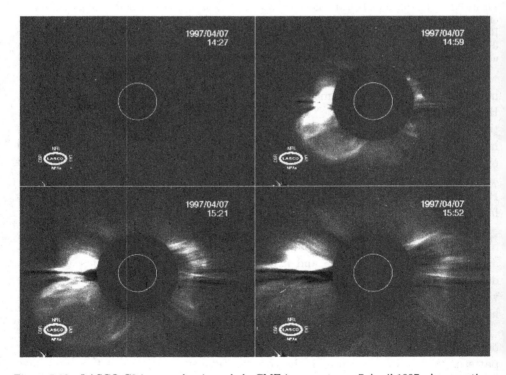

Figure 5.12   LASCO C2 images showing a halo CME in progress on 7 April 1997. A consortium of the Naval Research Laboratory (USA), the Max-Planck-Institut fuer Aeronomie (Germany), the Laboratoire d'Astronomie (France) and the University of Birmingham (UK) produce the SoHO/LASCO data used here. SoHO is a project of international co-operation between ESA and NASA.

large geomagnetic storms occurred and 15 of them could be fairly confidently associated with halo CME events. Three of the others occurred during *SOHO* data gaps, but that still leaves 28 halo events that were Earth-directed and yet did not produce major geomagnetic activity. Arguably, if B_z is southwardly directed 50% of the time, this number would be about expected. However, there is clearly still more work to be done.

The discovery of low coronal signatures of CME onset has made the possibility of developing prediction tools for potentially hazardous events a much more attainable goal, although we are still a long way away from understanding what triggers CMEs and predicting their occurrence. Observations with *Yohkoh* and *SOHO* in the EUV and X-ray have shown that the expulsion of mass associated with the CME can be seen as sudden subsequent depletions or dimmed regions in the low corona. The mass estimates for this plasma are typically an order of magnitude lower than those determined from white-light coronagraph measurements, clearly demonstrating that a significant amount of the CME mass comes from these low coronal regions. Recently, spectroscopic measurements in dimming regions have indicated significant outflow velocities, confirming that this plasma is indeed being expelled (Harra and Sterling, 2001).

Other signatures that seem to be associated with CME onset are coronal waves, seen in a few cases with the EIT instrument on *SOHO*. A recent study by Biesecker *et al.* (2002) found that 92% of 38 EIT waves had an associated CME over the period March 1997 to June 1998. These 38 waves were those that could be identified with the highest certainty to be waves. These waves propagate across the disc at average speeds of a few hundred km/s. They are by no means ubiquitous, however.

In terms of potential prediction methods the appearance of S-shaped or reverse S-shaped structures (Figure 5.13) in the corona has been a topic of particular interest recently. These sigmoidal structures have been suggested to have a greater probability of eruption than non-sigmoidal regions (Canfield *et al.*, 1999). The

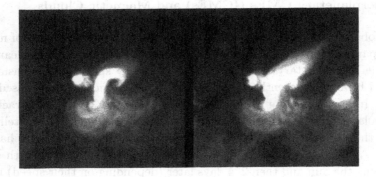

Figure 5.13   A sigmoid at S25E01 before (*left*, 8 June 1998, 15:19 UT) and after (*right*, with cusp, 9 June 1998, 16:17 UT) an eruption has taken place. Solar X-ray image from the *Yohkoh* Soft X-ray Telescope. The Soft X-ray Telescope (SXT) was prepared by the Lockheed Palo Alto Research Laboratory, the National Astronomical Observatory of Japan and the University of Tokyo with the support of NASA and ISAS. The image size is 491 × 491 arcsec.

S-shape suggests that these regions may possess a high level of magnetic shear, a factor that is important in determining the amount of free energy available in the magnetic field. However, Glover *et al.* (2000) have shown that while it appears that sigmoidal regions do have a greater probability of erupting, projection effects can lead to the mis-classification of regions as sigmoidal.

The processes that lead to the onset of a coronal mass ejection are still very unclear, although the earlier suggestion of a flare-related pressure pulse has been largely ruled out. There are currently two favoured classes of models: one has free magnetic energy building up in a structure as a result of shearing of its footpoints and when the shear reaches a critical value equilibrium is lost and the structure erupts. However, for simple bipolar structures calculations by Aly (1984) have shown that the energy associated with an open field in this case is greater than for any other configuration. As a result this model requires additional energy input in order to trigger the CME. The existence of additional sources of magnetic flux, or more complex non-bipolar configurations, can make the eruption more likely, allowing parts of the overlying field to be removed via reconnection and allowing the core field to erupt. This is the breakout model of Antiochos *et al.* (1999). Recent observations with the EUV Imaging Telescope (EIT) on *SOHO* and in H$_\alpha$ (Sterling *et al.*, 2001) have identified a phenomenon called crinkles, which show brightening prior to flare and CME onset at a location on field lines adjacent to those of the eventual flaring region. These authors have suggested that this may be evidence of the partial reconnection required by the breakout model to allow the underlying sheared field to erupt.

The other popular model involves driving of the CME by currents generated in the photosphere. This requires that a flux rope already exist in the corona. The idea is that magnetic flux is injected into the flux rope from below, increasing the pressure, and it responds by moving outwards away from the Sun.

5.13   Interplanetary CMEs (ICMEs) and Magnetic Clouds

CMEs are observed in interplanetary space using *in situ* measurements of magnetic fields and particles, and there are many cases where CMEs have been clearly identified in the interplanetary medium by interplanetary spacecraft. Measurements of this kind have been recently augmented by the existence of space-based instrumentation. In particular, the LASCO instrument on *SOHO* which can track CMEs out to 30 solar radii, and the *Advanced Composition Explorer* (*ACE*) satellite, also located at the L1 point, about 215 Earth radii upstream of the Earth, have been invaluable. *ACE* has given scientists the opportunity to observe Earth-directed CMEs leaving the Sun and then 2–4 days later (depending on their speed) measure their plasma and magnetic field properties at 1 AU.

The basic properties of an ICME are shown in Figure 5.14, which shows a leading forward shock, enhanced magnetic field and smooth rotation of the geocentric

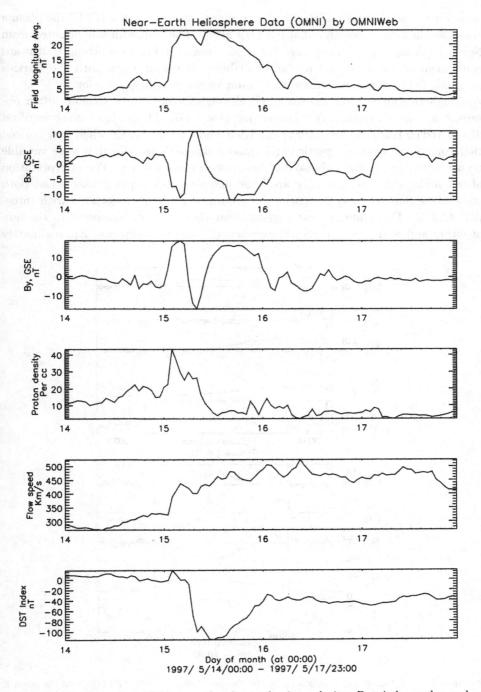

Figure 5.14 Total magnetic field strength, plasma density, velocity, D_{ST} index and y and z magnetic field components in GSE coordinates for an ICME in May 1997. Data from OMNIWEB.

solar Earth coordinates (GSE) y and z components. Within the ICME the plasma composition also shows an enhanced O/p ratio. The z direction is oriented from North to South, so a strong negative z component indicates a strong southward component of the field and a high probability of reconnection with the Earth's magnetic field. In this case these two examples are also magnetic clouds.

Magnetic clouds are large-scale helical magnetic structures that are often observed as the interplanetary consequence of a CME. Magnetic clouds were first discovered by Burlaga *et al.* (1981) and Klein and Burlaga (1982) when they noticed that some events showed particularly smooth magnetic profiles that were variable on a timescale of hours and also had low plasma temperatures. The interpretation of magnetic clouds is that they are large magnetic flux ropes connected at both ends to the Sun and may in fact be the prominence cavities associated with three-part CMEs. The evidence that suggests that they are still connected to the Sun at either end is the existence of bi-directional heat flux electrons. Approximately

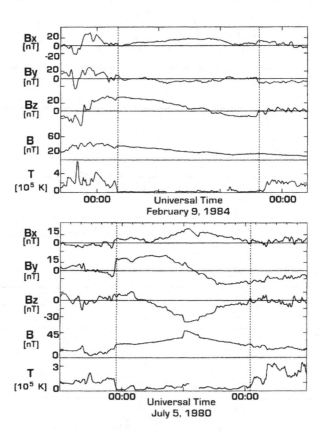

Figure 5.15   Total magnetic clouds seen by the *Pioneer Venus Orbitor* (*PVO*). Note the smooth variation in B_z. (From Mulligan, Russell and Luhmann, 1998; reproduced by permission of American Geophysical Union.)

30–40% of ICMEs fall into this category of magnetic clouds, their importance from a space weather point of view being down to the smoothly varying field — this can lead to the long periods of southwardly directed field necessary to produce geomagnetic storms. Typically this implies the need for $B_z < -10$ nT for a period > 3 h. Examples of two typical magnetic clouds are shown in Figure 5.15.

5.14 Magnetic Storms and Substorms

One of the principal manifestations of space weather is the geomagnetic storm and one of the first published works on magnetic storms was by von Humboldt in 1808. In this work he described the results of an experiment that measured magnetic declinations of small magnetic needles every half hour from midnight to early morning on 21 December 1806. During the experiment von Humboldt noted auroras overhead. When the auroras ceased, the magnetic deflections also ceased. He called the episode '*magnetisches ungewitter*', or a magnetic storm. The variations that he observed were associated with the auroral electrojet. Typically this is found at an altitude of approximately 100 km and about 65° magnetic latitude in the midnight sector, and the associated currents are normally about 10^6 A during substorms. In an intense magnetic storm the electrojet will move to lower latitudes and be much more intense, sometimes causing fields at the Earth to be several thousand nT. The principal defining feature of a magnetic storm is a decrease in horizontal magnetic field intensity followed by a subsequent recovery, as illustrated in Figure 5.16.

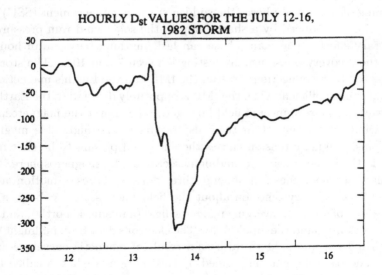

Figure 5.16 Great magnetic storm of 12–16 July 1982. (From Gonzales *et al.*, 1994; reproduced by permission of American Geophysical Union.)

The D_{ST} index is generally used to measure the intensity of magnetic storms. This indicates the hourly values of the average global variation of the low latitude horizontal component of the magnetic field. The average deflection from four or more ground-based magnetometers near the magnetic equator forms the basis of this index, first proposed by Chapman. There are many other indices used to measure geomagnetic activity, some of which are no longer used since developments in understanding the physical processes have led to the introduction of more appropriate ones. Some have been maintained, however, because of the consistent long term record that they provide. Among the other indices often used are:

- K_p — the K index is a quasi-logarithmic index of the three-hourly range in magnetic activity relative to an assumed quiet day for a single geomagnetic observatory site. The planetary three-hour-range index K_p is the mean standardised K index from 13 geomagnetic observatories and ranges between 0 and 9. Large storms have $K_p \geq 7$.
- A_p — the three-hourly A_p index maps the K_p index to a linear scale. The units are nT and the A_p value can be thought of as approximately half the range of the most disturbed horizontal magnetic components, with a range between 0 and 400.

'Superstorms' are those that have a $D_{ST} \leq -350$ nT, while major storms have $D_{ST} \leq -100$ nT. Magnetic storms have three phases: an initial phase during which the magnetic field strength increases from up to 10–50 nT, a main phase where the magnetic field decreases by ≥ 100 nT, and a recovery phase where the field returns to its pre-storm ambient value. The sudden jump in the magnetic field at the beginning of the initial phase (the sudden storm commencement (SSC)), which lasts < 5 min, is produced by a sharp jump in the solar wind ram pressure at the interplanetary shock. The main phase can last anything between an hour and a day, with the recovery phase usually lasting between 7 and 10 h. The storm main phase is caused by reconnection between the IMF and the Earth's magnetic field, a process that is most efficient when the IMF is oppositely directed to the Earth's field, i.e. southwards. The reconnected field lines get dragged over the magnetosphere by the solar wind and reconnection then also occurs in the night-side magnetotail. Following this, magnetic tension causes the entrained plasma to be shot from the tail towards the Earth, near the midnight sector of the magnetosphere. Plasma trapped on closed field lines can undergo three different types of motion, as shown in Figure 4.5. Firstly, gyromotion about the field lines; secondly, bounce motion between regions of strong converging magnetic field (a magnetic bottle); and, finally, azimuthal motion around the dipole field. The electrons drift from midnight to dawn and ions drift from midnight to dusk as a result of magnetic curvature and field gradients. The resulting current formed by this motion is what is called the *ring current* (Figure 5.6). The defining property of a magnetic storm is the creation of an enhanced ring current between 2 and 7 R_E that produces a magnetic field disturbance that, at the equator, is oppositely directed to the Earth's dipole field.

The decrease in the Earth's magnetic field as a result of the formation of this current manifests itself as a decrease in the D_{ST} index during the storm's main phase.

The recovery phase represents the loss of the ring current particles from the magnetosphere via convection, charge exchange with atmospheric neutrals, Coulomb collisions and wave–particle interactions.

A measure of the magnetic field disturbance produced by the formation of the ring current is given approximately by the Dessler–Parker–Sckopke relationship (Dessler and Parker, 1959; Scopke, 1966), viz.

$$\frac{D^*st(t)}{B_0} = \frac{2E(t)}{3E_{\text{m}}},$$

where D^*st is the field decrease caused by the ring current, B_0 is the average equatorial surface field, $E(t)$ is the total energy of the ring current particles and E_{m} is the total magnetic energy of the geomagnetic field outside of the Earth ($= 8 \times 10^{17}$ J). D^*st is the Dst value corrected for magnetopause currents. From this equation we can see that the decrease in the magnetic field is linearly related to the total energy of the ring current particles and thus D^*st is a measure of the total particle energy injected into the inner magnetosphere and the intensity of the magnetic storm.

The primary cause of geomagnetic storms is a strong dawn–dusk electric field associated with the passage of southward-directed IMFs for sufficiently long periods of time. Note that an IMF of the correct orientation is not sufficient to produce a geomagnetic storm; it must remain this way for a long enough period of time, approximately three hours or more. The energy transfer mechanism, as described above, is magnetic reconnection between the IMF and the Earth's magnetic field, a process that has an efficiency of approximately 10% during intense magnetic storms (Gonzales *et al.*, 1994). The electric field has two drivers: solar wind velocity and southward IMF. Gonzales and Tsurutani (1987) have shown that for intense storms with $Dst \leq -100$ nT $B_s \geq 10$ nT for longer than three hours.

There are two parts of ICMEs where the magnetic field can be strong enough to produce a magnetic storm. These are the magnetic cloud region and the interplanetary sheath, which is upstream of the ICME and behind the shock. In magnetic cloud events the Dst index decrease is coincident with the characteristic smoothly varying southward field component of these events. In the case of events caused by interplanetary sheaths the shock leads to the main storm phase. An increase in the southward magnetic field is observed behind the shock, which is most likely caused by shock compression of the B_s component of the slow stream ahead of the shock.

5.15 Very Intense Storms

The very intense storms that can lead to power outages and satellite losses are obviously of great interest. At present, although there are a number of possibilities that may account for these events, there is not yet a clear picture of their causes.

The current possibilities are:

(1) High velocity CMEs — there is a statistical relationship between the magnitude of the magnetic field of an ICME at 1 AU and its velocity. At present this is only a statistical relationship, but the cause is likely to be related to the CME onset and acceleration profile of the CME at the Sun.
(2) Multiple ICMEs — in this case the first shock due to an ICME would compress the ambient plasma, increasing the field intensity, and the second would increase it even further.
(3) Multiple magnetic storms — repeated storms that occur in quick succession can increase the total ring current to produce an effect that appears to be a particularly large storm.

Improving our understanding of these events will require more observations, simulations and modelling.

5.16   The Future

There is clearly a long way to go in raising our level of understanding of space weather to the point where we can predict the circumstances on the Sun that will lead to an event that will have geomagnetic consequences. Also, there are many other aspects of solar–terrestrial physics that play a role in space weather that have not been mentioned or have only been touched on here. These include: the relationship between CMEs and magnetic clouds, the relationship between substorms and major geomagnetic storms, the radiation belts, cosmic rays and the longer term but no less important effects on climate evolution. Of fundamental importance will be developing an understanding of what triggers CMEs and the factors that lead to the production of magnetic clouds. Clearly, in this case the association of the three-part CME with magnetic flux ropes is an important one, but we are not there yet. It is also becoming clear that perhaps we have to look further than the flare site and the CME shock in order to understand what produces SEP events. Big advances in our understanding of flare trigger and particle acceleration mechanisms on the Sun should come with observations from the recently launched *Reuven Ramaty High Energy Solar Spectroscopic Imager* (*RHESSI*), while the unique 3-D vantage point offered by the *STEREO* mission in 2005 will provide an unprecedented opportunity to track CMEs as they leave the Sun and propagate through the interplanetary medium, using both remote sensing and *in situ* measurements. On the same timescale, the Japan–UK–US mission *Solar-B* will provide extremely-high-resolution information on magnetic fields, electric currents and velocity fields, and will have the ability to link what is happening in the photosphere to the corona. *Solar-B* will provide the microscope to magnetic phenomena on the Sun and *STEREO* will provide the global picture. In addition the *Solar Orbiter* mission due for launch in the 2011–12 timeframe will provide a completely new vantage point out of the ecliptic, providing an opportunity to observe the transition between the fast and slow speed solar wind in detail. Its close proximity to the Sun, with a perigee of 45 solar radii, and co-rotational phase will allow the connection between solar events and *in situ* measurements to be made in a way not yet possible.

From a terrestrial point of view it is obvious that upstream monitoring of solar wind parameters provides a vital link between solar remote sensing and the actual arrival of potentially hazardous conditions in the Earth's vicinity. In particular, missions such as *Cluster* and other future multi-satellite missions in the inner and the outer magnetosphere will provide important information on the science of space weather, leading to enhanced operational prediction capabilities.

Chapter 6

The Physics of the Sun

Louise Harra

6.1 Why Do We Study the Sun?

The Sun has been a source of mystery to the human race for many thousands of years. One of the first pieces of evidence of this is demonstrated at Newgrange, outside Dublin. It is a passage tomb built more than 5000 years ago, and one of its most notable features is the small opening over the door. For several days around the winter solstice, sunlight enters the passage and creeps in a snake-like fashion to the back of the burial chamber. Another example of the dedication to Sun worshipping is at Stonehenge, in Wiltshire. This was built between 2800 and 1500 BC. At the centre of the henge there is a large stone (Heelstone), which marks the point of sunrise at the summer solstice. Both of these annual events are still held in awe and respect by everyone privileged enough to experience them. An experience that is accessible to everyone is witnessing the beauty of a solar eclipse. Many people across Europe had the pleasure of witnessing the eclipse on 11 August 1999.

The Sun is our closest star, and provides heat and light for our existence. It also has other effects, such as producing energetic particles, which interact with and can damage the artificial satellite systems that we are becoming so dependent on. There is also evidence that fluctuations in the Sun's output produce changes in the climate.

The Sun is the largest and most massive object in the solar system, and yet is an ordinary middle-aged star. It is expected that the Sun will still shine for more than twice its present age, and before it dies it will become so large and luminous that it will engulf the Earth completely. Its final destiny is predicted to be a very compact star — a white dwarf.

The Sun is important to us from day to day, but also provides a laboratory for understanding other, more distant objects in the Universe, such as other stars and active galactic nuclei.

In this chapter, the basic physics of the Sun will be described — starting with the energy source at the core, and working out to the surface and into the solar atmosphere.

6.2   The Structure of the Solar Interior

The model that has evolved over the years in order to understand the Sun is shown in Figure 6.1. In this cartoon, we see the Sun's energy source at the centre (the *core*), the inner two-thirds of the Sun (*radiative zone*), where the energy is transported from the core by radiation, and the outer one third of the Sun (*convection zone*), where the convection process takes over as the main energy transport mechanism.

The vital statistics of the Sun are given in Table 6.1. The Sun's size is inferred from its distance and angular extent, the Sun's mass can be determined accurately from the motions of the planets, and the Sun's age is determined by studying the abundance of heavy radioactive elements in meteorites. The Sun's age is the least well-determined quantity with an accuracy of 2%. These characteristics of the Sun can be used as input into a mathematical model (the 'standard model') in order to determine, for example, how the pressure, density and temperature vary from the centre of the Sun to the surface.

There are only two ways in which the interior of the Sun can be probed to determine whether the standard model is correct. The first is by measuring the neutrino flux and the second is by measuring solar oscillations. Both of these will be discussed in the next sections.

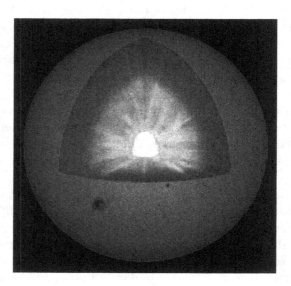

Figure 6.1   A cartoon illustrating the different layers of the Sun. Courtesy of Dr David Hathaway, Marshall Space Flight Center.

Table 6.1   Vital statistics of the Sun.

Mass	2×10^{30} kg (333,000 times the Earth's mass and 745 times the combined mass of all the planets)
Age	4.5 billion years
Radius	6.9×10^{8} m (109 times the Earth's radii)
Density (core)	151,300 kg m^{-3}
Temperature (core)	> 15 million K
Temperature (surface)	5700 K (16 times the temperature of boiling water)
Abundances	90% H and 10% He (with some trace heavier elements)
Spectral Type	G2V
Mean distance	1.5×10^{11} m (light takes 8 min 20s to reach the Earth)

6.3   The Energy Source of the Sun

The dominant source of energy is at the Sun's core. At the core a temperature of 15×10^{6} K and a density of 1.48×10^{5} kg m^{-3} are predicted. This region makes up a small percentage of the volume of the Sun.

The dominant energy generation process occurring in the core is thought to be the proton–proton (pp) chain. The first reaction is the fusion of two protons to form a deuteron, which is a nucleus of heavy hydrogen consisting of a proton and a neutron. One of the colliding protons thus converts to a neutron:

$$p + p \rightarrow {}^{2}D + e^{+} + \nu_e \,,$$

where e^{+} is a positron and ν_e is a neutrino, which are both emitted in order to maintain the balance in the reaction. The two colliding protons overcome the electrostatic force that tends to repel like charges and one of the protons decays to a neutron and a positron.

The next stage in the pp chain is the fusion of another proton with a deuteron, which produces an isotope of helium (consisting of two protons and one neutron), and a gamma ray:

$$^{2}D + p \rightarrow {}^{3}\text{He} + \gamma \,.$$

Finally, two helium nuclei fuse together to form another nucleus of helium (the more common alpha particle):

$$^{3}\text{He} + {}^{3}\text{He} \rightarrow {}^{4}\text{He} + p + p \,.$$

In summary, four protons fuse together to make a helium nucleus, and release energy through Einstein's equivalence of mass and energy ($E = mc^{2}$). The mass of an alpha particle is 4.00386 amu compared with the mass of four protons, which is 4.03252 amu. The energy released as a result of the pp chain is then 4.3×10^{-12} J.

A small fraction of this energy is removed by the neutrinos produced from the proton–proton interaction.

In addition to the main pp reactions listed above, there are reactions that produce ^{8}B neutrinos. First, an alpha particle (^{4}He) and a ^{3}He particle combine to form a beryllium nucleus:

$$^3\text{He} + {}^4\text{He} \rightarrow {}^7\text{Be} + \gamma \,.$$

The beryllium nucleus may capture a proton to form an ^{8}B boron nucleus:

$$p + {}^7\text{Be} \rightarrow {}^8\text{B} + \gamma \,.$$

The boron nucleus is unstable and can then decay, producing a beryllium nucleus, a positron and a neutrino:

$$^8\text{B} \rightarrow {}^8\text{Be} + e^+ + v_e \,.$$

Although these ^{8}B neutrinos are much rarer than those occurring in the pp chain, they are more sensitive to the Sun's internal temperature than the pp neutrinos (the typical (^{8}B) flux is proportional to T^{18}, compared with the flux (pp), which is proportional to $T^{-1/2}$) and have much greater energy, and hence it is likely they are easier to detect.

The pp interaction has a low probability, but since there are a large number of protons in the core, a large number of these interactions take place. In fact around four million tonnes of hydrogen is converted into helium every day. This large amount of hydrogen being converted on a daily basis prompts the question 'How much fuel is left in the Sun's core?'. The Sun's mass is being reduced annually by around 1.4×10^{17} kg. During its lifetime (4.6×10^9 years), the Sun has lost only 0.03% of its mass (approximately 3×10^{30} kg). Calculations have predicted that it will be another 5×10^9 years before the Sun starts on its next stage in the journey of a star — that of a red giant.

6.4   The Neutrino Problem

Studying solar neutrinos provides us with a direct test of the theory of stellar structure and evolution, and of the nuclear energy generation. In addition, the study of solar neutrinos can perform particle physics experiments, for example determining whether neutrinos have non-zero mass.

Neutrinos interact very weakly with matter, and have an optical depth that is approximately 20 orders of magnitude weaker than that of a typical optical photon. However, the probability of absorption does increase with energy.

Neutrinos were discovered 30 years ago and since then an enormous effort has been expended to produce an experiment to measure the solar neutrino flux accurately. The Brookhaven National Laboratory constructed one of the first experiments in the Homestake goldmine in South Dakota. The experiment is placed deep in the mine to avoid contamination by cosmic rays. Neutrinos with high enough energies (> 0.8 MeV) can interact with a common isotope of chlorine, which is used in cleaning fluid. This isotope accounts for a quarter of all natural chlorine and consists of 17 protons and 20 neutrons. The reaction involves the combination of chlorine with a neutrino to produce an isotope of argon:

$$^{37}\text{Cl} + v_e \rightarrow \,^{37}\text{Ar} + e^- \, .$$

The isotope of argon is unstable and it is this decay that can be detected:

$$^{37}\text{Ar} \rightarrow \,^{37}\text{Cl} + e^+ + v \, .$$

Due to the high energies involved, the observation is limited to predominantly the rare ^{8}B neutrinos (see Figure 6.2). Bahcall defined the unit of measurement of neutrinos to be the SNU (1 SNU $= 10^{-36}$ interactions s^{-1} atom^{-1}). The first theories in the 60s predicted a rate of 30 SNUs. Improvements in the modelling in the 90s have produced consistent results of 7 SNUs for the chlorine reaction. The current experimental result gives only 2.2 SNUs. The discrepancy is well outside of experimental and theoretical errors.

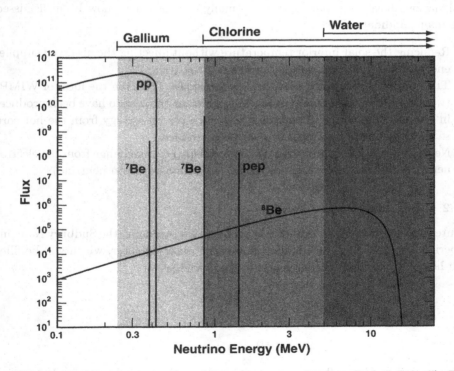

Figure 6.2 Distribution of neutrino flux with neutrino energy (courtesy Prof. J. Bahcall). The top of the figure shows the range over which the different available experiments are most sensitive.

Other experiments have been carried out making use of different reactions. The Kamiokande experiment in Japan was initially designed to study the decay of the proton using a 5 kilotonne tank of pure water located underground in a mine. Its first detection of neutrino was from the explosion of supernova 1987A. Since then, efforts have been made to improve the performance to detect solar neutrinos. Now known as Super Kamiokande, detection is by neutrino scattering from electrons. The energy of the neutrinos that can be detected is > 7.5 MeV, so again it is only sensitive to the ^{8}B neutrinos. The advantage of this experiment is that it is possible to measure the angular distribution of the scattered electrons in order to confirm that the neutrinos originate from the Sun. Again the measurements from Super Kamiokande are around a factor of 2 lower than the theoretical predictions. Two major gallium experiments are also being conducted — SAGE and GALLEX. More than half the neutrinos measured in these experiments have low energies in the range of the pp neutrinos. The measurements are still a factor of 2 lower than prediction.

6.4.1   *Possible explanations*

Since the discrepancy in the number of neutrinos was discovered, a number of explanations have been put forward — many of which have now been dismissed. The main candidates are:

(a) Reducing the solar interior temperature without affecting the observed supplied energy. This will reduce the production of neutrinos.
(b) The existence of a large amount of hidden mass. This is in the form of WIMPS (weakly interacting massive particles), which are thought to have been produced in the early Universe. They could carry some of the energy from the hot core, so that the number of neutrinos could be reduced.
(c) Neutrinos oscillate from one flavour to another (i.e. can change from the electron neutrino that we have been discussing to a muon or tauon neutrino).

6.4.2   *New experiments*

Future experiments will address these problems. For example, the Sudbury Neutrino Experiment (SNO) in Canada will use another water detector, which will be filled with heavy water (D_2O). The reactions produced are

$$v_e + {}^2H \rightarrow p + p + e^-$$

or

$$v + {}^2H \rightarrow n + p + v,$$

where neutrinos of any flavour can participate in the second process. If neutrino flavour changing does occur, the number of neutrinos which can produce the first reaction will be reduced. The first results from the SNO took place in 2001, and confirm that the electron neutrinos 'oscillate' or change flavour on their journey to the Earth. These new results are in excellent agreement with the predictions of the standard solar models, and we can finally lay this puzzle to rest.

6.5 Helioseismology

The Sun is now known to oscillate in a very complicated manner, and observational measurements of waves produced can provide another test for the standard model. The period of the waves allows us to probe the interior of the Sun to obtain information on the temperature, chemical composition and motions from the core right to the surface. Helioseismology (the study of seismic waves on the Sun) provides us with a means of measuring the invisible interior and dynamics of the Sun.

The solar oscillations are thought to be caused by the turbulent convection near the surface of the Sun. The temperature at the bottom of the convection zone (top of the radiative zone) is cool enough (around 2 MK) to allow some heavy nuclei to capture electrons. These heavier particles increase the opacity and cause the plasma to increase in temperature. This build-up of energy is transported by convection. Convection produces patterns of motion on the Sun's surface — which result from over 10^7 resonant modes of oscillation in the interior! Leighton, Noyes and Simon made the first observations of solar oscillations in the 1960s at the Mount Wilson Observatory. They measured very small Doppler shifts in the wavelengths of absorption lines in the photosphere's spectrum. This showed that patches that appear to oscillate intermittently with periods close to 5 min occupy approximately half of the Sun's surface. The oscillations have an amplitude of close to 1 km/s, and patches persist for around 30 min. Initially it was thought that this pattern was due to turbulent covection. It was later proposed that the oscillations are the superposition of the resonant acoustic modes of the interior.

The Sun acts as a resonant cavity because acoustic waves become trapped in a region bounded at the by a large density drop near the surface and bounded at the bottom by refraction due to an increase in sound speed. The speed of sound increase because the Sun is hotter at greater depths ($V_s \propto T^{1/2}$), and hence the wave front is refracted since the deepest part is travelling at greater speed than the shallowest part.

There are two different types of oscillation that are possible on the Sun — p-modes and g-modes. In the former the most important restoring force is pressure and in the latter the most important restoring force is gravity. The p-modes have frequencies between about 2 min and 1 h and the g-modes are predicted to have much longer periods, but have not yet been observed. The g-modes are trapped lower in the Sun's interior, below the convection zone.

Physically and mathematically, the ascillation modes can be understood as spherical harmonics l, m and n. The quantities l and m are quantum numbers

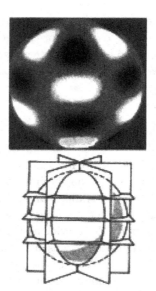

Figure 6.3   This shows spherical harmonics $l = 6$ and $m = 3$. The grey regions are the nodal boundaries.   The white regions are those that are moving radially outwards, while the black regions are those that are moving radially inwards. Credit: Noyes, R., 'The Sun', in *The New Solar System*, J. Jelly Beatty and A. Chaikin, eds., Sky Publishing Corporation, 1990, p. 23.

(where $l \leq m \leq +l$) that determine the angular behaviour of the oscillation over the surface of the Sun. The harmonic degree l indicates the number of node lines on the surface (i.e. the number of planes slicing through the Sun). The azimuthl number m describes the number of planes slicing through the Sun longitudinally. The order n is the number of nodes in the radial direction. Figure 6.3 shows spherical harmonic numbers $l = 6$ and $m = 3$. There are three planes in the longitudinal direction ($m = 3$) and three planes in the latitudinal direction ($l - m = 3$).

For small values of l the waves can penetrate almost to the centre of the Sun, whereas for high values of l the modes are trapped close to the surface. Figure 6.4 shows one set of the Sun's oscillations. The frequency of the mode shown is determined, from the Michelson–Doppler Interferometer (MDI) on board the Solar and Heliospheric Observatory (*SOHO*) spacecraft, to be $293.88 +/- 0.2$ uHz. The cut-away shows the wave motion propagating in the solar interior. It is also possible to observe the rotational splitting of the different l modes by large scale flows. This is used to determine how the Sun's rotation varies with the radius and latitude.

Solar rotation is found to be differential. It is dependent on both the heliocentric latitude and the radial distance to the solar centre. Carrington originally determined the rotation period from sunspot observations to be 27.725 days. Solar rotations are numbred, starting with Carrington totation number 1 on 9 November 1853. The rotation of the convection zone has also been found to show differential rotation similar to that observed by Carrington at the surface of the Sun. This rotation varies with latitude (see Figure 6.5). Below the convection zone there is a layer

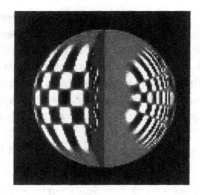

Figure 6.4   A Computer-generated model of one set of standing waves on the Sun's surface with radial order $m = 16$, angular degree $l = 20$ and angular order $m = 16$. Black and white denote areas that are displaced in opposite directions. This image is adapted from 'Structure and Rotation of the Solar Interior: Initial Results from the MDI Medium-L Program', by Kosovichev *et al.*, 1997, with kind permission of Kluwer Academic Publishers.

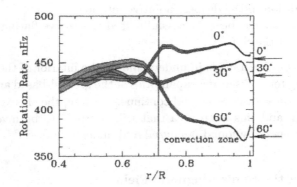

Figure 6.5   This shows the inferred rotation rate as a function of depth and latitude within the Sun. This has been determined at three different latitudes (equator, 30° and 60°). Courtesy of SOHO/MDI consortium. *SOHO* is a project of international cooperation between ESA and NASA.

of shear (named the tachocline), which is a narrow region connecting the radiative zone (which rotates rigidly) with the convection zone.

There have been a number of methods used to measure oscillations in the Sun. Networks of observatories around the world have been set up to allow extended observations to be made so as to search for long period oscillations. It is important to separate very low degree modes (i.e. $l = 0$), where the whole Sun will pulsate, from those with higher-level degrees. The University of Birmingham has developed a method of helioseismology that uses an optical spectrum of the Sun's integrated light. These observations are made at six observatories throughout the world, including Hawaii, Tenerife and Australia, and the network is known as the Birmingham Solar Oscillations Network (BiSON). There are also sites, including the Antarctic, Chile, California and Morocco, which have been developed by researchers at the University of Nice. These also observe the integrated Sun. This group of instruments is called IRIS (International Research on the Interior of the Sun). The Global Oscillations Network Group (GONG) involves six ground-based observatories which are capable of observing oscillations over the entire Sun. GONG consists of 67 collaborating institutes from around the world. The MDI instrument on *SOHO* already mentioned also has this capability. The advantage of *SOHO* is its location near the Earth–Sun Lagrangian point (L1), which allows continuous observations of the Sun.

These observations have led to a much deeper understanding of the solar interior. The discoveries include:

(a) The angular velocity observed at the Sun's surface extends right through the convection zone with little radial dependence.

(b) There is a region of strong shear called the tachocline, which is located at the base of the convection zone.

(c) The base of the convection zone is determined to be at 0.713 R.

(d) The tachocline is the region where the angular velocity adjusts to the apparent solid body rotation in the deeper radiative interior.

(e) Seismic holography has been successfully used to observe flaring active regions at the far side of the Sun.

Much progress has been made in understanding the interior of the Sun, and it is clear, with the long term observing capabilities of the ground-based and space-based observations, that these discoveries will continue. To date, the elusive g-modes have not been observed, and this is an area that will continue to be investigated until they have been successfully found, analysed and understood.

6.6   Creation of the Sun's Magnetic Field

Magnetic fields have long been known to be the source of solar and stellar activity. The solar interior is composed of highly conducting plasma. In this circumstance, the magnetic field is 'frozen' into the plasma. The solar dynamo is located some-where in the interior where the plasma, which is rotating, generates electric currents and converts the plasma motion to magnetic energy. Initial modelling work in the 1970s suggested that the location of the dynamo was in the convection zone. However, the recent observations from *SOHO* have shown that the likely location of the dynamo is in the region of high shear between the convection zone and the radiation zone (see Figure 6.5).

Several theories have been proposed to explain how the solar dynamo operates. Any theory that explains the solar dynamo must also explain why sunspots emerge at high latitude early in the solar cycle, and appear at lower latitudes later in the cycle (Spörer's law). This is illustrated by the well-known butterfly diagram shown in Figure 6.6, which clearly shows that sunspots do not appear randomly but are

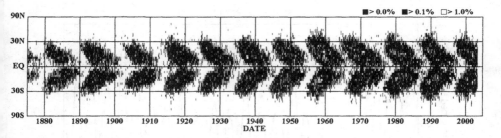

Figure 6.6 The location of sunspots is shown in this diagram from 1874, when the Royal Greewich Observatory began observing sunsports. Courtesy of NASA/MSFC.

concentrated in two latitude bands on either side of the equator. Any theory would also have to explain the periodic reversal of the field every 22 years.

Babcock published a semi-observational model of the variation of the Sun's magnetic field in the early 1960s. Initially the field lines on the Sun are poloidal, connecting the north and south poles. This occurs approximately three years before the onset of the new sunspot cycle. Differential rotation, with a rotation rate of 25 days at the equator and nearly 30 days at the poles, stretches the magnetic field horizontally around the Sun (this was originally suggested by Cowling in the 1950s). This takes the form of a spiral pattern in the north and south hemispheres. The field is most intense around latitudes $\theta = 30°$, due to the $\sin^2 \theta$ term in the Sun's differential rotation, and this is the location of the first active regions of the cycle (i.e. emergence of magnetic flux at the surface).

The magnetic field is embedded in the interior of the Sun, which can be viewed as a fluid with an associated pressure. The magnetic field also has an associated pressure given by $B^2/8\pi$. The magnetic pressure pushes the plasma out of the magnetic flux tube in order to achieve pressure balance with the outside plasma (p_0):

$$p_i + \frac{B^2}{8\pi} = p_0 \,.$$

The loss of plasma inside the flux tube gives rise to the buoyancy force which pushes the magnetic flux tube up to less dense regions in the interior. The flux tube will eventually penetrate through the surface of the Sun. The active latitudes will continue to move towards the equator, reaching $\theta = 8°$ after around eight years, with the emergence of active regions following Spörer's rule and Hale's law. Hale's law states that active regions on opposite sides of the hemisphere have opposite leading magnetic polarities, which alternate between successive sunspot cycles.

The final stage in Babcock's model describes the reversal of the poloidal field due to the tendency of active regions to have the leading polarity at lower latitude to the following polarity. The following polarity can migrate towards the nearest pole whereas the leading polarity has more chance of moving towards the equator and cancelling with opposite polarities from the opposite hemisphere. The following

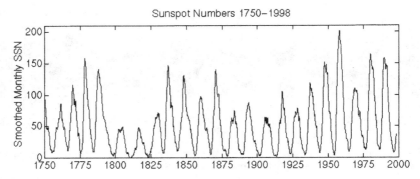

Figure 6.7   The sunspot cycle from 1750 to 1998. The data was obtained from the National Geophysical Data Center, U.S.A.

polarities that make it to the poles first cancel the existing magnetic field and then replace it with flux of the opposite polarity. After 11 years the field is poloidal again, and the process restarts until after 22 years the polarities return to the initial starting point.

Other models have been proposed over the years, but most use the idea of differential rotation to generate a toroidal field from a poloidal field. The original models assumed that the field lines lay just below the surface of the Sun. It is now known that this is not possible, and the location as shown in Figure 6.5 is at the region of high shear between the radiative zone and the convective zone.

Solar activity levels increase and decrease over an 11-year cycle. This cycle is illustrated by counting the number of sunspots. As can be seen in Figure 6.7, the amplitude of the sunspot number can vary by more than a factor of 4 between cycles in records dating back to 1750. The cycle period also varies from peak to peak, with intervals in the range of 8–15 years, making it difficult to predict when the peak of solar activity will occur. There have been occasions when no sunspots were seen at all — the most famous being described by Maunder, who found that during the period from 1645 to 1715 the sunspot cycle virtually disappeared. There can also be an imbalance between the activity levels in the northern and southern hemispheres. This was seen during the Maunder minimum when most of the spots that did appear were in the southern hemisphere. The recent sunspot cycles have shown an almost equal number of sunspots in the southern and northern hemispheres.

It has been argued that the phase of the solar cycle appears to be coupled to a periodic oscillation of the solar interior. Shorter cycles are generally followed by longer cycles. The dataset, however, covers only a few hundred years, and is not long enough to define the phase exactly.

6.7   The Photosphere

The photosphere was the first part of the Sun to be observed. Sunspots had been seen even before telescopes were invented in the early 1600s, especially in China.

With the improvements in ground-based telescopes and observations from space, many other features of the photosphere can now be observed also.

One of the amazing features of the photosphere is how opaque it is, so that to us it looks like a solid sphere. In fact, the density at the photosphere is 10,000 times less than that of the Earth's atmosphere. The reason for the high opacity of the photosphere is the combination of absorption at discrete wavelengths (this accounts for a small fraction of the opacity) and continuous absorption between ultraviolet and infrared wavelengths in the solar spectrum (this is the dominant source of the opacity). In the upper interior of the Sun and in the photosphere there are many free electrons that have been released from atoms by excitation. Hydrogen atoms require a relatively large amount of energy to release their one electron, but can more easily capture another electron to give H^-. This process of capture is rare but since there is a large abundance of hydrogen, and many free electrons, H^- is quite abundant. The main result of this is that photon absorption then occurs relatively easily, since photons with a wide range of energies can liberate the extra electron in H^-.

The other main feature of the overall photosphere is limb darkening. This is due to the temperature variation, which drops from 6400 K in the lower part of the photosphere to 4400 K in the upper part of the photosphere. This temperature drop occurs over a small height range of a couple of hundred kilometres. The upper part of the photosphere is referred to as a 'temperature minimum' and is the start of the chromosphere. Limb darkening can be observed in full disc images of the Sun.

From ground-based telescopes in 'good seeing' fine granules can be seen all over the disc. This is evident in Figure 6.8 outside the sunspot. Granules visible in this way have an average size scale of 1100 km (1.5 arcsec), although many smaller granules exist. Closer to the sunspot the appearance of the granules changes such that the polygonal shape becomes lengthened at the boundary. They occur as

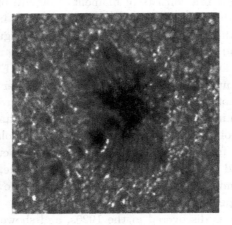

Figure 6.8   An image of a sunspot (4305 A G Band) taken by the Swedish Vacuum Solar Telescope in La Palma. The granulation is clearly seen outside of the sunspots as regular polygonal features surrounded by dark lanes.

umbral dots within the spot itself. The change in the shape of the granules is associated with a strong horizontal magnetic field. The granules are separated by intergranular lanes, which are darker than the granules themselves, indicating a temperature that is approximately 100 K cooler. The darker a feature is on the disc, the cooler it is relative to the surrounding material.

Granule lifetimes range between 10 and 20 min, with the largest granules lasting longest. Granules are associated with upward motions with a typical velocity of 2 kms^{-1}, and the intergranular lanes show downward motions. The characteristics of granules suggest that they are rising convection cells of hotter plasma and the intergranular lanes are cooler plasma that is falling. The granules decay by three methods: merging with other granules, fragmentation, and decay. New granules have been shown to arise predominantly from the fragmentation of old granules.

Another, much larger and deeper convection takes place on the Sun that is known as supergranulation. These have size scales of 30,000 km, with a typical velocity of 0.3 kms^{-1}. In each supergranulation cell, material flows outwards from the centre and sweeps magnetic field to the cell boundaries where downflow is observed. The supergranulation may be related to the chromospheric network, which will be discussed in the next section.

The largest features in the photosphere are sunspots and these have radii of between a few hundred km and 100,000 km. There have been occasions when larger spots have been observed although this is not common. Although sunspots are the largest feature on the photosphere, they cover a maximum of only about 1% of the Sun's disc at any time. Sunspots are darker than their surroundings, which means that they have a lower temperature. The overall photospheric emission fits a black body of around 5700 K, whereas the temperature within a sunspot is approximately 4000 K.

A sunspot has a dark core (umbra). As it becomes larger it also develops a penumbra. The penumbra consists of filaments that are brighter and approximately radial. Penumbras can have more complex structures in larger, non-circular sunspots. The umbra and penumbra can be clearly seen in Figure 6.8. Sunspots are dynamic structures, and horizontal flows of material are seen to move radially outwards from the umbra and penumbra. This is known as the Evershed flow. Larger spots tend to form groups, with the leading spot being closer to the equator than the following spot. Photospheric faculae are frequently seen surrounding sunspot groups. These are around 300 K hotter than their surroundings, and they consist of many small bright points. They have opposite magnetic polarity to the sunspot, and the unipolar field from the sunspot connects to the faculae. Material from the penumbra is pushed towards the opposite polarity faculae, giving the observed Evershed flow of around 2 kms^{-1}. The inverse Evershed flow is observed in the chromosphere with a speed of around 20 kms^{-1}.

The Wilson effect was discovered in the 1920s, and showed the sunspot to be depressed with respect to the surrounding photosphere when observed at the Sun's limb. This is thought to be due to the difference in temperature between the umbra and the surrounding photosphere. In the umbra the abundance of H^- is lower and hence the opacity is lower. Lower depths can then be seen in the umbra. The Wilson effect is somewhat controversial since it has not been consistently observed.

6.7.1 *The photospheric magnetic field*

Zeeman splitting of photospherically formed Fraunhofer lines measures the photospheric magnetic field. The Fraunhofer spectrum consists of a continuous background emission with numerous absorption lines imprinted on it. Kirchoff's law states that this absorption occurs when a cool gas is in front of a hot continuum source. The energy level diagram for hydrogen is shown in Figure 6.9. Emission lines are formed when photons are emitted at certain energies when an electron jumps from, for example, the $n = 2$ to $n = 1$ orbits. For absorption lines, as seen in the Fraunhofer spectrum, the photons are absorbed and push the electron from a lower to an upper orbit.

Fraunhofer originally identified the strongest lines in the spectrum. Although hydrogen is the most abundant element, the Lyman series (see Figure 6.9), which arises from the ground state, occurs in the ultraviolet. The hydrogen lines seen in the visible are from the Balmer series, which require excitation from the $n = 2$ level. Other strong lines are resonance lines of less abundant elements such as sodium and magnesium. Resonance lines have absorption from the highly populated ground state. The most famous examples are the Na I D lines at 5890 Å and the H and K lines of ionised calcium at 3900 Å. There are many elements which exist but are

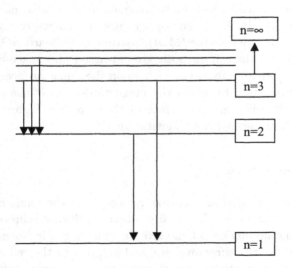

Figure 6.9 The energy levels in hydrogen. The Hα emission is produced by electrons moving from the $n = 3$ level to the $n = 2$ level. Transitions to the $n = 1$ level produce the Lyman series in the UV.

not observed, e.g. C, N, He. These elements have their excitation potentials above 10 V and hence their strongest lines are in the UV.

Hale discovered the Zeeman effect in sunspots in 1908, proving that the Sun is a magnetic star. Under the influence of a magnetic field a spectral line can be split into several spectral lines with slightly different frequencies (see Chapter 8). The reason why this occurs is that the electron orbits the atom as a negative electric charge, giving rise to a small current with an associated magnetic field. When an external field is added to the system, it interacts with the electron's magnetic field, hence adding or subtracting to the energy of the photon that forms the spectral line. For a three-component line, the shift in wavelength of the two outer components from the central wavelength is given by

$$\frac{\Delta\lambda}{nm} = 4.7 \times 10^{-8} g^* \left(\frac{\lambda}{nm}\right)^2 \left(\frac{B}{T}\right),$$

where n and m are the quantum numbers of the levels, and g^* is the modified Lande g factor. In this way the magnetic field can be determined. For feasible measurements, a large sensitivity of $\Delta\lambda/B$ is required. When the magnetic field is pointing towards the observer, the two outer components of the spectral line are circularly polarised and the undisplaced line disappears. When the magnetic field is perpendicular all three components will be visible.

Babcock developed the magnetograph in 1952, which utilises the circular polarisation of the Zeeman components of the spectral lines, and allows the measurement of magnetic fluxes in active regions and also quiet regions. The largest magnetic field strengths occur in sunspots with values of around 0.4 T. There are many ground-based observatories which make magnetograms (which measure the longitudinal field), e.g. Kitt Peak. Magnetograms are also currently available from the Michelson Doppler Interferometer (MDI) instrument onboard *SOHO*. The recent high-resolution observations from both ground-based and space-based telescopes have shown that magnetic flux is everywhere on the Sun's surface. Magnetograms from MDI have been used to model the magnetic loops. It has been found that magnetic fields exist all over the surface of the Sun (see Figure 6.10), and this phenomenon has been termed the magnetic carpet.

6.8   The Chromosphere

The chromosphere is defined as a narrow region above the temperature minimum (the top of the photosphere). It was first observed during eclipses as a red ring, which is due to the dominance of the red Hα emission. Sir Norman Lockyer described this region as 'the chromosphere' in 1869, due to the red colour observed. Detailed telescopic observations soon revealed that the chromosphere is irregular, with many hundreds of narrow jet-like features seen. These features are known as spicules and can exceed 10,000 km in length but are only around 1000 km wide.

When the chromosphere was observed with full disc spectroheliograms, its appearance was confirmed to be much more complex than the photosphere. The main chromospheric Fraunhofer lines that are used for diagnosing the atmosphere are the $H\alpha$ and Ca II H and K lines. These are superimposed on a continuum that stretches across the visible range down to a wavelength of around 1400 Å and is formed by scattering of photospheric light by free electrons in the chromosphere. The Balmer continuum is produced at wavelengths of less than 3646 Å, and is formed by the recombination of free electrons with protons. The electrons are captured into the $n = 2$ orbit and emit photons. The photon can have any energy above the amount corresponding to the difference between the $n = 2$ orbit and the energy required to just set the electron free.

Complexity is revealed in lines formed at different heights in the atmosphere. A few lines in the visible part of the spectrum have high opacity near the central line wavelength. Hence the line core is formed in the chromosphere above the temperature minimum, where the temperature starts to increase (see Figure 6.14). The $H\alpha$ line is photoelectrically controlled (i.e. the photons from the photospheric continuum create or destroy $H\alpha$ photons through the excitation or de-excitation of hydrogen atoms — see Chapter 8). Consequently, the $H\alpha$ line intensity is controlled more by the photospheric radiation and hence does not provide information on the temperature of the chromosphere. An example of a full disc image of the Sun in $H\alpha$ is shown in Figure 6.11. Sunspots are visible, surrounded by bright plage regions. Elongated dark features known as mottles and fibrils are seen all over the disc. The large, long, dark regions are known as prominences. The vertical extensions of these structures are apparent when they are at the limb. The entire chromosphere is made up of small bright cells that form the chromospheric network. Tuning the wavelength of narrow-band images from the core of the $H\alpha$ line to the wings reveals successively the mid-chromosphere down to the photosphere.

The Ca II H and K lines are collisionally controlled, and hence the line profiles clearly reflect the temperature of the region (see Figure 6.12 for an image in Ca II K). As shown in Figure 6.14, the temperature begins to rise with height from the temperature minimum out into the chromosphere. For example, the Ca II K line is observed in absorption on the disc with very broad wings. There is an increase of intensity towards the line centre, with a sharp dip at the centre. The opacity increases towards the centre of the line and hence the line profile can be roughly interpreted as a map of temperature with depth in the solar atmosphere. Around 0.3 Å (known as K1) from the line centre there is an intensity minimum, which corresponds to the temperature minimum region between the photosphere and the chromosphere. At 0.15 Å (known as K2) from the line centre the intensity rises again, which corresponds to the low chromosphere, and finally the line centre (known as K3) corresponds to the high chromosphere. The dip (or central reversal) at the line centre occurs because the opacity of this wavelength is so high that we see layers of very low density where the emission falls well below the level that plasma

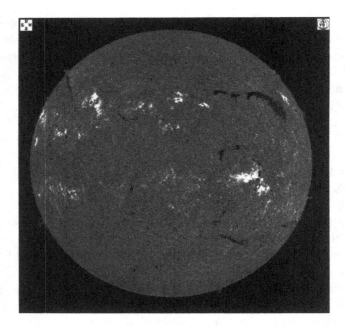

Figure 6.11   An image of the Sun in Hα on 4 January 2001. It has been contrast-enhanced. Sunspots are visible as dark spots, plage are visible as bright regions and filaments are visible as several long, dark features. Courtesy of the Big Bear Solar Observatory.

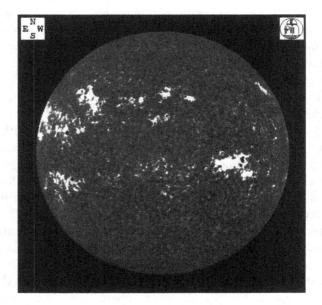

Figure 6.12   A full disc image of the Sun in CaII K taken on 4 January 2002. It has been contrast-enhanced. The chromospheric network and active regions show more clearly in this image than in Hα, whereas the filaments are barely visible. Image courtesy of the Big Bear Solar Observatory.

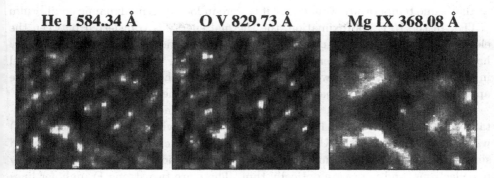

He I 584.34 Å   O V 829.73 Å   Mg IX 368.08 Å

Figure 6.13   Images of the quiet Sun clearly showing the network in the chromosphere (He I) and the transition region (O V). At coronal temperatures (Mg IX) the network is very diffuse and loop-like structures dominate.

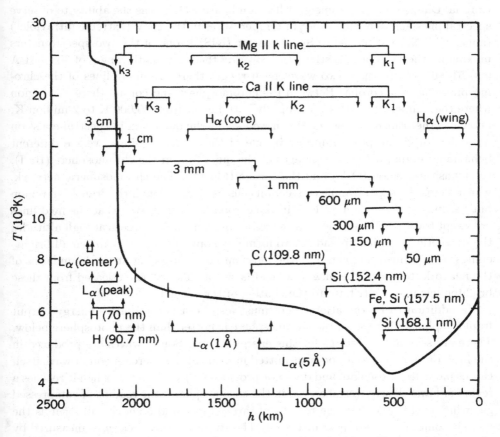

Figure 6.14   Temperature as a function of height in the quiet Sun from Vernazza, Avrett and Loeser 1981 (*Astrophysical Journal*). The approximate heights where various lines and continua are formed are indicated.

at that same temperature would have if it were in local thermodynamic equilibrium (LTE). In the LTE approximation, the thermal energy is transferred between the electrons and the atoms by collisions between the particles, and hence the density has to be large enough to ensure a high collision rate. This assumption is valid in the photosphere, and in weak lines in the chromosphere. Figure 6.14 shows the approximate regions of formation of various chromospheric radiations.

Further information can be obtained on the chromosphere, and more importantly its connection to the higher corona, from observing in the ultraviolet (UV) wavelength range. In the UV band, the photospheric continuum has decreased substantially. The spectral lines in this range are emission lines that are excited by free electrons, and many are optically thin. There are two strong Fraunhofer lines in the UV range, which are the Mg II H and K lines at 2802 Å and 2795 Å respectively. The strong Lyman α line appears at 1215 Å, and is formed by free electrons exciting atoms from their ground state to their $n = 2$ state (see Figure 6.9). The Coronal Diagnostic Spectrometer (CDS) on board *SOHO* has the ability to observe spectral lines across a wide range of temperatures in the EUV (extreme ultraviolet) range. The Normal Incidence Spectrometer (NIS) is one of the two spectrometers making up the CDS, and is stigmatic, covering the wavelength ranges of 308–381 Å and 513–633 Å. In these two wavelength ranges there are many lines of the chromosphere and the corona. Between the chromosphere and corona there is a region where the temperature rises very rapidly with height from 20,000 K to 2 million K, known as the transition region. When we look at the quiet Sun in spectral emission lines that originate predominantly in one of these three regions, we see different behaviour. Figure 6.13 shows images of the quiet Sun in the chromosphere (He I), the transition region (O V) and the corona (Mg IX). The chromospheric network, which aligns the supergranule convection cells, is clear through the transition region but becomes diffuse when we look in the corona. It is suggested that the field lines are swept towards the supergranule edges at low altitudes, whereas at high altitudes the magnetic field lines spread out to form a canopy. There are many small brightenings seen (named 'blinkers') — the most intense ones being at the boundaries of the network, to which the magnetic field is swept. The energy released from these brightenings may contribute to the heating of the corona.

In addition to its radiation, there must also be a non-radiative energy output from the photosphere since the chromosphere is hotter than the photosphere below. There are several suggestions for this energy source. Sound waves are produced in the interior, and hence can be transmitted into the atmosphere. A sound wave itself causes particles to oscillate and does not provide energy. However, when it reaches a region of lower density (as observed in the chromosphere) the regions of compressed gas will travel faster than the regions of rarefied gas, and a shock will form as the wave becomes more and more distorted. The strength of a shock, η, is measured by the fractional change in density in the undisturbed and shocked regions of plasma:

$$\eta = \frac{\rho_{\text{undisturbed}} - \rho_{\text{shocked}}}{\rho_{\text{shocked}}}.$$

A shock front will give rise to heating and it is thought to be responsible for the heating in the lower chromosphere.

Higher up in the chromosphere, Alfvén waves are thought to be responsible for heating. Alfvén waves are formed when a magnetic field line is displaced and released. The displacement is thought to be due to the convective motions below the surface. Alfvén waves are also one of the mechanisms thought to heat the corona and will be discussed in the next section.

Spicules have widths that are close to the spatial resolution limit of most current instrumentation. The lifetime of a spicule can range between 1 and 10 min, with an average upward velocity of 25 kms^{-1} and temperatures that range between 5000 and 15,000 K. Although the spicules are surrounded by hotter coronal plasma, their temperatures never reach coronal temperatures. It is not clear whether the spicules fade away or fall back to the surface. The outflow derived from the density of spicules is a factor of 100 higher than the average solar wind flux, and hence the maximum outflow from spicules is 1%. There have also been indications that

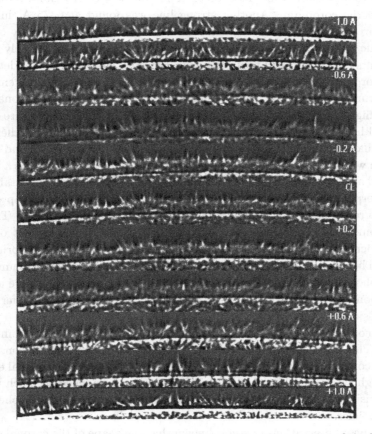

Figure 6.15 This figure shows the limb of the Sun in different wavelengths of the Hα spectral line (from the blue wing to the red wing). Beyond 0.6 Å the general chromosphere disappears and only the spicules remain. Image courtesy of the Big Bear Solar Observatory.

spicules can have a rotational motion. It is now recognised that spicules delineate the magnetic network, and that they cover only a few per cent of the solar surface. Images of spicules at different wavelengths within Hα are shown in Figure 6.15. Spicules are extremely dynamic and this is evidenced by the fact that they are seen in both blue- and red-shifted material.

6.9   The Corona

The outer atmosphere of the Sun, which was named the corona (from the Latin, meaning 'crown'), is seen as an extended white halo during an eclipse. The reason why we only see the corona during an eclipse is that it is a factor of 10^6 fainter than the photosphere and the sky brightness is normally larger than the coronal brightness by 3–5 orders of magnitude. Although eclipses had been seen for many hundreds of years it was during an eclipse in 1869 that Young and Harkness discovered the coronal emission line at 5303 Å. This line was then unknown and hence was named the coronium. Another strong line was found at 6375 Å, in the red wavelength range. Sixteen other unknown 'elements' were subsequently observed. The periodic table at this time was nearly complete, and it seemed unlikely that any new elements could be added to it. It was in 1939 that Grotrian finally determined that the coronal red line was due to FeIX, which he determined was emitted at a temperature of 500,000 K. This was the first indication that the corona had an extremely high temperature relative to the photosphere — and hence there emerged the coronal heating problem, which is still unresolved today. In 1942, Edlén identified the remaining coronal lines as forbidden transitions of iron, nickel and calcium. The reason why these highly ionised ions can exist is that the corona has a very low density and a high temperature. Forbidden transitions occur when metastable levels become overpopulated, because collisional de-excitation is rare (see Chapter 8).

Coronagraphs have been used to simulate eclipses for many years. There are several ground-based observatories, as well as the LASCO instrument on board *SOHO*, which use the technique of an occulting disc to block out the bright photosphere. The spectrum obtained during an eclipse consists of a continuum due to scattered photospheric light with bright emission lines superposed. There are three main components to coronal light: K(Kontinuierlich)-corona, F(Fraunhofer)-corona and E(emission)-corona.

The K-corona is formed from photospheric light being scattered from coronal electrons. This is the dominant component out to two solar radii. It consists of a featureless continuum with no Fraunhofer lines. The speeds of the coronal electrons are very high (around 0.03 c) and hence the lines are broadened so much they are no longer visible. The F-corona is the result of the scattering of photospheric light by dust particles.

Due to these observations it is now known that the shape of the coronal emission gradually changes through the solar cycle. At solar minimum, the corona has long symmetrical streamers at the equator, and plumes at the poles. The plumes at the poles are suggestive of a large-scale dipole magnetic field. During solar maximum there are many more structures, indicative of magnetic complexity, which eliminate the symmetry of the corona.

The outer atmosphere of the Sun seems to consist of hot, magnetised plasma. The theory that deals with the interaction of hot plasma and magnetic fields is known as magnetohydrodynamics (MHD; see Chapter 9). By combining Faraday's law, Ampére's law and Ohm's law, the time dependence of the magnetic field can be determined. This is known as the induction equation:

$$\frac{\partial B}{\partial t} = \nabla \times (v \times B) + \eta \nabla^2 B,$$

where B is the magnetic field strength, v is the velocity of the fluid, and the magnetic diffusivity η is given by $1/\mu_0\sigma$, where μ_0 is the permeability of free space and σ is the conductivity. The first term on the right indicates that the magnetic field changes with time due to plasma transporting the magnetic field, and the second term on the right describes how the magnetic field diffuses through the plasma.

For a fluid at rest, or a situation where the conductivity is low, we obtain the standard diffusion equation:

$$\frac{\partial B}{\partial t} = \eta \nabla^2 B,$$

which implies that magnetic field located in a area given by L will diffuse away in a time given by $\tau_d = L^2/\eta$. For a typical length scale of 10^5–10^6 km, the diffusion timescale is millions of years. It takes a long time for magnetic fields to leak out of any star by diffusion. If, on the other hand, the conductivity is very high then the diffusive term is small, and the rate of change of the magnetic field is given by

$$\frac{\partial B}{\partial t} = \nabla \times (v \times B).$$

This equation describes how the magnetic field is 'frozen into' the plasma and hence moves with it. The timescale for significant motions within the fluid is given by $\tau_v = L/v$. The ratio of the two numbers is the magnetic Reynolds number:

$$R_M = \frac{\tau_d}{\tau_v} = \frac{\tau_d v}{L}.$$

In the solar atmosphere R_M is very large, since the diffusion times are long, and the transport of magnetic field lines by the plasma dominates. In fact, this assumption is valid for most of the Universe. One exception is the case of current sheets, where the magnetic gradients are large and magnetic reconnection can take place, releasing energy. This process is thought to be important in solar flares and coronal heating.

6.9.1 *Coronal heating*

Understanding why the corona is so much hotter than the surface of the Sun has been one of the main goals of solar physicists since the problem was discovered more than 50 years ago. Although the energy that is required to heat the corona is only around 0.01% of the Sun's total luminous output, the actual mechanism is still not known. The first suggestion of heating by sound waves was made in the 1940s by Biermann, Schwarzchild and Schatz. Since the convection zone generates sounds waves, they suggested that the sound waves, which are longitudinal, would strengthen into shocks as they moved into a region of lower density, hence releasing energy. The energy flux of a sound wave is constant until the energy is dissipated through a shock, and is given by

$$E = \frac{1}{2}\rho v_s^2 c,$$

where c is the sound speed, ρ is the mass density, v_s is the velocity of the plasma caused by the passage of the sound wave and c is the sound speed given by $(\gamma p/\rho)^{1/2}$, where γ is the ratio of specific heats. The ratio of densities in the photosphere and corona is approximately 10^8 and the sound speed is around 10 kms^{-1} in the photosphere and 200 kms^{-1} in the corona. Hence, the velocity of the acoustic wave is increasing as the wave moves away from the photosphere and develops into a shock in the lower chromosphere. The energy flux of acoustic waves has been measured in the photosphere and chromosphere and the transition region and has been found to be short of the energy needed to heat the corona by a factor of 100–1000. It was thus accepted that the corona must be heated by a magnetically dominated mechanism. Figure 6.16 (Plate 9) shows a soft X-ray image of the Sun demonstrating the magnetic complexity of the coronal emission. The brightest structures are located in active regions. However, loops are seen which connect active regions to other active regions and there are large diffuse loops, which exist in their own right.

There are two types of mechanisms proposed to heat the corona:

(1) Alternating current (AC) models, i.e. the dissipation of waves;
(2) Direct current (DC) models, i.e. the dissipation of stressed magnetic fields.

The source of both of these is believed to be in the convection zone. Since the magnetic field is frozen into the plasma, the convective motions push flux tubes around.

6.9.1.1   Wave models

Magnetic waves are created through periodic motions of the flux tubes, which can propagate upwards and potentially dissipate their energy in the corona. There are several types of magnetic waves: Alfvén waves, slow mode MHD waves and fast mode MHD waves.

Alfvén waves are transverse or torsional waves, which travel along the field lines in the same way that a wave can travel along a rope. The wave propagates with a velocity that was determined by Alfvén to be

$$V_A = \frac{B}{\sqrt{4\pi\rho}},$$

where B is the magnetic field and ρ is the mass density. The average magnetic field in the corona is approximately 10 Gauss (10^{-3} Tesla), whereas in the photosphere it is around 1000 Gauss (0.1 Tesla). The Alfvén velocity is thus only around 10 kms^{-1} in the photosphere whereas it reaches a velocity of 300 kms^{-1} in the corona. The Alfvén velocity is the main parameter that determines how quickly coronal structures can change. Alfvén waves can propagate easily into the corona — the problem is how to extract the energy in order to heat the corona. Since they are transverse, they can grow in amplitude as the density decreases without a shock forming. In order to dissipate the energy strong gradients in the magnetic field or velocity are required.

The other slow and fast mode MHD waves are longitudinal waves that propagate across the magnetic field. These have the same Alfvén velocity, and since they are longitudinal, they can steepen into shocks and dissipate energy easily.

Waves have been searched for in different types of structures in the solar atmosphere. Intensity and velocity fluctuations in chromospheric and transition region spectral lines have been found in the quiet Sun. These provide evidence for upward-propagating waves from the photosphere right to the upper transition region. Flare-related waves have also been observed which propagate across the disc away from flare sites with a velocity of around 300 kms^{-1}. These are observed using the EUV imaging telescope onboard *SOHO*. Eclipse observations, which have the advantage of high time cadence, have been searched for short period Alfvén waves, but have not yielded convincing results. Spectral lines in active regions have been analysed from the early 1970s in order to search for a relationship between the line width and temperature through the atmosphere. The relationship found is consistent up to the high transition region, but there are few spectral lines that can be used for analysis in the corona. The high-resolution TRACE images have shown evidence for oscillations in loops, which are shaken by the impact of a flare (see Figure 6.17). These waves damp over a period of around 10–12 min. The observed decay of the loop oscillation amplitude demonstrates the presence of strong dissipation of the wave energy. As the oscillation amplitude decreases, the temperature of the loops is observed to rise.

It is clear that observations are showing many occasions when MHD waves exist in all regions of the corona — from quiet regions to flaring regions. The difficulty is in determining whether these waves have sufficient energy to heat the corona.

6.9.1.2 *Microflare and nanoflare heating*

Coronal heating can also occur by reconnection of magnetic field lines in narrow sheets. E. Parker suggested that this can occur by the random shuffling of flux

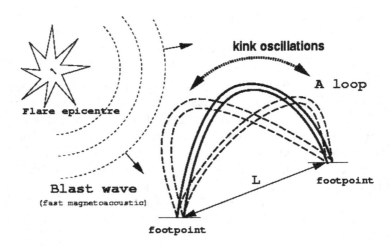

Figure 6.17   A cartoon showing how a flare can excite oscillations (from Nakariakov, V.M., in *Waves in Dusty, Solar and Space Plasmas*, eds. F. Verheest *et al.* (AIP CP537, 2000)).

tubes which causes twisting, and subsequently generates field-aligned currents. The energy released by a single nanoflare would be 10^{24} ergs. The details of magnetic reconnection will be discussed in the next section on solar flares.

The dissipation of the current can take the form of small impulses. Observations were made of frequent hard X-ray impulses during solar maximum (1981) from a balloon flight. Since then, observations have been made in active regions and quiet Sun regions. The frequency distribution of energies of the flares can be fitted with a power law distribution:

$$dN = AE^{-\gamma}dE\,,$$

where E is the flare energy, and γ is the value of the power law. To obtain enough energy from the flares to heat the corona, γ must be greater than 2, and the energy range which is integrated must be large enough. This means that there must be more events in the lower energy regime. There has been much recent work, which concentrates on determining the value of the power law. Using the soft X-ray telescope on board *Yohkoh*, Shimizu studied small flare events in active regions. He found that the power law value was only 1.7, which does not provide enough energy to heat the corona. Small flares have also been found to occur in the Sun — both in the network and inter-network regions. The steepness of the power law is in dispute, with values ranging between 1.6 and 2.6 from different authors. Different methods, which include using spectrometers and imagers in different temperature emission lines, provide different results, making it difficult to determine whether small scale flaring is a dominant heating mechanism. It is, however, clear that there are numerous flare events that occur all over the Sun, in active regions, network and even in the inter-network regions. Figure 6.18 shows a surface plot demonstrating

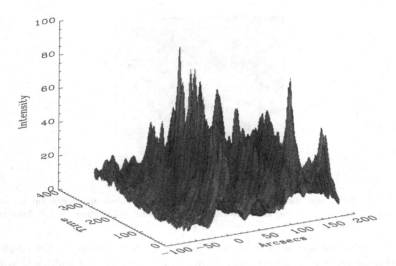

Figure 6.18   A surface plot of small flares in the quiet SUN. The intensities are derived from the O V line, which has a temperature of 250,000 K and is formed in the mid-transition region. The CDS instrument on board *SOHO* was used for the observation. The surface plot shows the intensity that fluctuates all over the quiet region — in both the cell and network regions. The *x*-axis shows the position of the slit in the *y* direction, and the *y*-axis shows time and the *z*-axis shows intensity.

the universal nature of flaring events. These flares occur right through the solar cycle, from minimum to maximum. This is also apparent from the work by Carolus Schrijver, Alan Title and colleagues, which showed that bipolar magnetic fields rise up into the photosphere and disappear within a few days, and are being continuously replenished on a timescale of 40 h (see Figure 6.10; Plate 8).

In summary, there is evidence for the presence of both waves and flares, which can in principle provide the energy necessary to heat the corona. The current work is concentrating on distinguishing which mechanism dominates in each of the widely different types of structure in the corona.

6.10   Solar Flares

Solar flares are the most powerful and explosive events in the solar system, releasing up to 10^{32} ergs in less than an hour. Flares generally cover only 0.1% of the visible hemisphere at any one time. They release energy over a wide range of the electromagnetic spectrum, and in the form of mass, particles and shock-wave motions. Flares can produce noticeable effects on the Earth — including aurorae that can be seen far from the Earth's magnetic poles and even towards mid-latitudes. Electrical transformers can explode as a result of the high voltages induced and passengers on high-flying aircraft can experience large radiation dosages. The study of these types of effects on the Earth is known as space weather (Chapter 5).

Figure 6.19   A white light flare that was observed by the *Yohkoh* spacecraft on 15 November 1991. The sunspots are dark, and the flare can be seen in white. Courtesy of S.A. Matthews.

The first observation of a flare took place in 1859, made by Carrington and Hodge. During regular observations of sunspots, a brightening suddenly appeared, which lasted only around five minutes. This was the first white light flare to be observed. An example of a white light flare is shown in Figure 6.19. They are extremely energetic events and are now known to be rare. Soon after this observation there were magnetic disturbances recorded at Kew Observatory, and less than a day later aurorae were seen at mid-latitudes. At that stage, although it was not clear how the Sun, which is so far away, could have an effect on the Earth, it was beginning to be accepted and investigated in earnest.

Flares have a wide range of sizes and energies, and various classification schemes have been developed over the years. The first scheme was developed from ground-based observations and is based on the area of the flare and the brightness as seen in the Hα spectral line. This is illustrated in Table 6.2 In addition, there is a qualitative measurement of the intensity that ranges from faint (f), normal (n) to brilliant (b). So a small flare covering less than 2 square degrees, which is very bright, will be classified as Sb.

A classification scheme to describe the X-ray intensity of flares (see Table 6.3) was introduced when the *GOES* (*Geostationary Operational Environmental Satellite*) series of satellites were launched. These record the total intensity in soft X-rays covering a wavelength range of 1–8 Å. A new *GOES* satellite (*Soft X-ray Imager — SXI*) was launched in 2001, which has a spatial resolution of 5 arcsec.

Flares have a very different appearance, depending on which wavelength you observe them in. Flares observed in hard X-rays are impulsive and spiky and last for short periods of time. However, flares observed in soft X-rays are gradual, smooth and last for longer. The EUV emission shows impulsive spikes at the very start of the flare, but more gradual emission later in the flare. The different appearance in these wavelengths is indicative of different phases of a flare. In the early impulsive phase, electrons are accelerated to high energy and velocities and emit hard X-rays and radio emission. The gradual phase is detected in soft X-rays and is a

Table 6.2 The Hα classification scheme for flares.

Importance	Flare area (square degrees)
S	< 2
1	2.1–5.1
2	5.2–12.4
3	12.5–24.7
4	> 24.7

Table 6.3 The soft X-ray classification of solar flares.

GOES class	Intensity (ergs cm^{-2} s^{-1})
B	10^{-4}
C	10^{-3}
M	10^{-2}
X	10^{-1}

response to the plasma heating (via Bremsstrahlung) which is a consequence of the high-speed electrons impacting the denser chromosphere. There has been much discussion about a preflare phase, as sometimes a small increase in soft X-rays is seen beforehand. It is also apparent that soft X-ray emission lines show a large increase in their width before the main flare impulsive phase begins. Occasionally jets of plasma are also seen.

The Hα flare patrols from a number of ground-based observatories provided insight into the cause of flares. It soon became evident that flares occur close to sunspots (see Figure 6.19), and that the more complex the sunspot, the greater the flare occurrence in that active region. The Hα eruptions tend to be located between regions of opposite polarity. This suggested strongly that the cause of flares was magnetic in origin.

Once it was realised that the stored magnetic energy is the most likely source of energy during the flare, the problem became 'how is so much energy released in such a short time?'. At the start of Section 6.7, the magnetic Reynold's number was discussed. We showed that it takes millions of years for the magnetic field in the Sun to diffuse. However, either reducing the length scale or enhancing the resistivity can shorten the diffusion timescale. The most likely process to reduce the diffusion timescale is magnetic reconnection. In the 1950s and 1960s, Sweet and Parker investigated what happens when two oppositely directed field lines are brought together. As the distance between the field lines is reduced, the current becomes large and hence ohmic dissipation can be large. The field is said to reconnect at the point where diffusion occurs. All the plasma travels through the diffusion region. Their work showed how the diffusion time could be reduced by a factor of $\sqrt{R_M}$ by enhancing the diffusion of the magnetic field. This is still too low to account for explosive flares, and reconnection was then dismissed as a viable mechanism. However, at this time the possibility of enhanced resistivity was not considered. In the case of a flare the resistivity could be greatly increased, which would lead to a much faster energy release. Another model which was developed by Petschek showed that field lines are not required to reconnect along the whole boundary layer but merge over a much shorter length. For this to happen it is necessary, from continuity, for the boundary layer to consist of slow shocks that can accelerate matter, which does not pass through the diffusion region. With this method a very rapid reconnection can be achieved.

Various observations have been made which are in agreement with predictions from theory of magnetic reconnection. A cartoon of one possible scenario is shown in Figure 6.20. In this example the reconnection occurs high in the corona. Hard X-rays produced by electron-ion Bremsstrahlung are observed in the initial impulsive

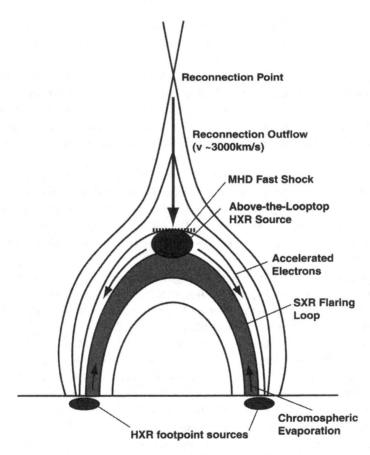

Figure 6.20   A cartoon of the reconnection configuration of a flare as described by Masuda *et al.*, 1995, *Proc. Astr. Soc. Jl.* 47, 677–689.

phase of flares. It is common for hard X-ray sources with photon energies of greater than 30 keV to be located at the footpoints of flare loops. No significant time lag is found between the appearance of hard X-ray emission at the two footpoints and hence it is reasonable to assume that the energetic electrons are streaming down the loop legs from the reconnection point to the footpoints. These high-energy electrons heat the dense chromosphere and plasma is evaporated into the loops. This is observed as thermal emission in soft X-rays. In addition to the footpoint hard X-ray sources, occasionally a looptop hard X-ray source is observed. This is an indication that the reconnection point is high above the source and that the reconnection outflow is impinging on the higher density plasma in the soft X-ray loops and causing a shock.

Other types of reconnection can also occur — for example, an emerging bipole can reconnect with magnetic loops. Although there is much evidence for reconnection, there are still some basic problems. For example, the inflow and outflow jets necessary for reconnection to occur are not regularly observed. This could be due to line-of-sight difficulties or insufficient resolution. Future instrumentation should enable us to provide the necessary observational constraints for the theory.

6.10.1   *Coronal mass ejections*

Ejection of material from the Sun has been observed in the white light corona for many years. Speeds of more than $1000 \, \mathrm{kms^{-1}}$ have been observed, with around 10^{13} kg of mass being lost. Coronal mass ejections (CMEs) have comparable energies to flares. Coronal mass ejections have been associated with long duration flares and erupting filaments, which can occur both inside and outside active regions. Due to the latter association, CMEs occur during the whole solar cycle. There are roughly two CMEs a day during solar minimum, and on order of tens per day occur during solar maximum. An example of a CME is shown in Figure 6.21 (Plate 10). The twisted structure of this particular CME is clear, and indicates a relationship to a prominence eruption.

Since the first observation of a flare in 1859, there has been an association between flares and effects on the Earth. Initially it was thought that only flares had the energy to produce an effect. However, after CMEs were observed it became clear that they could also produce space weather effects. Then ensued a long debate on whether flares caused CMEs or vice versa. Observationally, large flares are not necessarily associated with a CME, as was previously thought. B class flares have been found to be associated with *geoeffective CMEs*. When a flare is seen to be eruptive (i.e. plasma ejections are observed), a CME is observed. If a flare is confined (i.e. no plasma ejection is observed), no CME is observed. Flare and CME onset times have been found to occur any time within tens of minutes of each other. Currently, there is widespread agreement that both flares and CMEs are caused by a global rearrangement of the magnetic field.

In addition to eruptive flare and prominences, various other CME indicators have been defined. These are important for determining the location, and hence the source of CMEs. One new technique is based on a reduction in intensity of the emission observed, or 'dimming', and this is one of the best signatures of a CME. The first example of dimming was found above the limb, and was related to mass loss in an eruptive flare. Since then various examples have been found on the disc. It is possible to estimate the mass of the dimmed area, and in one case this was determined to be around 10^{14} g, which is equivalent to the mass lost in a CME. Making use of spectroscopic observations from the CDS instrument onboard *SOHO*, it was found that the bulk of the mass was lost from plasma at around 1 MK. Trans-equatorial loops which join two active regions have also been observed to be related to CMEs. Coronal waves, which are related to flares, are also thought to trigger CMEs and flares. It has been found that sigmoids (S-shaped structures) have a higher probability of erupting since they are twisted and sheared structures. The main problem with all the signatures is: How can we predict when the CME will occur? For example, sigmoidal structures can survive for a couple of rotations without erupting.

Emphasis is currently being placed on predicting when and where CMEs will occur. This is important because of the effects that high-speed particles have when they impact the Earth's magnetosphere. The most spectacular product is an aurora. The more intense the geomagnetic storm, the further the aurora extends towards the equator. The high-speed electrons excite oxygen and nitrogen atoms, emitting the well-known green and red light. However, these storms can affect our lives in a different way. We are currently dependent on satellite communication for television, defence and telephone communication. Spurious commands can be given to the spacecraft because of particles impacting the electronics. The enhanced radiation heats the Earth's atmosphere, causing it to expand outwards. This will increase the drag on satellites and pull them into a lower orbit. Ironically this had terminal consequences for two spacecraft designed to study the Sun — the *Solar Maximum Mission* and the *Skylab Space Station*.

6.11   Solar Wind

The existence of a solar wind was first proposed to explain several observational facts: the orientation of comet tails (this cannot be explained by the radiation pressure from the Sun), persistent and faint aurorae near the Earth's poles, and a strong 27-day period for strong aurorae. It was predicted from the orientation of comet tails that there has to be a continuous outflow of material from the Sun with a speed of the order of hundreds of kms^{-1}.

Two theories were developed to explain outflow from the Sun. The first was by Chapman in 1957, who showed that the corona must extend into space rather than being confined, as with most planetary atmospheres. Parker then showed an inconsistency in this work, and deduced that the corona has to actually expand into space.

The existence of solar wind was shown a couple of years after Parker's derivation by using ion traps on board rockets, and also the *Mariner 2* spacecraft. These data showed that there were two components to the solar wind. One was a fast component with speeds of around 700 kms^{-1} and one was a slow component with speeds of around 300 kms^{-1}. Images from the *Skylab* mission in the 1970s showed that some of the high-speed wind originates from the low-density regions known as coronal holes (see Figure 6.22). Another method of measuring the solar wind speed made use of remote cosmic radio sources that fluctuate due to the solar wind. This method showed that the slow solar wind is not steady, and close to minimum of the cycle it is confined to mid-solar latitudes.

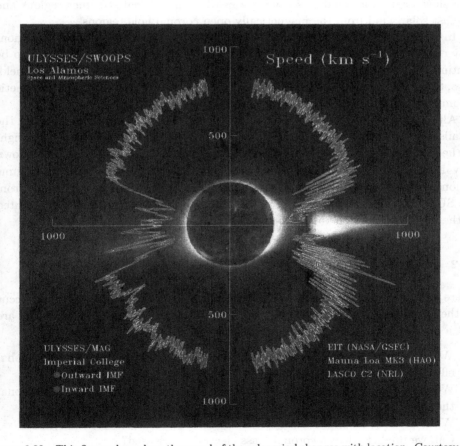

Figure 6.22   This figure shows how the speed of the solar wind changes with location. Courtesy of McComas *et al.*, 1998, *GRL* 25, 1–4; copyright (1998) American Geophysical Union; reproduced by permission of American Geophysical Union. The solar wind speeds are shown at different latitudes. EIT, Mauna Loa and LASCO images are overlaid to illustrate the sources of the slow and the fast solar wind.

The spatial distribution of the solar wind varies dramatically during the solar cycle. At the minimum of the cycle, the north and south poles have a different magnetic polarity. The negative and positive field lines meet at the equator and form a current sheet, which is based in the streamer belt. The streamer belt is the source of the sporadic slow solar wind, and the coronal holes are the source of the fast solar wind. At solar maximum the picture becomes a lot more complicated. The global magnetic structure no longer takes the form of a magnetic dipole. The slow winds can originate from anywhere, and the polar coronal holes reduce in size.

The *Ulysses* spacecraft travelled over the poles of the Sun during solar minimum in 1994/95. The observations provided a clear picture of the solar wind speed at all latitudes. This is shown in Figure 6.22, where the high-speed wind is confined to the high latitude region and the low speed wind is confined to the equatorial regions. The slow solar wind emanates from the nearly symmetrical streamer regions and the fast solar wind from the magnetically open coronal hole regions.

In addition to the streamers, it has been found in some cases that active regions may expand with speeds of around 40 kms^{-1}. The expansion rate is found to be continuous, and the loops will eventually reconnect with the global magnetic field. Also, there are many noted cases of jets of material being released following magnetic reconnection induced by, for example, emerging flux.

Although it is known that the source of the high-speed wind is coronal holes, the details are still unclear. One possibility is polar plumes. These features are bright in the UV and are long and thin structures. However, measurements have shown very little difference in the speed of the wind between plume and inter-plume regions. There is, however, some evidence from spectroscopic observations from the SUMER instrument on board *SOHO* that the high-speed flow is concentrated at the boundaries of the magnetic network.

6.12   Where to Next?

There have been huge leaps in our understanding of the many processes that occur on the Sun. However, there are still many questions to address. The following are just a few — any of them could be the making of many PhD theses!

(a) What is the relationship between the waves that occur inside the Sun and global effects seen outside the Sun?
(b) What processes dominate in the heating of active regions, the diffuse corona, the quiet Sun?
(c) What actually triggers a flare?
(d) Can we predict when a coronal mass ejection is released?

There is much more work to be done, and some of these questions will be addressed by future space missions and experiments, for example *STEREO*, the *Solar Orbiter*, *Solar-B* and *SDO*.

Further Reading

The Sun as a Star, by Roger J. Taylor, Cambridge University Press, 1997.

Guide to the Sun, by Kenneth J.H. Phillips, Cambridge University Press, 1992.

The Sun from Space, by Kenneth R. Lang, *Astronomy and Astrophysics Library*, Springer-Verlag, Berlin, Heidelberg, 2000.

Astrophysics of the Sun, by H. Zirin, Cambridge University Press, 1988.

Spectroscopy of Astrophysical Plasmas, edited by A. Dalgarno and D. Layzer, Cambridge Astrophysics Series, 1987.

The Atmosphere of the Sun, by C.J. Durrant, Adam Hilger, Bristol, Philadelphia, 1988.

Solar Astrophysics, by Peter V Foukal, John Wiley and Sons, 1990.

Solar Magnetohydrodynamics, by E.R. Priest, D. Reidel, 1982.

The Physics of Solar Flares, by Einar Tandberg-Hanssen and A. Gordon Emslie, Cambridge University Press, 1988.

Chapter 7

X-Ray Astronomy

Keith Mason

7.1 Astronomy from Space

The ability to fly astronomical instruments in space has been a major factor in the spectacular advances that have taken place in our understanding of the Universe within the past half-century. The main advantage of astronomy from space is access to the full range of the electromagnetic spectrum, and freedom from the censoring effects of the Earth's atmosphere. Less that a century ago our view of the heavens was limited to the narrow 'optical' window within which human eyes function. Technological advances extended our view from the ground to radio wavelengths, and more recently infrared radiation. Space observatories have given us access to gamma rays, X-rays, extreme ultraviolet, and ultraviolet light, as well as a much more complete view of the infrared band.

This chapter concentrates on X-ray astronomy. The X-ray band was one of the first to be opened up by space travel, and X-ray astronomy has by now reached a level of sophistication and utility to rival optical astronomy as a tool for exploring the Universe. The hottest and most energetic regions in the Universe emit X-rays (Figure 7.1). Objects that we can detect in this band include comets moving in the solar wind (Chapter 3), the hot outer coronae of the Sun and nearby stars, and the remnants of dead stars. Quasars are prodigious emitters of X-rays, with luminosities so great that some can be seen at a time when the Universe was only 10% of its present age. X-rays also reveal the hot gas within clusters of galaxies, which pervades the space between the galaxies themselves. X-ray astronomy has taught us the importance of accretion power as a source of energy, and it is currently playing a crucial role in efforts to understand how galaxies form. The origins of this new astronomy are explored briefly in the next section, setting the scene for an examination of some of the key topics in the field as they currently stand.

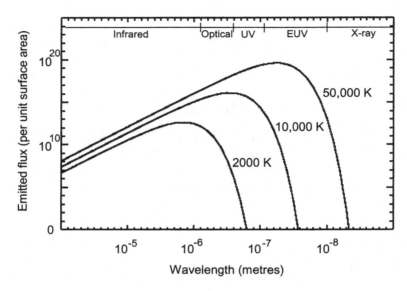

Figure 7.1   An illustration of blackbody radiation. As we raise the temperature of a blackbody, the spectrum of the energy it emits moves to shorter and shorter wavelengths. At a temperature of 2000 K a blackbody will emit primarily in the red part of the optical range and in the infrared. To our eyes it would have a dull reddish colour. When the temperature is raised to 10,000 K, the emission extends throughout the optical range and into the ultraviolet. We would perceive this as a bluish-white colour. At temperatures of 50,000 K the emission begins to extend all the way up to X-rays.

7.2   The Origins of X-Ray Astronomy

The brightest source of X-rays in the sky is, as in visible light, the Sun (see Chapter 6). The X-radiation comes from the hot and tenuous solar corona which has a temperature in excess of 1 million degrees. In 1949 Herbert Friedman and co-workers at the Naval Research Laboratory demonstrated unequivocally that the Sun was an X-ray source using rocket-borne Geiger counters, confirming Burnight's photographic evidence from the previous year (Burnight 1949; Friedman, Lichtman and Byram, 1951).

An X-ray source similar in brightness to the Sun at the distance of the nearest stars would require a detector many orders of magnitude more sensitive than available in the 1950s. Thus, at that time, the prospects of detecting X-rays from anything outside the Solar System seemed remote indeed. Nevertheless, a group of physicists, including Rossi, Giacconi, Clark and Olbert, working at a company called American Science and Engineering, Inc., thought it might be interesting to try. As Rossi (1990) has recalled, it was probably a good thing that these were physicists and not astronomers; otherwise they would have rapidly been discouraged by the apparent futility of their endeavour. Indeed, their initial proposal to NASA was turned down precisely because there seemed to be so little prospect of their being able to detect anything. In a change of tactic, a rocket experiment was eventually put together under a U.S. Air Force programme to attempt the apparently much more tractable task of detecting scattered solar X-rays from the Moon. After two failed attempts, a sounding rocket payload was successfully flown on 8 June 1962. The team did not detect the Moon. However, they did amaze themselves and the world by detecting a very bright source in the constellation of Scorpius. They christened it Sco X-1, and the science of cosmic X-ray astronomy was born.

Many other sub-orbital rocket flights followed as research groups around the world pursued the mystery of the origin of the cosmic X-ray emission. By 1971 a total of about 50 discrete X-ray sources had been detected, though some were highly variable and seen for only a brief period of time. Sco X-1 itself was found to vary in flux by a factor of about 2, and its optical counterpart, a 13th magnitude blue variable star, was located. A source in Taurus, Tau X-1, was identified with the Crab Nebula — the remnant of the supernova of 1054 AD. Both the nebula and the central pulsar were X-ray emitters, the X-ray emission from the latter being pulsed in the same way as its optical and radio output. X-ray sources had also been found to be associated with extragalactic objects, including the galaxy M87 in the Virgo cluster and the quasar 3C 273.

The next big breakthrough in X-ray astronomy came when X-ray detectors were flown on satellites, greatly increasing the exposure and sensitivity achievable compared to the few minutes of observing time realised by a sounding rocket. Some observations of cosmic X-ray sources were made fortuitously from satellites designed for other purposes, for example from the Orbiting Solar Observatories. The first satellite-borne X-ray detectors designed explicitly to study cosmic sources were on the *Vela* series of satellites in the late 1960s, but the primary purpose of these missions was to monitor compliance with the nuclear test ban treaty. The first satellite dedicated to X-ray astronomy was *Uhuru*, launched in December 1970. This achieved the first systematic survey of the X-ray sky, yielding a final catalogue of 339 objects. *Uhuru* revealed X-ray pulsars orbiting about binary companions, results that demonstrated that these are neutron stars, which generate their power by accreting matter from their companion. *Uhuru* also found extended X-ray emission from clusters of galaxies, the first evidence for hot gas between the galaxies.

Many other X-ray satellites followed. Nevertheless X-ray astronomy remained largely a niche science until the advent of X-ray telescopes that used grazing incidence imaging optics. The first of these was *Einstein*, launched in November 1978.

Up until *Einstein*, most X-ray instruments were simple proportional counters, with no intrinsic image resolution, whose viewing direction was mechanically collimated. The angular resolution of collimated proportional counters was typically in the range of 0.5–3 degrees, and with resolutions like this only the brightest sources could be distinguished above the general background X-ray emission. Special add-ons such as rotation modulation collimators could improve the ability of these instruments to position sources, but still it was difficult to achieve accuracy better than about 1 arc min.

Ordinary mirrors such as those widely used in optical astronomy are ineffective in the X-ray band because the X-rays simply penetrate the mirror and are absorbed. However, if X-rays graze the surface of a mirror at a very shallow angle, significant reflectivity can be achieved. As early as 1972, grazing incidence 'flux collectors' had been pioneered on the *Copernicus* satellite (Figure 1.2) to focus X-rays onto small proportional counters, thereby improving the signal-to-background ratio. This was also the first time that cosmic X-ray instruments had flown on a three-axis stabilised satellite, allowing extended continuous observations of sources.

However, *Einstein* was the first cosmic X-ray observatory to use large mirrors to focus X-rays onto a detector that had imaging capability. *Einstein* was a three-axis stabilised platform and employed an arrangement of grazing incidence mirrors in a configuration that was originally developed by Woltjer for use as an X-ray microscope. The telescope, involving successive reflections from a hyperboloid and paraboloid cylindrical mirror, was capable of focussing low-energy X-rays from a point source within an area only a few arc sec across (Figure 7.2). Much higher contrast was thus achieved compared to the background, allowing much fainter sources to be revealed. The underlying richness of the X-ray sky then soon became apparent, with emission being detected from a wide variety of stars and galaxies.

Telescopes with improved capability followed, including the hugely successful German-led *ROSAT* mission, which among its many achievements surveyed the whole sky with an X-ray telescope for the first time (Figure 7.3; Plate 11). The Japanese *ASCA* observatory, while having poorer image quality than *ROSAT*, was able to focus X-rays up to about 10 keV, compared to only 2.5 keV for *ROSAT* and about 4 keV for *Einstein*. The higher energy X-rays are more penetrating, and can reveal objects whose low-energy emission is extinguished by absorption in overlying material. *ASCA* also gave access to the important complex of iron emission lines between 6 and 7 keV. The recently launched *Chandra* and *XMM-Newton*

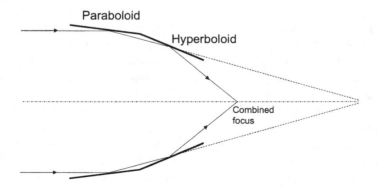

Figure 7.2   Principle of a grazing incidence X-ray telescope.

observatories represent the current state of the art, and dwarf *ASCA, ROSAT* and *Einstein* in their capabilities. *Chandra* emphasises image quality, and is able to concentrate X-rays within a circle that is a fraction of an arc sec across. *XMM-Newton*, on the other hand, emphasises collecting power, with each of its three X-ray telescopes being made up of 58 closely packed mirror shells to maximise the mirror area. The two new observatories use advanced X-ray CCD detectors to record the images produced by the telescopes, and arrays of gratings to form dispersed spectra of X-ray sources. *Chandra* is producing X-ray images that match those from ground-based optical telescopes in their quality and detail. *XMM-Newton*, on the other hand, is able to measure the X-ray spectra of faint objects such as X-ray sources in nearby galaxies, which previous observatories could only barely detect.

Even larger observatories are currently in the planning stage. These will use detectors that operate at temperatures that are a fraction of a degree above absolute zero to achieve greatly improved discrimination of the energy of the X-rays detected, coupled with high throughput. The current best CCD detectors achieve an energy resolution of about 100 eV at 6 keV. New cryogenic detectors, for example superconducting tunnel junction (STJ) detectors or those that work basically as bolometers, detecting the heat input of individual X-ray photons absorbed by the detector material, should be capable of resolving a few eV at similar energies, allowing individual atomic transitions to be separated (see Chapter 13). In order to provide enough photons to make use of this superb energy resolution, very large X-ray mirrors or mirror arrays will be required, dwarfing those on even *XMM-Newton*. The *Constellation-X* mission, under development in the USA, achieves this with an array of spacecraft working in unison. In Europe an alternative concept involving a single large mirror that is assembled on the International Space Station is being studied (called *Xeus*). Missions such as these can be expected to take X-ray astronomy to new levels of sophistication, permitting detailed studies of physical processes occurring deep in the Universe.

Finally, as a footnote to the birth of X-ray astronomy, the sunlit portion of the Moon is indeed an X-ray source, as shown to good effect by a *ROSAT* satellite image taken in the early 1990s!

7.3 Binary X-Ray Sources — The Power of Accretion

The origin of the many bright, often highly variable X-ray sources discovered by early sounding rocket surveys was initially a puzzle. As a group, these sources were concentrated in the plane of the Galaxy, with a further concentration towards the galactic centre (Figure 7.4; Plate 12). This meant that some at least were as far away as the centre of our Galaxy, about 10 kpc, and therefore had a very high luminosity of order 10^{31} W. Evidence that this energy was being generated by accretion of matter onto compact stars became conclusive with the discovery, using the Uhuru satellite, of pulsating X-ray sources in binary systems. The pulsation

Particle of mass m attracted towards star of mass M due to gravity

Gravitational potential energy is converted into kinetic energy of motion

$$\frac{1}{2}mv^2 = \frac{GMm}{r} \qquad (1)$$

The energy is liberated when the particle strikes the star's surface and comes to rest

periods recorded were as short as a few seconds, and varied periodically because of the Doppler shift induced as the pulsating source orbited about a companion object. The underlying stability and short period of the pulsation could only be explained as the rotation of a very compact neutron star. Models were thus developed of X-ray binary systems that contained a neutron star accreting material from a relatively normal companion.

The amount of energy that can potentially be liberated in this way can be easily calculated. Consider an element of matter of mass m falling from infinity onto a star of mass M (see inset). To conserve energy, the infalling matter travels faster and faster as it approaches the accreting star, as gravitational potential energy is converted into kinetic energy of motion. If the accreting star has a 'solid' surface, the inflowing matter must come to rest, liberating its excess kinetic energy through radiation. The more compact a star is, for a given mass, the greater the luminosity that can be liberated per unit mass accreted. For a solar mass white dwarf with a radius of 5×10^6 m, the fraction of the gravitational potential energy that can be liberated is about 3×10^{-4}. For a neutron star of the same mass and a radius of 1.5×10^4 m, the fraction increases to about 0.1. For comparison the equivalent number for nuclear burning, which powers a normal star such as our Sun, is about 7×10^{-3}. Thus accretion onto a neutron star has the potential to generate a very high luminosity. Accretion onto a black hole can be similarly effective. Black holes do not have a solid surface on which the accreting material can be brought to rest. Nevertheless, provided the density in the accreting flow is high enough, viscosity and the need to dissipate angular momentum can ensure that a high fraction of the gravitational potential energy is released before the accreting matter disappears beyond the event horizon of the black hole.

A nice illustration of the relative effectiveness of accretion and nuclear burning in producing energy occurs in a class of X-ray source known as X-ray bursters (Figure 7.5). These generate energy continuously via accretion. The accreted material settles onto the neutron star surface until, after perhaps a few hours, it reaches a critical density, sufficient to trigger a nuclear reaction. This causes a short 'burst' of additional radiation, typically lasting for a minute or so, during which time the

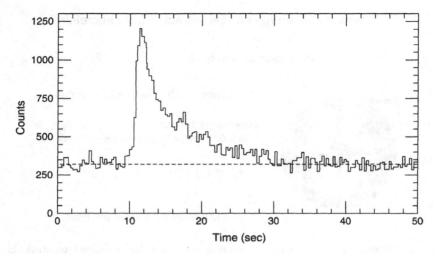

Figure 7.5   A plot of count rate versus time for an accreting neutron star X-ray source showing an X-ray burst. The 'steady' emission is caused by accretion onto the neutron star, whereas a nuclear explosion on the star's surface produces the burst.

nuclear fuel that has built up on the neutron star surface is consumed. Because the burst of nuclear burning has a very different temporal characteristic to the accretion energy release, it is possible to measure the fraction of the total energy generated by the two mechanisms averaged over a long time period. A ratio of about 100 to 1 in favour of accretion is found. The magnitude of this ratio points to the burst being caused by the burning of helium rather than hydrogen.

A similar effect occurs in some accreting white dwarf systems. They accrete continuously, accumulating unburnt nuclear material on the surface of the white dwarf. Because a white dwarf is much larger than a neutron star, it takes much longer for the accumulated material to reach the critical density needed to start a nuclear reaction. This is both because the surface area is larger, and because the gravitational field is less intense. Eventually the critical point is reached, and a nuclear explosion occurs. Such events are known as novae. Nova explosions appear so spectacular compared to the normal brightness of the underlying system because of the large interval between explosions (years to centuries), which means that a large reservoir of material can be accumulated, and because accretion, which provides the only energy output for the bulk of the time, is ineffective compared to nuclear burning in a white dwarf.

It follows from Equation (1) that the luminosity emitted by accretion is proportional to the rate at which matter is being accreted onto the star. Thus, the higher the rate at which matter is accreted, the higher the luminosity of the source. The mass accretion rate and the luminosity cannot be arbitrarily high, however. The limit is set by radiation pressure. When the luminosity generated is high enough, the pressure exerted on the accreting matter by the radiation (by means of

How much energy can be extracted from accretion?

The source luminosity $L = \dfrac{GM\dot{m}}{R}$ (2)

where $\dot{m}$ is the mass accretion rate

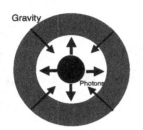

However, if the accretion rate is too high, radiation pressure suppresses the infall

$$L_E = 1.3 \times 10^{31} \frac{M}{M_{Sun}} \quad \text{watt}$$

This is known as the **Eddington limit.**

Thomson scattering of the electrons, for example) will be sufficient to stop the inflow. This is known as the Eddington limit (see inset). For pure electron scattering in a spherically symmetric flow, the Eddington luminosity for a star of mass equal to that of the Sun is 1.38×10^{31} W.

7.3.1 *Thin accretion discs*

Angular momentum plays an important role in determining how accreting sources behave and appear. Let us take the simple case of a compact star accreting material from its binary companion. This will happen if, for example, the companion fills its Roche lobe, the gravitational equipotential surface that encompasses both stars. In this situation matter at the surface of the companion that is directly 'underneath' the compact star experiences as high a gravitational attraction from the compact star as from its parent, and material can flow from one to the other.

This material still carries with it the orbital angular momentum it had when it was part of the companion, however. It cannot fall directly onto the compact star, and is forced instead into an orbit about the latter appropriate to its angular momentum. If matter were coming off the companion in a continuous stream, it would tend to form a ring about the compact star in the plane of the companion's orbit at a radius corresponding to its specific orbital angular momentum. In practice, viscous interaction between the particles will tend to smear the ring out into a disc. As they lose angular momentum and energy through viscous interaction, individual particles spiral closer and closer to the compact star, while angular momentum is transported outward through the disc.

We can consider particles within the disc to be travelling in circular Keplerian orbits at a given radius. To move into an orbit with a smaller radius, the disc material must lose energy. Viscous interactions heat the disc, which dissipates the excess energy through radiation. The smaller the radius, the higher the Keplerian velocity and the greater the rate of energy release. Thus we expect the temperature

Formation of a thin accretion disc in a compact binary star system

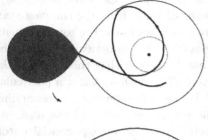

Consider an idealised situation in which a star orbiting a compact companion just begins to overflow its Roche lobe. The gas stream from the companion cannot fall directly on to the compact star because it still has the angular momentum it possessed when it was part of the companion. Instead it swings around the compact star and eventually intersects its original path. The interactions will tend to circularise the orbit at a radius where the angular momentum is the same as material at the overflow point of the Roche lobe.

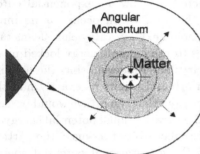

When the accreting matter settles into a ring at the cicularisation radius differential rotation will result in viscous interaction between particles in neighbouring radial orbits. Slower-moving outer particles will gain energy at the expense of faster-moving inner particles, and the matter will spread radially to form a disc. There is a net flow of matter inwards, and a net transfer of angular momentum outwards.

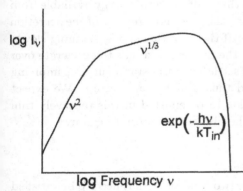

The viscous interactions heat the disc which radiates away the energy. The smaller the radius, the higher the Keplerian velocity, and the greater the rate of energy release. The temperature of the disc therefore increases with decreasing radius. For a steady state thin disc we expect T to vary as $r^{-3/4}$. If we make the assumption that each annulus in the disc is radiating as a blackbody, we can calculate the expected radiation spectrum of the disc by weighting the emission of each annulus by its area. We find that the spectrum should have a slope of 1/3 in log frequency units. At high and low frequencies, the spectrum should approximate the blackbody spectrum of the innermost and outermost disc annuli respectively.

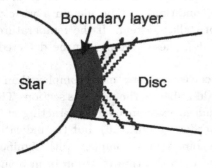

Only half of the original gravitational potential energy is dissipated by particles in reaching a given radius in the disc, the rest being converted to the kinetic energy of orbital motion. When disc material finally encounters the central accreting star, it has to lose this remaining kinetic energy in order to settle onto the star's surface (assuming the latter is not spinning rapidly). This results in a small boundary layer region where matter is heated to high temperature by shocks.

of the accretion disc to rise as we approach the compact star. Allowing for the energy associated with the transport of angular momentum outwards as well as the release of gravitational binding energy, it is found that the temperature in a steady-state thin disc should vary as $r^{-3/4}$, where r is the distance from the compact star. If each annulus in the disc is assumed to radiate as a black body, the total intensity of the disc is proportional to the surface area that emits at temperature T times the black body intensity at that temperature. It can be shown that this leads to the simple result that the intensity of radiation given off by such a disc at a particular frequency, ν, should be proportional to $\nu^{1/3}$ between frequencies corresponding to the peak emission of the innermost and the outermost radius of the disc. At higher frequencies, the disc spectrum is expected to show an exponential cutoff corresponding to the highest temperature in the disc, which occurs at its inner edge, whereas at low frequencies the spectrum should exhibit a Rayleigh–Jeans tail.

A further property of accretion discs relates to the kinetic energy locked up in the orbital motion of the disc particles. In a circular Keplerian orbit, the amount of kinetic energy locked up in a particle at an orbital radius r is exactly half the difference between the gravitational potential energy that the particle would possess at infinite distance from the attracting body and its gravitational potential energy at distance r. In other words, in order to enter a circular orbit of radius r, the particle must lose an amount of energy equal to half the gravitational potential energy difference between infinity and r. It follows that half the total energy available from accretion is still locked up in orbital motion in the innermost regions of the accretion disc, just above the accreting star's surface. If the accreting star is spinning slowly compared to the inner disc orbital velocity, the disc material has to decelerate over a very small physical distance in order to settle onto the star's surface, implying large energy release over a small area, and thus a high temperature. We expect half the total energy available from accretion to be emitted in this relatively thin 'boundary layer' at the interface between the disc and the accreting star.

7.3.2   *Real life*

From the above discussion it is clear that the two conditions that need to be satisfied for accretion energy to be important are a compact star and a supply of fuel. Observationally we have found that these conditions can be met in a variety of circumstances and by a number of types of stellar system. If the temperature in the emission region is high enough ($> 10^5$ K), these systems can be detected as X-ray sources.

The situation found in most X-ray sources is, however, more complex than the idealised thin, optically thick accretion disc described in the previous section. There is good evidence that accretion discs do indeed exist in many interacting stellar binary systems as a result of Roche lobe overflow (Figure 7.6), but the indications in many systems are that these discs are geometrically complex, with significant structure and thickness. It is also clear that optically thin emission from tenuous material is important in accounting for many X-ray sources as well as optically thick

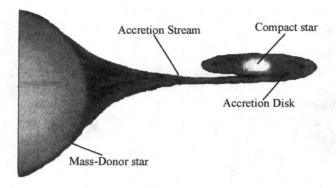

Figure 7.6   Schematic of a disc-accreting X-ray binary.

emission from a disc or boundary layer. In many systems the compact accreting star, a neutron star or a white dwarf, has a strong magnetic field, which can disrupt the accretion flow. Mass transfer can also occur by other than Roche lobe overflow. For example, compact objects in orbit about early type stars can accrete substantial amounts of material direct from the stellar wind of that star. In these circumstances the accretion flow can be much more chaotic than the relatively ordered geometry of Roche lobe overflow. There is even evidence, for example, that isolated neutron stars can accrete sufficient material from the interstellar medium to be detected as X-ray sources, dispensing with the need for a binary companion.

7.3.3   *The X-ray binary zoo*

It has become clear, therefore, that compact X-ray sources are a heterogeneous group, and that X-ray binaries come in a variety of shapes and sizes. More importantly, these different types occur in several different branches of stellar and binary evolution. Among X-ray binaries, one distinction is based on the type of companion star that orbits the X-ray source. In 'high-mass' X-ray binaries, the companion is a massive star that dominates the optical light of the system. In low-mass X-ray binaries the companion is much fainter and less massive, and its intrinsic optical light is often swamped by that generated by reprocessing of the X-ray emission.

7.3.4   *High-mass X-ray binaries*

Many of the pulsating X-ray sources that originally demonstrated the importance of accretion energy are found in orbit about bright, early-type supergiant stars (Figure 7.7). They have orbital periods typically measured in days ($P \leq 10$ days) and accrete matter for the most part direct from the stellar wind of their companion. The neutron star accretor often has a significant magnetic field whose axis is offset from the rotation axis of the star. The magnetic field lines channel the accretion flow into particular areas on the neutron star surface, resulting in an

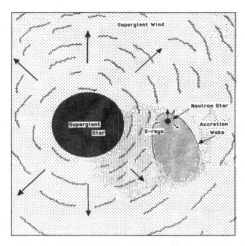
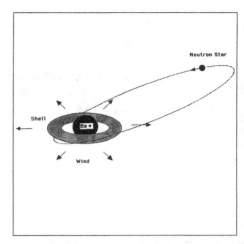

Supergiant companion   **Be star companion**

Figure 7.7   Two types of high-mass X-ray binary.

intense energy release within a small volume, and therefore high temperatures and a bright X-ray source. The viewing aspect of the X-ray emission region is modulated by the rotation of the star, producing the characteristic pulsed X-ray emission.

Other pulsating X-ray sources are in orbit about lower-luminosity B stars, many of which show Be characteristics (Figure 7.7). Be stars, or shell stars, are usually of spectral type B and one of their distinguishing characteristics is that they occasionally eject an equatorial ring of material. These ejection episodes typically have timescales of years and occur irregularly, accompanied by characteristic changes in the appearance of their optical spectral lines due to emission and absorption in the ring, or 'shell'. The orbital periods of X-ray pulsars in orbit about Be companions are generally longer ($P \geq 20$ days) than those in the supergiant binaries, and many of these orbits have a substantial eccentricity. The X-ray flux of Be star X-ray binaries can vary substantially as the neutron star passes through regions of different density resulting from the shell episodes, and particularly when the neutron star makes its closest approach to the Be star in an eccentric orbit (periastron).

Taken as a group, the spin period of the neutron star in these high-mass X-ray binaries ranges from less than a tenth of a second to over 10 min.

7.3.5   *Low-mass X-ray binaries*

The nature of the low-mass X-ray binaries has proved much more difficult to unravel than that of their high-mass counterparts, even though many of the brightest and most luminous X-ray sources in the Galaxy are probably of this type. This is because there are no bright optical stars in these systems, and because few X-ray sources of this type exhibit X-ray pulsations. Thus the two major diagnostics of binarity used to probe the high-mass systems are absent.

A significant fraction of the low-mass X-ray binaries should nevertheless be seen at an angle close to their orbital plane. These high-inclination systems should betray their binary nature when the mass-donor companion star passes in front of the X-ray-emitting star causing an eclipse of the X-rays. For many years it was a puzzle that no such eclipses were readily apparent, leading some authors even to doubt the binary hypothesis. This is currently understood as a selection effect caused by the fact that the accretion discs in these systems have significant thickness, or at least have thick rims (Figure 7.8). Thus X-rays are completely blocked from view when one of these is observed in the orbital plane. Work with the *EXOSAT* satellite, which benefited from long continuous observations because of its extended Earth orbit, revealed many low-mass X-ray binaries that do exhibit orbital modulation among fainter systems, either due to eclipses or, more frequently, orbital-phase-dependent obscuration of the X-ray source by disc material. The latter

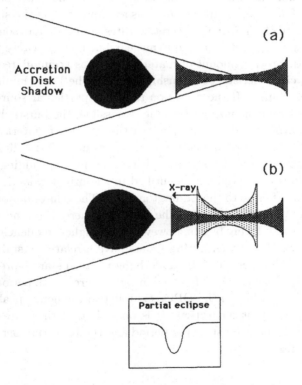

Figure 7.8 A thick accretion disc rim can cause a selection effect whereby sources viewed in the orbital plane are never detected in X-rays. This can explain why eclipses by the companion star are rarely seen in low-mass X-ray binaries. However, a scattering corona above the disc plane can render such a source visible. In this case the eclipse is partial and gradual because the corona is large.

phenomenon results in irregular 'dips' in the X-ray flux which favour particular orbital phases, and indicates that the disc or disc rim thickness changes with azimuth around the disc. A few systems, the best example of which is the source X1822-371 (V691 CrA), show partial X-ray eclipses, indicative of a large X-ray source. What we are seeing in such systems is probably the effects of a large scattering corona about the X-ray source. No X-rays are received from the accreting star directly in X1822-371, either because the scattering corona is optically thick, or because the disc always blocks the direct line of sight. Very low level ($< 1\%$) coherent pulsations with a period of 0.59 s have recently been detected in X1822-371 in high-sensitivity observations made with the satellite *Rossi X-ray Timing Explorer* (*RXTE*) (Jonker and van der Klis, 2001). This shows that the accreting object in the system is a neutron star.

Sensitive observations have by now in fact revealed a rich vein of phenomenology within low-mass X-ray binaries. In addition to the orbital dips and eclipses mentioned above, these include X-ray bursts and quasi-periodic oscillations (QPOs).

X-ray bursts were mentioned in an earlier section and represent the episodic nuclear burning of accreting material that has accumulated on the surface of a neutron star. This occurs for particular accretion rates, and therefore source brightness, which are too low to permit continuous nuclear burning. Because the luminosity in a burst depends on the amount of material that has accumulated since the last burst, there is a characteristic relationship between the burst luminosity and the time since the last burst. If the accretion rate in a particular source increases to the point where continuous nuclear burning is possible, the bursts disappear.

QPOs, on the other hand, probably reflect the motion of material orbiting close to the neutron star. They are manifest as oscillations in the X-ray flux whose period is unstable. Various types of QPO have been identified, including QPOs within bursts. X-ray QPOs were originally identified in the bright galactic bulge sources, which are accreting at close to the Eddington limit. These have periods in the range of 6–60 Hz and may well originate in the magnetosphere of the accreting neutron star. The QPOs found during bursts, however, have higher frequencies, in the range of 300–600 Hz. The QPOs seen during bursts have a relatively stable period, and may reflect the rotation period of the underlying neutron star. More recently, even higher frequency QPOs have been detected in some source, with periods from 500 Hz to in excess of 1 kHz. These 'kHz' QPOs exhibit two frequency peaks, which have an approximately constant separation. It is possible that these are caused by the interplay between the neutron star spin period and the Keplerian period of material in the innermost disc.

7.3.6 *Cataclysmic variables*

Accretion onto a white dwarf is also important despite the relatively small amounts of energy that can be extracted compared to a neutron star or black hole. Indeed, an important class of variable star, the cataclysmic variables (CVs), generate most of their energy in this way. CVs have on average much lower luminosities than neutron star or black hole binaries, but their space density is much higher so that there are many close enough to us to be studied in detail. Indeed, because they are close by and relatively unaffected by the obscuring effects of the interstellar medium, CVs offer significant advantages as laboratories for studying accretion. A further advantage is that it is easier to study the intrinsic properties of accretion discs in these systems. The light of the disc is invariably swamped by reprocessed X-radiation in neutron star or black hole binaries, but this is not the case in CVs.

As with neutron star and black hole binaries, there is a variety of diagnostically rich phenomenology associated with CVs. This includes eclipses, orbital modulation due to disc structure, quasi-periodic oscillations, and the opportunity to study the effects of accretion instabilities in discs. The latter result in spectacular outbursts in a subset of CVs known as dwarf novae, during which their brightness increases by as much as a factor of 100. The orbital periods of CVs are rarely more than a few hours, and can be as short as a few minutes in systems where the mass donor star, like the accreting star, is a degenerate dwarf.

Magnetic fields can also play a significant role in CVs. An important subset have an accreting white dwarf that has a magnetic field strong enough to channel the accretion flow, overcoming the effects of angular momentum and creating a situation where matter falls quasi-radially onto the surface of the white dwarf near the magnetic poles. As discussed earlier, this is a favourable situation for maximising the accretion energy released. If the magnetic and rotational axes of the white dwarf are offset from one another, the emitted radiation will be modulated on the rotation period to produce the white dwarf analogue of an X-ray pulsar. These are known as intermediate polars and are usually characterised by relatively strong bremmstrahlung emission in the X-ray band coming from the gas that has been heated in a shock above the white dwarf surface.

On the other hand, in the subset of magnetic CVs known as polars, or AM Her stars, the drag due to the magnetic field has caused the white dwarf rotation to become phase-locked to the orbital motion (Figure 7.9). This results in the accretion stream becoming entrained in the magnetic field at large distances above the white dwarf. The accretion flow is thus focussed on to a very small spot on the white dwarf's surface. Such systems are characterised by three main emission components: thermal emission from gas that is shock-heated above the surface of the white dwarf — this has a temperature of typically a few tens of keV, and is bright in X-rays; emission from the heated photosphere of the white dwarf beneath the accretion column, which typically has an equivalent black body temperature of a few tens of eV — this is bright in the soft X-ray and EUV bands; and emission resulting from cyclotron cooling in the accretion column, which usually peaks in the far-red and infrared portion of the spectrum.

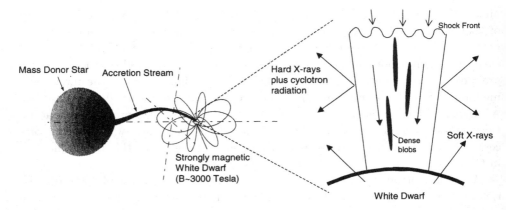

Figure 7.9　Schematic of an AM Her star or polar. The accreting white dwarf in these binaries has a strong magnetic field, the effects of which are to lock the white dwarf's rotation to the orbital motion. Accreting matter is threaded by the field lines and channelled directly onto the white dwarf. A shock forms above the white dwarf photosphere as the material is brought to rest, as depicted in the detail on the right. The accreting material is heated by the shock, and cools by emitting hard X-rays and cyclotron radiation as it settles onto the photosphere. The photosphere below the accretion column is heated by the radiation from the column, and by dense blobs of material that penetrate it directly, causing it to emit in the soft X-ray and extreme ultraviolet bands.

7.3.7　*Black hole or neutron star?*

Both black holes and neutron stars have a deep enough gravitational well to be able to generate the luminosity of an X-ray binary through accretion. The question then naturally arises as to which type of compact star is powering a given source.

In some cases there is unequivocal evidence for the presence of a neutron star. X-ray pulsations, X-ray bursts and certainly the more stable types of QPOs would be difficult if not impossible to account for if the accreting star was a black hole. In contrast, positive evidence *for* a black hole is more difficult to come by. By association we have come to recognise certain state transitions in the X-ray spectrum as being characteristic of accretion discs around black holes. However, the most compelling evidence for black holes comes from measurements of the orbital velocity of the companion star in the binary, and the demonstration that the mass of the compact object in certain systems exceeds the mass limit of a neutron star.

The original, and for many years the best, black hole candidate among X-ray binaries is Cygnus X-1. This X-ray source is in orbit about a bright supergiant companion, and measurements of the orbital velocity of this star suggested that the mass of the X-ray emitter was indeed in excess of that which is plausible for a neutron star. Cygnus X-1, moreover, does not show any of the positive indicators of a neutron star, and does exhibit the spectral state transitions associated with a black hole. However, the mass estimates for the X-ray star do depend on knowing the mass of its much heavier companion, and uncertainties in the latter, driven by poor knowledge of its evolutionary history, are sufficient to engender some doubt.

More secure mass estimates have been made in recent years for certain transient X-ray sources, i.e. those that flare up for a period of a few months before the X-ray emission decays back to its quiescent level. The advantage of such systems is that the companion star is a relatively dim dwarf, and much less massive than its X-ray-emitting partner. In these circumstances, uncertainties in the mass of the companion star are less critical in determining the mass of the X-ray source. Further, once the transient emission has decayed, the companion can be studied without the disturbing effects of the bright X-ray source, which would otherwise swamp its emission. Studies of a number of these systems suggest a minimum mass for the compact star that is significantly above the theoretical limit for a neutron star. Thus we can be confident that at least some X-ray binaries are indeed powered by accreting stellar-mass black holes.

7.3.8   ...or white dwarf?

The question also arises as to whether the accreting star in an individual system might be a white dwarf. One can argue statistically that on luminosity grounds the majority of the bright galactic X-ray binaries are not accreting white dwarfs, because of their much lower accretion efficiency compared to neutron stars or black holes. Moreover, the characteristics of accreting white dwarfs as a class are well established through the study of cataclysmic variables. X-ray sources that show very rapid pulsations, with periods less than about 1 s, cannot be rotating white dwarfs because the velocities at the surface of the star would exceed the break-up speed of such an object. Further, X-ray pulsars generally exhibit long-term fluctuations in their pulse periods that are too fast to be consistent with the moment of inertia of a white dwarf.

There is a class of luminous X-ray source that may well be due to emission from a white dwarf, however. These are the so-called 'supersoft' sources, such as LHG 83 and LHG 87 identified in the Large Magellanic Cloud. As the name implies, the X-ray spectrum of these sources is very soft, with an equivalent blackbody temperature of no more than a few hundred eV. Their luminosities are also high — in fact exceeding the Eddington luminosity for a solar mass star. Rather than being due directly to accretion, it is likely that the luminosity of these objects is being generated by continuous nuclear burning of accreted matter on the surface of a white dwarf.

7.4   Supernova Remnants

The vast majority of elements in the Universe heavier than helium are believed to have been formed by nuclear reactions in the centres of stars. The way in which these elements find their way into the interstellar medium to be recycled into new stars and planets is by means of supernova explosions. These disrupt the parent star and release the elements that would otherwise remain locked deep in its interior.

Two main types of supernova explosion have been identified. A Type 1a supernova is believed to result when the white dwarf star in a cataclysmic variable is pushed over its stability limit by the mass of accreted matter it has accumulated from its companion. Beyond a certain point the pressure in the star's interior is insufficient to prevent its implosion under gravity and it collapses to form a neutron star, ejecting its outer layers in a supernova explosion. A Type II supernova, on the other hand, is believed to occur in a much more massive star when the rate of energy generation in its interior falls to the point where it can no longer prevent the collapse of the star's core. Again the end product may be a neutron star with the outer layers of the original star being blasted off into the surrounding medium.

The result of a supernova explosion is a supernova remnant, which can be a bright X-ray source. There are two main types. A central pulsar, the neutron star formed in the supernova explosion, powers the *filled-centre* remnants. These are comparatively rare, but a good example is the Crab Nebula, which is the remnant of a supernova explosion that took place in AD1054 (Figure 7.10; Plate 13). The central pulsar is the power source for highly energetic electrons that radiate away energy by means of the synchrotron process (Chapter 8) resulting in the extended nebula .emission. In contrast, the *shell-like* remnants are thermal emission from gas heated by the supernova blast wave propagating into the interstellar medium (Figure 7.11; Plate 14).

Shell-like supernova remnants are of interest because they allow us to study the process by which the interstellar medium becomes enriched with heavy elements. Recent X-ray images taken with, for example, the *Chandra* satellite trace the distribution of elements through their line emission, showing that the distribution of particular species is clumpy. Supernova remnants are also a major source of heating of the interstellar medium.

The evolution of a shell-like supernova remnant can be described in four stages:

In *Phase I* the outer envelope of the exploding star is expelled into the surrounding low-density interstellar medium at velocities of 10–20 thousand kilometres per second. This acts as a supersonic ram, heating the interstellar matter in a shock wave which projects ahead of the expanding ejecta. The ejected material itself cools adiabatically as it expands (see inset).

By *Phase II* the expanding shell has swept up a significant mass of interstellar matter, which causes a 'drag' on the expansion of the shell. The outer parts of the expanding shell are decelerated, but the ejecta in the interior of the expanding bubble continue to expand at high velocity until they encounter the decelerating outer layers. The result is a reverse shock that propagates back into the centre of the remnant, reheating the ejected gas and converting a significant portion of the energy of the explosion back into heat (see inset).

<u>Illustration of the first two phases in the production of a shell-like supernova remnant.</u>
<u>Phase 1</u>: Outer layers of star ejected at velocities of 10-20 thousand km/s

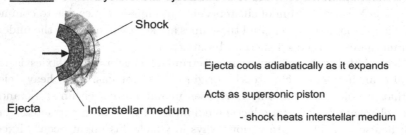

Ejecta cools adiabatically as it expands

Acts as supersonic piston

- shock heats interstellar medium

<u>Phase 2</u>: Blast wave sweeps up significant mass of interstellar matter

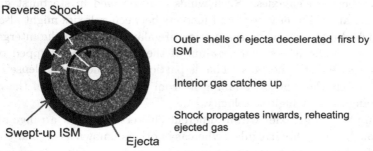

Outer shells of ejecta decelerated first by ISM

Interior gas catches up

Shock propagates inwards, reheating ejected gas

Phase III is marked by the rapid cooling of material in the shell by means of line emission, which becomes important at a temperature of below about 1 million degrees. To preserve pressure balance, the density rapidly increases, and the shell forms a dense 'snowplough'. A good example of a remnant in this stage is the Cygnus loop, which is about 50,000 years old and is characterised in the optical band by dense cooling filaments that are bright line emitters.

Phase IV is the final stage in the evolution, where the expansion velocity of the shell becomes sub-sonic (< 20 kms^{-1}) and the remnant loses its identity.

7.5   Clusters of Galaxies

The detection of hot diffuse gas in clusters of galaxies ranks as one of the major discoveries of X-ray astronomy. The majority of galaxies in the present day Universe occur in clusters or groups, and the study of clusters is a key element in tracing the ongoing competition between gravity and the Hubble expansion to determine the large-scale structure of the Universe.

The discovery of diffuse X-ray emission in clusters of galaxies was the first indication that the space between galaxies is pervaded by hot gas. The luminous content of galaxies makes up only about 5% of the total mass of clusters, which is inferred from dynamical studies. The hot intergalactic gas is at least as important a contributor to the total mass as the galaxies, alongside the unseen dark matter that makes up the remainder. The temperature of the X-ray-emitting gas can be as high as 100 million degrees and it is estimated to have a density that is of order 10^3 m^{-3}. The X-ray spectrum of clusters contains emission lines due to common elements such as iron and oxygen, and these lines are key evidence that the underlying emission mechanism is indeed thermal bremsstrahlung.

The fact that the X-ray emission spectrum of clusters of galaxies is rich in lines is also an important clue to the origin of the hot gas. The heavy element species that are observed can only have been manufactured within stars, and thus the heavy elements in the intergalactic medium must have been extracted from the galaxies themselves. There are various ways in which this might occur. Excessive supernova activity within a galaxy might drive a wind of hot-metal-enriched gas into the space between galaxies. Such winds are observed in starburst galaxies like M82. Gravitational interactions and mergers between galaxies might also drive off metal-enriched material, while the motion of galaxies through the intergalactic medium may cause the interstellar medium of the galaxy to be stripped off due to the ram pressure of the hot gas. This is particularly true in the dense central regions of a cluster where galaxies falling in from the outer regions of the cluster will be travelling at their highest velocity.

The distribution of galaxies within different clusters can show a range of morphologies, some being highly irregular, whereas in others, known as regular clusters, the density of galaxies rises towards the centre of the cluster. Often there is a very large and dominant galaxy (or pair of galaxies) at the centre of a regular cluster (see Figure 7.12). The distribution of X-ray-emitting gas follows a similar pattern. In irregular clusters the hot gas tends to be clumpy, whereas in regular clusters the hot gas peaks towards the cluster centre and the distribution of X-ray emission is smooth.

The range of cluster morphology can be understood in terms of the dynamical relaxation of the cluster material. In the early Universe clusters would begin to form first in regions where the mean density of matter is highest, and grow under the influence of gravity. The more massive the proto-cluster, the more rapidly it would grow. Similarly, the deeper the gravitational potential well, the more rapidly galaxies will move within the cluster and the easier it will be for them to interact. Thus in this picture regular clusters are those whose dynamical timescales are short enough for them to have been able to achieve a 'relaxed' state within the current age of the Universe. In contrast, the dynamical timescale of clusters that currently appear irregular is longer than the present age of the Universe and they will not achieve a relaxed state until some time in the future. In practice this simple picture is likely to be modified by sub-clustering and merging between individual clusters reflecting a distribution in the scale size as well as the amplitude of matter fluctuations in the early Universe.

Some clusters exhibit enhanced X-ray emission towards their centres, often associated with a dominant central galaxy. These central enhancements can be understood as 'cooling flows' (Figure 7.12), which can form when the timescale for

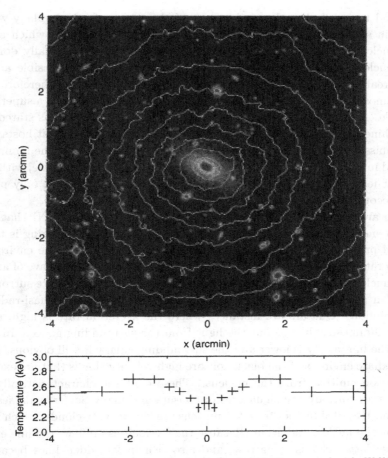

Figure 7.12   X-ray contours of the cluster of galaxies Sersic 159-03 measured with *XMM-Newton* superimposed on an optical image of the cluster taken from the Digital Sky Survey. This is a regular cluster, and the X-ray emission is symmetric and centred on the bright central (CD) galaxy. The lower plot shows the temperature of the X-ray spectrum measured across the cluster. The temperature is lower in the central regions, suggestive of a cooling flow.

the gas to lose energy due to the bremsstrahlung process is short compared to the dynamical timescale. As the gas emits energy and cools, it is compressed in the gravitational potential. Since the efficiency of the bremsstrahlung process increases with density, the rate of cooling increases, and the net result is a flow of cooling material towards the centre of the gravitational potential. Such flows can contribute significantly to the mass of the central galaxy.

7.6   Active Galactic Nuclei

Carl Seyfert recognised the first examples of active galactic nuclei in the 1940s. He examined galaxies with a bright star-like nucleus and found that they exhibited strong and unusually broad emission lines. Other types of active galaxy were discovered in subsequent decades, including radio galaxies, some of which also had bright nuclei (N-galaxies), and quasars, where the emission is totally dominated by the nucleus and the surrounding galaxy is very faint if it is visible at all. It was the realisation that quasars were at enormous distances and therefore of very high luminosity that propelled to the fore the idea of accretion onto a supermassive black hole. A quasar can have a luminosity up to 10^{45} erg/s and, as stated above, can outshine the summed emission of all the stars in the galaxy that hosts it. The quasar emission can also vary on short timescales, indicating that the luminosity is generated in a small volume (Figure 7.13). Accretion onto a black hole, with a mass that is at least one million times that of the Sun, seems to be the only plausible way to account for these properties.

There are many different types of active galactic nucleus (AGN) that can be linked to energy generation by a massive black hole. Current thinking is that the variety of properties observed is a consequence of differences in the environment, accretion rate and viewing angle (see inset). The basic concept we have of an active galactic nucleus is of a black hole accreting matter from its immediate surroundings through an accretion disc or, in low-luminosity examples, via a quasi-radial low-viscosity inflow. Quasars are instances where the accretion rate is high, and the accretion luminosity drowns out the light from the surrounding galaxy. In Seyfert galaxies the luminosity is lower and the surrounding galaxy is still obvious. Beyond the immediate environs of the black hole are high velocity clouds that are excited by the radiation emitted from the nucleus. These emit the characteristically broad emission lines that are the signatures of quasars and many Seyfert galaxies in the optical and UV. Still further from the nucleus are low-velocity clouds, which also are excited by the nuclear radiation. Because their velocities are low, the line emission from these latter clouds is narrow, and also rich in forbidden lines because the density of these clouds is relatively low (Chapter 8). Between the broad and narrow

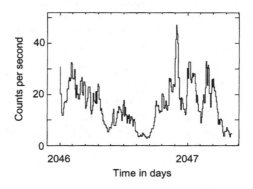

Figure 7.13   The X-ray light curve of the Seyfert 1 nucleus in the galaxy NGC 4051 measured in the 0.1–2 keV energy range with the *XMM-Newton* satellite. The time is measured in days since Julian date 2,450,000.0. Each data point is an accumulation over 500 s.

line regions is a cold dense torus of gas. If the galaxy is viewed such that the torus is seen edge-on, it can block the line of sight to the accreting black hole. In these circumstances the only obvious evidence of the active nucleus is the narrow emission lines, which come from beyond the torus. Such systems are known as Seyfert 2 galaxies. In fact there is an observational gradation between a full Seyfert 1 galaxy and a Seyfert 2, which is characterised by the ratio of the broad to narrow line strength (thus Seyfert 1, 1.5, 1.8, 1.9 and 2).

Direct evidence that AGNs can emit relativistic jets comes from observations of radio galaxies where there are numerous examples of collimated emission extending either side of a galactic nucleus. The consequences of a jet being seen face-on, i.e. travelling directly into the line of sight, are believed to explain the Blazar phenomenon, which is seen in a small but important subclass of AGN. Because the jet is travelling towards us at relativistic speeds, its emission can be boosted by several orders of magnitude (the Lorentz factor) and the observed timescale of variability can be very short. The jet emission is non-thermal, which results in a continuum that is free from emission lines. One kind of Blazar is named after the system BL Lac, where the emission is always dominated by emission from the jet. In the optically violent variables (OVVs), however, the emission comes and goes, causing large changes in apparent brightness. In their low state, when the emission from the jet is low, OVVs characteristically reveal the thermal spectral lines of a normal AGN.

7.6.1   *X-ray emission from AGNs*

A characteristic of AGNs is that they can emit across the entire electromagnetic spectrum. Thus multiwavelength observations are essential to fully understanding the energetics of these objects. The X-ray band is particularly important, however, since it allows us to probe the regions very close to the black hole. The best X-ray data currently available concerns Blazars, many of which are bright in the X-ray band due to Doppler boosting of non-thermal emission from a jet, and relatively nearby Seyfert galaxies, which are bright because of their relative proximity.

Blazars are characterised by a high degree of variability in their X-ray flux on timescales of days and weeks. Their spectrum extends to above 1 MeV and has a simple power law form. Variations in their flux are accompanied by variations in spectral slope, generally in the sense that the objects have a harder spectrum when they are brighter. The high-energy emission of Blazars almost certainly arises due to inverse Compton scattering of photons by a power law distribution of relativistic electrons. Similarly, it is difficult to account for the high luminosities and short timescale variability seen without invoking relativistic beaming of the emission in a jet.

The spectrum of nearby Seyfert galaxies in the X-ray range can, on the other hand, be decomposed into three main components: an underlying power law spectral distribution, a reflection component and a soft X-ray excess.

Unified Quasar model

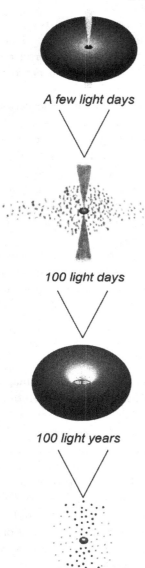

A few light days

100 light days

100 light years

Quasars exhibit a number of observational characteristics many of which can be explianed in terms of the 'unified' model illustrated here. At the centre of a quasar is a supermassive black hole, with a mass at least 1 million times that of Sun. The luminosity of the quasar comes about because the black hole is accreting matter from the surrounding galaxy, probably via some sort of accretion disk which is a few light days accross. X-ray emission comes from the inner part of the accretion disk, close to the black hole event horizon. There may be a relativistic jet emanating perpendicular to the disk plane. If the jet is viewed face on, its emission is boosted by the Lorentz factor, $\gamma=[1-v^2/c^2]^{-1/2}$, which can be very large when the velocity v of the jet is close to the speed of light, c. In these circumstances the emission we see is dominated by the non-thermal jet radiation. This is known as a BL Lac system. Surrounding the accretion disk on a scale of about 100 light days are a series of clouds which are irradiated by the central X-ray source. These clouds emit line radiation in the optical and ultraviolet bands and, because the they are circulating at speeds of as much as several thousand km/s in the strong gravitational field of the black hole, the lines they emit are broadened, by the Doppler effect, giving rise to the characteristic broad emission lines in the spectrum of a quasar, and their low-luminosity counter-parts, the Seyfert 1 galaxies. Beyond the Broad Line Region, on a scale of about 100 light years, is an optically thick torus, sometimes known as the molecular torus because it is a region where molecular lines can be seen. In a Seyfert 2 galaxy, the molecular torus blocks our view of the accretion disk and the broad-line clouds, and all we see is narrow emission lines from slow-moving clouds that are further from the black hole. In a Seyfert 1 galaxy, we are looking from a direction that is higher above the plane of the torus, and we see both the broad and narrow emission lines. As with the broad-line clouds, the narrow-line clouds are excited by radiation fromthe central black hole. Shielding from the torus results in a cone-shaped distribution of emission, which can be resolved in nearby Seyferts. The density of gas in the narrow-line clouds is sufficiently low to produce forbidden emission lines.

The power law distribution has an index of 0.9 when the flux is plotted in energy units, with, unlike Blazars, a relatively small dispersion from object to object. The spectrum is cutoff at high photon energies at a point that is generally not yet well determined but is of order 100 keV. The physics underlying this emission is still controversial. Thermal bremsstrahlung emission, Compton scattering and electron–positron pair production are all mechanisms that have been considered. The shape of the spectrum can be produced naturally by thermal Comptonisation models where hot electrons overlie a source of cool photons. The photons are scattered several times by the electrons and boosted in energy to produce the X-ray spectrum.

The second major X-ray spectral component in Seyfert galaxies can be explained in terms of Compton reflection of the primary power law by, for example, an underlying accretion disc. This component peaks somewhere between about 20 and 50 keV, which can be explained as a consequence of photoelectric absorption which progressively reduces the reflection efficiency towards low energy. The spectrum is often accompanied by relatively strong fluorescent emission from iron at a rest wavelength of about 6.4 keV.

Finally, many Seyfert galaxies have been found to exhibit excess emission below about 1 keV. It is tempting to see this soft X-ray excess as an extension of thermal emission from the accretion disc, which is usually found to rise through the ultraviolet spectral region (the big blue bump). The soft X-rays may again be due to Compton scattering of photons, but this time by a warm layer overlying the disc.

Some Seyfert X-ray spectra also exhibit the characteristic signature of absorption in 'warm' ionised gas that is interposed in the line of sight to the central source (Figure 7.14). Sharp absorption lines due to elements such as oxygen, nitrogen, carbon and iron betray the presence of such gas, while emission lines from forbidden transitions arise in gas that is distributed around the nucleus, and not projected against the bright central source. The state of ionisation of the gas can

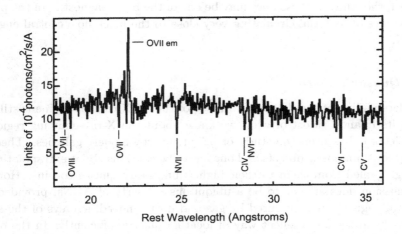

Figure 7.14 A high wavelength resolution X-ray spectrum of the Seyfert 1 nucleus of the galaxy NGC 4051 made with the *XMM-Newton* observatory shows numerous narrow absorption lines from warm ionised gas surrounding the source and projected onto the bright nucleus. Various absorption lines due to oxygen (O), nitrogen (N) and carbon (C) are marked. There is also an oxygen emission line at about 22 angstroms, which again comes from the tenuous gas surrounding the galaxy nucleus. The emission comes from a transition that is 'forbidden' except at low densities.

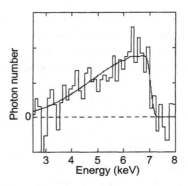

Figure 7.15 The X-ray spectrum of the Seyfert 1 nucleus Mrk 766 measured with the *XMM-Newton* observatory. The histogram shows the residuals from a simple power law fit to the continuum. There is excess emission believed to be due to a relativistically broadened iron line. A theoretical profile for a line produced in an accretion disc surrounding a spinning black hole is shown as the continuous curve, the parameters having been chosen to best match the data. The asymmetric profile results from the combination of beaming of emission from material circulating at close to the speed of light, and redshifting of the emerging photons due to the intense gravitational field of the black hole.

be determined from the energy and depth of the lines. In some galaxies there is evidence that the overlying gas is patchy and does not fully cover the underlying X-ray source, resulting in a complex spectral shape. Evidence is also accumulating that some Seyfert galaxies exhibit very broad and redshifted iron lines whose shape is the result of relativistic motion near the massive black hole (Figure 7.15). If confirmed, the study of these lines may be one of the best diagnostics of the physics and dynamics of material circulating very close to the black hole central engine of the AGN.

7.6.2 *Quasar evolution*

Our understanding of the detailed workings of the AGN central engine is still very much in its infancy. Most of what we know about the X-ray-emitting regions in these objects is based on the study of bright, nearby Seyfert galaxies. These are relatively low-luminosity objects and one has to be cautious about extrapolating the knowledge gained from them to their higher luminosity quasar cousins. However, X-ray emission does appear to be a ubiquitous property of AGNs, provided that the central engine is not obscured by gas and dust. Indeed, surveys of the sky in the X-ray band are an excellent way of locating quasars efficiently. In the optical band quasars must be picked out from the myriad of foreground stars and galaxies. By contrast, in X-rays most of the objects detected, at least away from the plane of our Galaxy, are quasars or other types of active galactic nucleus.

It is well known that quasars at a redshift of about 2 are on average much brighter optical emitters than they are in the local Universe. The same is true in

X-rays. At a redshift of 2 the Universe was about half its present age, and quasars as a whole were about a factor of 100 more luminous in both bands. The change in mean quasar luminosity represents an evolution in the mean properties of the quasar population over that time (see also Chapter 11). The most likely cause of this reduction in luminosity is a decrease in the rate at which matter is being accreted by the central black hole. It is not clear, however, whether each quasar has declined smoothly in luminosity over that interval, or whether the quasar phase is short-lived in individual galaxies, and that it is the mean intensity of such episodes that has declined with time.

Ideas are currently emerging that link the evolution of quasars with the evolution of galaxies in general. It is possible that all galaxies have a black hole at their centre. Indeed, the black hole might be an essential prerequisite for the formation of a large galaxy. In the early Universe the black hole might be enshrouded in gas and dust from the galaxy forming around it. The dust would absorb any radiation produced by accretion onto the black hole and re-radiate it at low frequencies. Thus these protogalaxies might be bright sources in the far infrared or sub-millimetre bands. Over time, radiation pressure from the black hole might drive a wind, which clears out the central regions of the galaxy. The nucleus emerges from its absorbing cocoon and shines as a bright quasar. As the wind clears out the centre of the galaxy, it also deprives the quasar of its accretion fuel, and the nucleus progressively dims, resulting in the reduction in measured quasar luminosities between redshift 2 and the present epoch.

Chapter 8

Using Quantum Physics and Spectroscopy to Probe the Physical Universe

Louise Harra and Keith Mason

8.1 Introduction

Quantum mechanics dates from the 1920s. It led to earth-shattering discoveries — for example, that Newton's laws of physics were completely wrong at the small-scale size of atoms. In quantum physics nothing can be measured or observed without disturbing it. It is fundamentally impossible to predict exactly what will happen to any particular particle in the Universe and therefore the future is not deterministic. If we set up an experiment in exactly the same way and under exactly the same conditions as yesterday, we will not necessarily get the same result today! Another consequence of quantum physics is that nothing can have a definite location *and* a definite speed! The field of quantum physics has widespread philosophical and scientific impacts.

Knowledge of the workings of atoms has provided us with a unique tool with which to probe quantitatively any object in the universe. This chapter will provide a glimpse of what can be achieved using quantum physics. We will begin with an overview of the basics.

8.2 Quantum Theory of an Atom

Many of the initial experiments leading to quantum physics were related to understanding the properties of light. The next section will detail some of the major discoveries and discuss the Jekyll-and-Hyde behaviour of light: that is, it can behave both as a particle and as a wave.

8.2.1   *Blackbody radiation*

It is commonly observed that when a solid, for example an iron poker, is heated its colour will change from a dark red to orange and yellow and then to a bluish white. This observation in the visible part of the electromagnetic spectrum is a simple demonstration of the fact that the peak wavelength of the continuous spectrum that is emitted by an object decreases as the temperature increases (Figure 7.1). Although this phenomenon is commonly observed, it cannot be explained easily by classical physics. Classical physics predicts that the amount of energy radiated at each frequency is proportional to the frequency. So this means that the higher the frequency the more radiation there would be. At the shortest wavelengths in the ultraviolet and beyond, infinite amounts of energy should be produced. This was known as the '*ultraviolet catastrophe*'.

After a determined struggle, Max Planck discovered a solution to this problem. He initially assumed that the walls of a blackbody contained oscillating electrons of all frequencies, and that emission and absorption of any radiation was through the oscillators. In order to obtain a good fit to the observations it was necessary for Planck to constrain the energy to be in chunks or quanta. Thus, the energy of an oscillator can no longer vary between zero and infinity, but must exist in a discrete set of energy levels. This is given by the equation $E = h\nu$, where the Planck constant, h, is determined by experiment and E is a quantum of energy. For high frequencies, the energy needed to emit one quantum of energy is very large, and only a few atoms (oscillators) of the blackbody will have enough to achieve this. Hence the puzzle of the ultraviolet catastrophe was solved. This was the first step towards quantum physics.

8.2.2   *Is light made up of waves or particles?*

The debate as to whether light consists of particles or waves has been going on for hundreds of years. Newton explained light in terms of particles since, for example, light travels in straight lines and can bounce off a mirror. Around the same time (17th century) Huygens developed the idea that light was in the form of a wave. Wave theory could also explain the reflection and refraction of light. Few people supported this idea until at the beginning of the 19th century Young carried out his famous 'two-slit' experiment. In this experiment he shone light upon a screen in which there were two narrow slits. A white screen was placed behind the slits and an interference pattern of alternating bands of light and dark was observed (see Figure 8.1). The light bands are where the waves from the slits add and reinforce (constructive interference) and the dark bands are where the waves from the slits cancel each other out (destructive interference). If light were in the form of particles we would not expect to see any interference. Instead we would see the simple sum of each slit opened individually. The natural explanation of Young's experiments was that light is made up of waves.

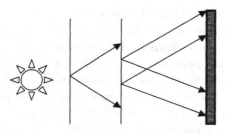

Figure 8.1   An illustration of Young's two-slit experiment.

8.2.3   *The photoelectric effect*

However, the issue was further confused by the discovery of another phenomenon, the photoelectric effect. The experiments of Lenard and Thomson showed how cathode rays (electrons) could be produced by shining light onto a metal surface in a vacuum. One of the most surprising results was that the intensity of the light source, or the distance from the light source to the metal, made absolutely no difference to the energy of the electrons leaving the metal. The only thing that could change the energy of the electrons was the frequency of the light. The energy of each electron is given by $h\nu$, where ν is the frequency of the light wave and h is Planck's constant. For light of a particular frequency the number of electrons emitted will increase for an increased light intensity, but the energy given to each electron will always equal $h\nu$. These results inspired Einstein to produce a piece of work that would later lead to a Nobel prize in physics. He concluded that the energy in light was propagated in discrete localised packets, known as photons or quanta.

So there is direct proof that light can act like a wave and can also act like a discrete particle — after centuries of debate we must conclude that both are correct! This behaviour is known as *wave–particle duality*, and is a general property of quantum physics. The same wave behaviour shown in Young's two-slit experiment can also be seen when electrons (particles!) are fired at two slits. So electrons are also both waves and particles. Which of these two guises comes to the fore is dependent on what sort of observation is being made — light can show properties of a wave when it travels through two slits, but acts like a particle when observed to bombard a metal plate.

8.2.4   *Heisenberg's uncertainty principle*

In 1927 Heisenberg presented the uncertainty principle, which was inspired by the puzzle of wave–particle duality. The uncertainty principle can be stated as

$$\Delta x \Delta p \geq \frac{h}{2\pi}.$$

The product of the uncertainty Δx in the position x of a particle, and the uncertainty Δp in the momentum of a particle must be greater than or equal to Planck's constant divided by 2π. This means that the greater precision that can be achieved in the measurement of the position of a particle, the greater the uncertainty in the determination of the momentum, and vice versa. In his classic thought experiment, Heisenberg imagined a gamma ray microscope being used to measure the position of an electron. Because an electron is so small, the microscope has to operate with the highest possible frequency radiation to have sufficient spatial resolution. However, gamma rays have so much energy they disturb the position of the electron when they interact with it, which is to say that when we try to use gamma rays to determine the position of the electron we succeed only in changing the electron's position! Thus there is a limit on how well we can ever measure where the electron is. In terms of wave–particle duality, the gamma ray, a high-frequency electromagnetic wave, acts as a particle, imparting momentum to the electron.

An alternative form to the uncertainty principle is

$$\Delta E \Delta t \geq \frac{h}{2\pi}\,.$$

This relationship states that energy and time cannot be accurately measured simultaneously. The uncertainty principle has a major impact on the way we view the physical world. It means that there is a fundamental limit to the accuracy of our measurements of any two physical parameters that no leap forward in detector technology will ever be able to overcome. Before we become despondent about our observations, we must remember that Planck's constant is extremely small, and our normal instrumental errors are much larger. It is only when we are considering measuring parameters on the atomic level that the uncertainty principle impacts what we are doing. It is one of the many strange and wonderful consequences of quantum physics that we are still grappling with, even decades after the theory of quantum physics was first formulated.

8.2.5   Bohr's atom

The main development of quantum physics occurred in the early part of the 20th century. Bohr made use of the various pieces of evidence already described above (e.g. the photoelectric effect) along with the results of work by Rutherford.

Rutherford proposed that the atom consists of an extremely small nucleus carrying most of the mass and that it has positive charge. Negatively charged electrons orbit the nucleus in a similar way to planets orbiting the Sun. However, classical theory predicts that the electrons would be accelerated continuously towards the nucleus, and hence the atom would radiate energy continuously. Ultimately the system would break down as the electrons spiral into the nucleus. Neither of these predictions was consistent with observations — hydrogen is observed to be stable, and the radiation from atoms takes the form of discrete lines.

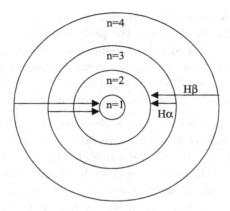

Figure 8.2   An example of the Bohr atom for hydrogen.

To get around these problems, Bohr used a cocktail of classical and quantum ideas to produce a model of the hydrogen atom with the electron orbits restricted. Electrons can only occupy certain orbits around the nucleus that correspond to a fixed amount of energy. An electron can move from one of these discrete orbits to another. It must emit energy if it travels towards the nucleus and it needs to absorb energy in order to move away from the nucleus. To explain why all the electrons would not end up heading straight for the nucleus, Bohr added the criterion that only a certain number of electrons are allowed in each orbit. If an orbit is full then an electron cannot jump there, no matter how much energy it has. All electron movements have to span a whole number of 'orbits'. Bohr's model explains the observation that the emission spectrum of an element occurs at the same discrete wavelengths as its absorption spectrum (see Figure 8.2 for an illustration of how the well-observed Hα and Hβ lines in hydrogen are formed).

Bohr's model was a brave and excellent beginning to understanding the quantum behaviour of an atom. The image of his model is still the one that many people have in their minds of how an atom works. However, there has been much further work carried out to achieve full quantum-mechanical solutions to the problem of understanding atoms.

8.2.6   *Visualising a real atom*

Many modifications to the Bohr model were proposed in the years following its publication. De Broglie put forward the concept of 'electron waves' in orbit around a nucleus. Since the electrons were in the form of waves, each orbit had to be an integer number of wavelengths in length. This led Schrödinger to use the mathematics of waves to describe the atomic energy levels allowed. Schrödinger's wave equation incorporates both the wave and particle behaviour of electrons, resulting in a series of wave functions that are represented by Ψ. The intensity of an electron 'wave' at any point determines the probability that there is an electron 'particle' at that point. The value Ψ^2 describes the probability distribution of an electron. Electrons do not have a defined orbit around the nucleus as in the Bohr model. Instead we can know only that the electron is somewhere within a certain volume.

Schrödinger's aim was to reproduce the actual energies observed using his theory. He had initially described an atom by three quantum numbers:

(a) Principal quantum number, n — this describes the size of an orbit, and can have any value as long as it is a positive (and non-zero) integer. This was the only quantum number used in Bohr's model. When n increases, the electrons are further away from the nucleus and they have larger energy. It is much easier for electrons existing in the higher n orbitals to escape from the nucleus. In the 'classical' view n determines the semi-major axis of an elliptical orbit.

(b) Angular momentum (or azimuthal) quantum number, l — this describes the angular momentum due to the electron's motion around the nucleus. It has integral values from 0 to $> n-1$ for each value of n. This differs from a planet's motion around the Sun since in the quantum world an angular momentum of zero is allowed. In the 'classical' view, l determines the semi-minor axis of an elliptical orbit.

(c) Magnetic quantum number, m — this was introduced because of the important effect that a magnetic field can have on an atom. An orbiting electron itself creates a tiny current, and acts like a magnetic dipole. In the 'classical' view this quantum number determines the orientation of the plane of the orbit in space. This number can have any integral value between l and $-l$.

The classification of orbitals (or energy levels) of electrons used a 'pre-quantum theory' terminology. The spectra of alkali metals were classified by their appearance into the sharp, principal, diffuse and fundamental series. The designation of the angular momentum of an electron $l = 0, 1, 2, 3$ now uses the first letters of these series (see Table 8.1).

Schrödinger's work still did not reproduce the spectroscopic results exactly, and it was necessary to consider a fourth quantum number.

The fourth quantum number was initially a mystery — but again represents a physical property of the electron. This was discovered due to the observed fact that spectral lines can exist extremely close together. For example, the well-known sodium yellow line (which has an electron changing from the 3s level to the 3p level and vice versa) was found to consist of two lines of wavelengths 589.0 nm and

Table 8.1   The universally used spectral notation for the n and l quantum numbers.

l \ n	1	2	3
0	1s	2s	3s
1		2p	3p
2			3d

589.6 nm. The 'new' quantum number hence had to exist in two different values to explain the line splitting. In 1926 Uhlenbeck and Goudsmit made the assumption that the fourth quantum number describes the electron's spin. The spin has two values that point in opposite directions and hence are referred to as 'up' or 'down'. The actual directions are parallel and anti-parallel to the local magnetic field, and the spin quantum number has a value of $+/- 1/2$ for one electron.

We thus should not now view the atom as consisting simply of a nucleus with electrons spinning around it in orbits similar to planets. It is difficult to visualise an atom where the orbits are ruled by the mysterious quantum numbers that have just been described. We also know that it is impossible, through Heisenberg's uncertainty principle, to know where an electron is and where it is going to next simultaneously. We do, however, know the probability of finding an electron in a particular area, and hence we can think of each orbital as the region of space that the electron is allowed to exist in. In Figure 8.3, we show a simple hydrogen atom that has one electron. The probability of finding an electron is shown in the 1s orbit ($n = 1, l = 0$). Orbitals with $l = 0$ are always spherically symmetrical. So if you have an electron in a 2s orbital, it would have a probability distribution that is spherical but is further away from the nucleus than the 1s orbital. If the electron has a higher angular momentum quantum number, for example $l = 1$, then the volume where the electron exists looks very different (Figure 8.3). From Table 1, we know that this case would be termed 2p. Remember that a 1p level cannot exist because of the rule that l ranges between 0 to $> n - 1$. So for quantum number $n = 2$, an electron has the choice of two very different orbitals — the 2s and 2p orbitals. As the quantum numbers become higher, the more complex the probability function, and hence the stranger the orbitals will look!

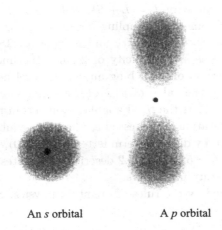

An *s* orbital   A *p* orbital

Figure 8.3   A representation of the orbitals in a hydrogen atom. This represents the region around the atom where an electron is most likely to be found if it has enough energy to be in the 1s and 1p orbital respectively.

Table 8.2   Illustration of the multiplet structure of an atom with two electrons and of the nomenclature used to describe each orbital.

Electron quantum numbers	L	S	J	States $^{2S+1}L_J$
$l_1 = 0,\ l_2 = 1,\ s_1 = 1/2,\ s_2 = 1/2$	0	0	0	$^1S_0$
		1	1	$^3S_1$
	1	0	1	$^1P_1$
		1	0, 1, 2	$^3P_0,\ ^3P_1,\ ^3P_2$

8.3   Rules of the Atom

Electrons are not trapped within one orbital, but can move between different levels. If the electron moves from one orbital to another one with lower energy it will radiate a single quantum producing a spectral line. By combining measurements of spectra and quantum theory a good understanding of what is permitted within an atom has been achieved.

Over the years a method of describing orbitals based on the quantum numbers has been developed. In order to describe the energy state and transitions within an atom, it is important to understand the nomenclature. The description is similar to the quantum numbers used in the previous section to describe a single electron ($nlms$). The total orbital angular momentum, L, is derived from the orbital angular momenta of the individual electrons in the atom, l. So for two electrons with orbital angular momentum l_1 and l_2, the total orbital angular momentum, L, can take values $(l_1 + l_2), (l_1 + l_2 - 1) \cdots (l_1 - l_2)$. Similarly, the total spin angular momentum for two electrons will be $S = (s_1 + s_2), (s_1 + s_2 - 1), \ldots, (s_1 - s_2)$. The spin and the orbital angular momentum will couple via their associated magnetic fields, to give the *total angular momentum* $J = (L + S), (L + S - 1), \ldots, (L - S)$. This type of coupling is known as spin–orbit coupling, or LS coupling. Another important quantity is referred to as the *multiplicity* and is equal to $2S + 1$; so for a single electron ($s = 1/2$), there is a multiplicity of 2 since the lines can split into two levels. The standard method of describing an energy level uses three numbers — the multiplicity ($2S + 1$), the value of the total angular momentum (J) and the angular momentum (L). Thus the total angular momentum for an atom or ion in a given electronic configuration is expressed as the term symbol $^{2S+1}L_J$. Different values of L are referred to by different term letters: $S(L = 0)$, $P(L = 1)$, $D(L = 2)$, $F(L = 3)$, $G(L = 4)$, $H(L = 5)$. Table 8.2 describes the states for two electrons for example values of the quantum numbers.

All atoms are governed by the rules of quantum physics. Some basic questions we can address are:

(i) *How many electrons can exist in one orbital?* For each value of n there are n possible values of $l(0, \ldots, n - 1)$ and for each value of l there are $2l + 1$ values of $m(-1 \cdots +1)$. Hence, using the first three quantum numbers, for each value of n it is possible to have $\sum_{l=0}^{l=n-1}(2l+1) = n^2$ different combinations of nlm. If we then include the two possible values of spin, the total possible 'states' that an electron can have for each orbital, n, is $2n^2$. Hence the lowest orbital, $n = 1$, which is normally referred to as the ground state, can only have two electrons. The next orbital, $n = 2$, can have eight electrons and so on.

(ii) *Can electrons move between orbitals?* There are a number of 'selection' rules that define which transitions are allowed (i.e. which orbitals the electrons are allowed to travel to). There is no restriction on the principal quantum number, n. However, for the angular momentum L it is only possible for an electron to move between two orbits that have a difference, $\Delta L = 0 + / - 1$. In addition, for the total angular momentum, J, only a transition with $\Delta J = 0 + / - 1$ is allowed. The final major selection rule is that a change in S is not allowed between transitions ($\Delta S = 0$). An example is the $^2S_{1/2} - {}^2P_{1/2}$ transition seen in He II. In this case $\Delta L = 1$, $\Delta J = 0$, $\Delta S = 0$, and hence does not break any of the rules. This emission line is very strong. These rules are not intuitive and have been derived over the years from experiment and theory. The selection rules do on the whole work for strong lines in typical laboratory plasmas. However, the description 'rule' is a misnomer. There are exceptions to these rules that will be discussed later. These 'rule-breakers' are extremely useful in determining plasma parameters accurately.

So far we have concentrated on the basic mechanisms within an atom. The next sections will illustrate how we observe various processes that occur within an atom, and how we can use this information to probe the basic characteristics of objects as disparate as the Sun and active galactic nuclei. These examples use the latest spectral information obtained from ground-based observatories and space instrumentation.

8.4 Spectroscopy — The Basic Processes

Spectroscopy is used extensively in all areas of astronomy to investigate places that we can only dream of visiting ourselves. We can determine the temperatures, densities, types of elements that exist, and the velocity of the plasma, all of which can be used to understand the basic physics of the objects that we are observing. In this section we shall, with examples, describe the basic processes that can occur in the Universe. These have been separated into thermal and non-thermal processes.

8.4.1 *Thermal processes*

8.4.1.1 *Blackbody radiation*

Blackbody radiation has been described earlier in the brief history of the development of quantum physics. The most important point about blackbody radiation is

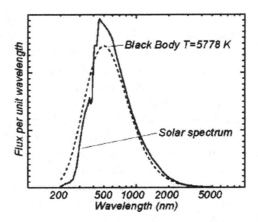

Figure 8.4   The smoothed spectrum of the Sun compared with a blackbody spectrum with a temperature of 5778 K.

that its behaviour depends only on temperature. This follows the law developed by Stefan and Boltzmann which states that the amount of energy radiated per unit area per second is proportional to the temperature of the object raised to the fourth power. Strictly, this law applies *only* to radiation from blackbodies (i.e. a perfectly absorbing body). In reality it is useful for understanding the approximate behaviour of the continuous energy spectrum emitted from many hot, optically thick plasmas (i.e. such as active galactic nuclei, the Sun, other stars, cataclysmic variables and compact X-ray binaries).

The Stefan–Boltzmann law made it possible in the late 1800s to determine accurately for the first time the temperature at the surface of the Sun. The temperature obtained was around 6600 K, which is close to the values derived in the present day. As an example, the actual spectrum of the Sun is compared with the best fitting blackbody curve in Figure 8.4.

8.4.1.2   *Discrete line emission*

Understanding the discrete line emission that had been observed for many years was one of the triumphs of quantum physics. Line emission is caused by the transition of an electron from one bound state of an atom to another. Figure 8.2 shows an idealised Bohr atom, which explains how an Hα emission line is formed. This would require an electron to drop from the $n = 3$ orbital to the $n = 2$ orbital. This is relatively easy to visualise for a hydrogen nucleus. However, for more complex atoms it is common to make use of '*energy level diagrams*'. These allow us to simply visualise how electrons can jump between the different orbitals that are defined by the four quantum numbers discussed previously. So, for example, in a He atom the strongest emission lines will be those that adhere to the selection rules described in the previous section. The so-called resonance line is produced when an electron

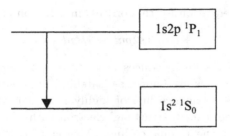

Figure 8.5   A basic energy level diagram of the He atom showing how one of the strongest emission lines is found.

jumps from the $1s2p^1P_1$ level to the $1s^{2\,1}S_0$ level. This is illustrated in Figure 8.5. This transition is also seen in what are known as 'He-like' ions. These are atoms that have been ionised to such an extent that they are left with only two electrons and hence their behaviour copies that of helium.

Line emission due to thermal processes appears predominantly in plasmas that have a temperature of less than 50 million K. The spectral lines result from collisions between atoms. These excite the electron from a lower energy orbital to a higher energy orbital, followed by emission of radiation as the electron reverts to its original state. It is then possible to determine various plasma parameters from these spectra, and this will be discussed in a later section.

8.4.1.3   *Radiative recombination continuum*

Radiative recombination is one of the processes that can produce a continuum. A free electron is captured by an ion into an unoccupied orbit. An electron can carry any amount of energy and in general when it is captured it will have more energy than it needs to be bound to the atom. This excess energy has to be lost and it appears as a photon. The equation for this process for He-like ions is given as

$$1s^2 + e \rightarrow 1s^2 2p + h\nu\,.$$

This is also known as a 'free-bound' process, and tends to dominate the continuum for plasmas with a temperature less than 1 million K.

8.4.1.4   *Dielectronic recombination*

This is quite an involved process and is different to radiative recombination because the incoming electron is required to have an exact amount of energy with no excess. This sounds highly unlikely, but it happens frequently in, for example, the solar corona. The incoming electron, nl (often referred to as a spectator electron), is captured by the ion, and part of its energy is transferred to an existing electron. The resulting configuration is highly unstable and can decay either by exactly the reverse process, which is known as *autoionisation*, or by the emission of radiation to a stable state. The equation for this process in a He ion is given by

$$1s^2 + e \leftrightarrow 1s2pnl \rightarrow 1s^2nl + h\nu \, .$$

Dielectronic recombination produces a series of emission lines with different values of n and l. The lines are known as 'satellite' lines, since they appear close to or blended with the resonance line of highly ionised plasmas. The strongest satellite lines have a spectator 'perturbing' electron with $n = 2$ and appear on the long wavelength side of the resonance line. This can be seen in Figure 8.13 for the He-like calcium spectra observed by the *Yohkoh* Bragg Crystal Spectrometer during a solar flare. The $n = 3$ satellite lines also appear close to the resonance line, with some being blended with the resonance line. The satellite lines become important for a large atomic number, and they should be included in calculations for any atoms with $Z \geq 12$.

An electron impacting an inner shell electron can also produce satellite lines. This is known as *inner shell electron excitation*. The equation for this is given by

$$1s^22s + e \rightarrow 1s2s2p + e' \rightarrow 1s^22s + h\nu + e' \, .$$

This process is not as effective as dielectronic recombination for He-like ions since it requires the simultaneous excitation of two electrons, which is less probable. It is most effective when there is a very high temperature and a low ion stage.

8.4.1.5   *Thermal Bremsstrahlung*

This process is also known as 'braking' radiation and produces continuum emission. For highly ionised plasma there are many free electrons present and these can be easily accelerated in the Coulomb field of another charge. The acceleration causes the electrons to radiate energy and this is known as 'free-free emission'. The emission produces an energy spectrum that is proportional to $T^{-1/2} \exp(-E/kT)$, and a luminosity that is proportional to $N_i N_e T^{1/2}$, where N_i and N_e are the number densities of ions and electrons respectively. Thermal Bremsstrahlung dominates over line emission at temperatures greater than 100 MK. It is for example seen in solar flare plasmas, but is not observed during quiet times.

8.4.1.6   *Two-photon continuum*

This can occur especially in He- and H-like ions. A demonstration of this process is the $1s2s\,^1S_0$ level in He. From the selection rules, it can be seen that the $1s2s$ electron cannot move to the lower level by a standard transition (i.e. it is not 'allowed'). This orbit is called metastable, since it is a long-lived state for an electron — it is not easy for it to fall into the lower level. At lower density the only way to decay from the metastable level is by the simultaneous emission of two photons that have a combined energy equal to the excitation energy of the metastable level.

8.4.2 *Non-thermal processes*

The term 'non-thermal continuum' generally refers to radiation from particles whose energy spectrum is not Maxwellian. In other words, the shape of the resulting spectrum is not simply a reflection of the thermal distribution of the particles. Non-thermal processes are potentially important in any environment where there are high-energy particles. Conditions suitable for accelerating particles to high energies occur widely in the Universe; for example, in the magnetopheres of planets, in flaring regions on the Sun, in supernovae, and in the vicinity of the massive black hole at the centre of quasars and other active galaxies.

8.4.2.1 *Cyclotron radiation*

As described in the section on thermal Bremsstrahlung radiation, accelerating an electron will cause it to emit radiation. If a plasma is permeated by a magnetic field, the electrons will be forced to gyrate about the field lines, and the radiation that is emitted as a result of this acceleration is known as cyclotron radiation when conditions are non-relativistic. The radiation is emitted at the gyrofrequency of the electron, i.e. the number of times per second that the electron revolves about the field lines. The frequency is proportional to the magnetic field strength, B, and is given by $\nu_g = eB/2\pi m_e$. The radiation emitted is linearly polarised when viewed perpendicular to the direction of the field lines, and circularly polarised when the field is viewed end-on. With electrons of higher velocity, there is significant radiation at higher harmonics of the gyrofrequency. For the mildly relativistic case, $\nu > 0.1c$, Doppler shifts become important due to the translational velocity of the electron along the field lines. The translational velocity is measured by the 'pitch angle', which is the angle between the component of velocity parallel and perpendicular to the field. For a typical plasma, the electrons will have a range of pitch angles, and this will cause the emission at each cyclotron harmonic to be spread out in frequency, and even to merge together to form a single continuum at higher harmonics.

8.4.2.2 *Synchrotron radiation*

Synchrotron radiation is similar to cyclotron radiation except that the electrons are relativistic, i.e. moving with velocities close to the speed of light. In this case, the emission produced is tightly beamed in the direction of motion, by an amount determined by the Lorentz factor, $\gamma = (1 - \nu^2/c^2)^{-1/2}$. The direction of motion is changing periodically due to the circulation of the electron about the magnetic field, and because the emitted beam is narrow, it sweeps across the observer in only a small fraction of the gyroperiod. This results in a narrow 'spike' of radiation every cycle, which is approximately $1/\gamma^2$ times shorter in length than the gyroperiod. The resulting radiation spectrum is therefore dominated by high harmonics of the gyrofrequency, ν_g, and peaks at $\nu \approx \gamma^2 \nu_g \sin\theta$, where θ is the pitch angle. As with cyclotron radiation, synchrotron emission is highly polarised. However, because of the effects of the relativistic beaming discussed above, circular polarisation is suppressed and the polarisation is primarily linear.

The synchrotron process is important in many astrophysical contexts, including the emission from extragalactic radio sources and radio and optical emission from the Crab Nebula supernova remnant. In practice the population of radiating electrons has a spectrum of energies. If the energy spectrum of the electrons can be described by a power law with index, p, the resulting synchrotron emission spectrum will have an index $\alpha = (p-1)/2$.

8.4.2.3   *Inverse Compton radiation*

If a high-energy photon collides with a stationary electron, it can transfer some of its energy to the latter, so that the frequency of the photon decreases and its wavelength increases. This process is known as Compton scattering. If, on the other hand, a low-energy photon comes across a high-energy electron, the photon can gain energy at the expense of the electron. This is known as inverse Compton scattering. An important illustration of this phenomenon is the Sunyaev–Zeldovich effect, where photons from the cosmic microwave background are scattered to higher energies by the hot gas that pervades clusters of galaxies (Chapter 7). This results in a deficit of low-energy (radio) photons in the direction of a cluster. Inverse Compton scattering of photons may also be important in accounting for the X-ray emission spectrum of quasars, together with the related process of synchrotron self-Compton radiation. In the latter, radio photons produced by the synchrotron process can be scattered up to X-ray energies by the energetic electrons that give rise to the synchrotron radiation in the first place. In an environment that is optically thick to Compton scattering, there can be efficient coupling between the radiation and particle fields that can, for example, significantly distort the thermal Bremsstrahlung emission from the plasma. This process is known as 'Comptonisation'.

8.5   Environmental Influences on Spectra

8.5.1   *Broadening of spectral lines*

8.5.1.1   *Natural width*

Heisenberg's uncertainty principle places a fundamental constraint on how well we can know the product of energy and time (Section 8.2.4). Thus, the uncertainty in the energy of the quantum emitted as a result of an electron transition within an atom depends on how long the electron has been in its excited state (Δt_1), and how long it will remain in its de-excited state (Δt_2). The uncertainty is consequently

$$\Delta E = \frac{h}{2\pi} \left(\frac{1}{\Delta t_1} + \frac{1}{\Delta t_2} \right)$$

or, in terms of wavelength, since $E = h\nu$ and $\lambda = c/\nu$, we have

$$\Delta\lambda = \frac{\lambda^2}{2\pi c}\left(\frac{1}{\Delta t_1} + \frac{1}{\Delta t_2}\right).$$

For normal ('allowed') transitions, the excited states can survive for typically about 10^{-8} s, which implies a natural width of a few tenths of a nanometre in the optical range. The lifetimes of metastable and forbidden states can be much longer, however — typically from seconds to hours — so these lines are intrinsically much narrower.

8.5.1.2 *Collisional broadening*

In higher density environments, the time for which an electron can remain in a given energy level is determined by collisions between atoms. The higher the density, the greater the rate of collisions and the broader the lines. This is known as collisional, or pressure, broadening. The mean time between collisions, Δt, can be expressed simply in terms of the mean distance between particles, l, and their velocity, v; so

$$\Delta t = l/v = \left(n\sigma\left(\frac{2kT}{m}\right)^{1/2}\right)^{-1},$$

where n is the number density, σ the collision cross-section and m the particle mass. An extreme illustration of the effects of collisional broadening can be seen in the spectra of white dwarf stars. The pressure at the photosphere of such stars is very high due to the intense gravitational field, and the absorption lines in their optical spectrum consequently have widths that are typically tens of nanometres.

8.5.1.3 *Doppler broadening*

Another source of line broadening is due to the Doppler motion of individual atoms, which is a function of temperature. If the gas is in thermal equilibrium, the atoms will be moving with a Maxwellian distribution, with a most probable line-of-sight velocity of $V = (2/3)^{1/2}V_{RMS}$. The line will thus have a typical width (two times Gaussian σ) of

$$d\lambda = \frac{2\lambda}{c}\left(\frac{2kT}{m}\right)^{\frac{1}{2}}.$$

Note that, because of the inverse dependence on mass, at a given temperature higher atomic weight elements will have narrower lines than those of lower atomic weight. Often, Doppler (or thermal) broadening is difficult to distinguish from random bulk motions due to turbulence (see below).

8.5.1.4   *Zeeman effect*

The effect of a magnetic field on an emitting atom is to split the energy levels by an amount that is proportional to the magnetic field strength. One can imagine that the energy of the magnetic field adds vectorially to the orbital and spin angular momentum of the electron, separating the $2j + 1$ degenerate levels because of the different orientations of the magnetic dipoles in the external field. For a singlet line ($S = 0$), the magnetic field will split any term into $2J + 1$ levels ($J = L + S$). The selection rules for transitions between different terms are $\Delta M = 0, \pm 1$. The frequency of a singlet line of frequency, ν, will be split into three components — the unshifted π component, and two σ components with frequencies separated by

$$\nu_\sigma = \nu \pm \frac{eH}{4\pi mc} = \nu \pm 1.4 \times 10^{10} H \text{ Hz} ,$$

where the magnetic field strength, H, is expressed in Tesla. The π and σ components will have different polarisation properties that depend on the orientation of the observer to the magnetic field. If the Zeeman components are not resolved, we see a broadened spectral line. When the field is strong and/or the intrinsic line width is narrow, we may resolve the Zeeman components in the spectrum, which allow us to deduce the magnetic field strength and orientation.

8.5.2   *Dynamical effects*

Under the section on line broadening, we noted that the thermal motion of individual atoms would cause the lines they emit to be shifted in wavelength because of the Doppler effect. The same principle applies to bulk motion of the emitting gas, which could be caused for example by turbulence, rotation, or motion of the whole emitting volume towards or away from the observer. This can further affect the appearance of spectral lines, blurring them and causing adjacent lines to merge, but it can also provide an important diagnostic of the dynamical environment of the emitter. By tracking the position of lines emitted by a star in a binary system, we can measure the orbital velocity, and hence gain some insight into the mass of the component stars in the binary. Similarly, we can measure the ejection velocity of material in a solar flare, or outflow velocities in galactic nuclei. The velocity- or red-shift of distant galaxies and quasars allows us to measure the expansion of the Universe. The broadening of spectral lines due to bulk motion allows us to measure the rotation rate of stars, and the circulation velocity of an accretion disc.

8.5.3   *Photoelectric absorption*

If a photon has an energy that exceeds the energy binding an electron within an atom, ion or molecule, it can be absorbed, ejecting the electron. This is the photoelectric effect, famously first explained by Einstein, and the loss of photons due to this process is known as photoelectric absorption. Photoelectric, or bound-free, absorption is a major source of opacity in stellar atmospheres and interiors, while photoelectric absorption in the interstellar medium has a major effect on the observed spectrum of distant X-ray sources. The absorption cross-section for a given electron within the atom peaks when the photon energy, ν, is exactly equal to the binding energy of the electron. Below this critical energy, or edge, the cross-section is zero, since the photon has insufficient energy to dislodge the electron. Above the edge, the absorption cross-section falls as approximately ν^{-3}.

In the X-ray band between about 0.1 and 10 keV, the dominant source of photoelectric absorption is the K-shell electrons of medium atomic weight elements such as carbon, nitrogen, oxygen, neon, magnesium, silicon, sulphur and iron. The summed photoelectric absorption cross-section increases to lower energies with the ν^{-3} dependence noted above, with discontinuities corresponding to the K-shell binding energies of the different atomic species. Thus the opacity of the interstellar medium increases as the photon energy decreases, leading to a marked turnover in the apparent spectrum of distant X-ray sources, often referred to as low-energy absorption. If the absorbing atom is partially ionised, the binding energy of the remaining electrons in the atom is affected, modifying the energy of their absorption edges.

8.5.4 *Electron scattering*

As discussed above, if a photon is scattered by an electron there will be an interchange of energy and momentum. In the case of a photon scattering off a stationary electron, the photon wavelength will be shifted to lower values, and this is known as Compton scattering. The simpler case of Thomson scattering assumes that there is no change in the wavelength of the photon, and this remains an adequate approximation in most circumstances provided that the energy of the photon is less than about $m_e c^2 \approx 0.5$ MeV, where m_e is the electron mass. In the low energy regime, the cross-section for scattering by electrons is given by the Thomson cross-section, $\sigma_T = 6.653 \times 10^{-29}$ m^2. At higher energies the scattering is described by the Klein–Nishina cross-section, which at energies above $m_e c^2$ tends to $\sigma_{KN} \propto (h\nu)^{-1}$.

In the X-ray band, for a gas with approximately solar abundance in which the medium atomic weight elements responsible for photoelectric absorption are in their neutral (un-ionised) state, photoelectric absorption will dominate over electron scattering at photon energies below about 10 keV even if all the hydrogen atoms in the gas, which dominate the electron number, are completely ionised. This balance will shift in favour of electron scattering as the degree of ionisation of the gas increases.

8.5.5 *Pair production*

For photon energies greater than twice the rest energy of the electron, $2\,m_e c^2$, electron–positron pair production is possible. This is an important source of opacity for gamma-ray photons. Annihilation of positrons and electrons can result in an emission 'line' at $m_e c^2 = 512$ keV.

8.6   Spectroscopy in Action

Spectroscopy is a powerful tool for making direct measurements of plasma parameters. Some examples are listed below illustrating various methods of probing the characteristics of different plasmas.

8.6.1   *How dense is it?*

In some cases, one can tell a great deal about how dense a plasma is just by observing the kind of spectral lines it emits. An illustration of this is the optical spectrum of Seyfert galaxies (see Chapter 7), an example of which is shown in Figure 8.6. These spectra typically exhibit both broad and narrow lines. The Balmer lines in Figure 8.6, for example, have both broad and narrow components. The broad lines are believed to come from relatively close to the massive black hole that is powering the quasar nucleus, their width being due to the high velocity of the emitting clouds as they circulate about the black hole. The narrow line spectrum, however, has substantial emission in the form of forbidden lines (Section 8.5.1.1) such as [OIII] (note the square bracket notation denoting a forbidden line). The presence of forbidden lines immediately tells us that the density in the narrow-line

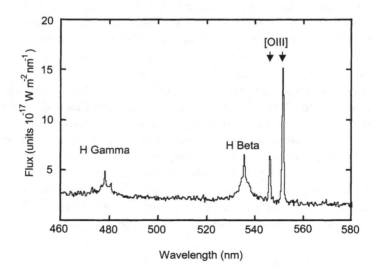

Figure 8.6   Optical spectrum of the redshift 0.1 Seyfert galaxy RE J2248-511. Emission due to the Balmer series lines Hβ and Hγ is marked, along with the pair of narrow forbidden lines due to [OIII] 495.9 nm and [OIII] 500.7 nm. Note that Hβ and Hγ have both narrow and broad components. A narrow line due to [OIII] 436.3 nm can be resolved on the red wing of Hγ.

region is low, typically about one atom per cubic metre, and about a million times more rarefied than the gas producing the broad lines. The reason for this sensitivity to density is that the transition probability for forbidden lines is very low, and at high densities the atom will lose energy through collisions before it gets a chance to radiate. The narrow-line-emitting clouds are thought to lie at large distances from the black hole (see Chapter 7) and to be moving relatively slowly.

We can also derive information on density by measuring of the strength of individual lines. In the coronal case (for plasmas with low densities and high temperatures), the intensity of a spectral line is proportional to the quantity known as 'emission measure'. The emission measure of the plasma at a particular electron temperature is proportional to the square of the density of electrons (N_e^2) at that temperature, T_e, averaged along a line of sight through the corona. Thus electron densities can be determined from the intensity of spectral lines if we can estimate the emitting volume.

A method has also been developed which can probe the densities of coronal plasma without making any assumptions on the volume of the emitting plasma. This entails making use of emission lines originating from 'metastable' levels. Metastable levels are long-lived because all the available methods of decay have a very low probability of transition. As noted above, 'forbidden transitions' are not totally forbidden — they merely have a very small transition probability. They do break the 'selection rules' stated earlier, but in particular circumstances these transitions can become quite strong. Since a metastable level has a low transition probability, an electron in that level will stay there for a long enough time to be affected by collisions from other levels. The collisional depopulation will naturally dominate at higher density, and will be proportional to the electron density, N_e. As mentioned earlier, 'allowed' transitions have an intensity that is proportional to N_e^2, and hence

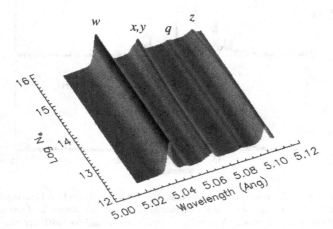

Figure 8.7 A theoretical model of intensities of the He-like sulphur emission lines as a function of density. As the density increases the population in the 'metastable' level decreases and hence the 'forbidden' line z decreases in strength.

the ratio of an allowed to a metastable level will provide an accurate determination of the electron density. Figure 8.7 shows a spectral model developed for the He-like sulphur ion as a function of density. It can be clearly seen that as the density increases the intensity of the forbidden (know as the 'z' line) begins to decrease. This is because, as the density increases, so does the number of collisions that knock electrons out of the metastable level.

Another type of transition that has a low transition probability is one that gives rise to an 'intercombination' line. This is because the transition requires a

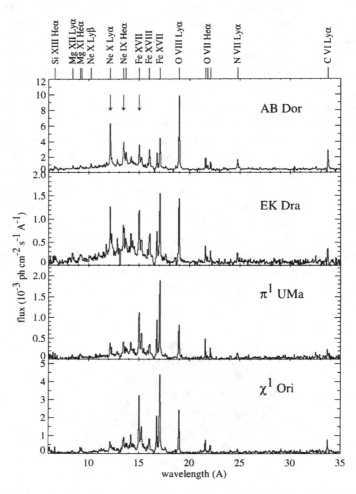

Figure 8.8   X-ray spectra from four solar analogues taken by the Reflection Grating Spectrometer on board *XMM-Newton*. The age of the stars increases from top to bottom, and the activity in the corona decreases from top to bottom. A number of ions are highlighted with arrows — the flux in the highly ionised ions decreases with a decrease in activity, indicating a reduction in temperature (Guedel *et al.*, 2002, *The Twelfth Cool Stars, Stellar Systems and the Sun*, eds. A. Brown, T.R. Ayres and G.M. Harper (Boulder: University of Colorado), in press).

spin change. For example, $^3P - {}^1S$ in He-like ions has a change in spin $\Delta S = 1$; the selection rules state that this transition is not 'easily' allowed. An example of intercombination lines is shown in Figure 8.13.

8.6.2   *How hot is it?*

Most objects in the universe do not have a single well-defined temperature. This is certainly the case in the solar corona, where the plasma is continuously changing. The rates of most of the processes described earlier are dependent on the kinetic temperature of the ions or electrons. In coronal plasma, ionisation equilibrium is assumed, i.e. coronal ions lose electrons predominantly by electron impact, and capture them by radiative and dielectronic recombination. Calculations of ionisation equilibrium are made that show which ionisation stage is most abundant at different temperatures. So, for example, Fe XII is abundant over the temperature range $T_e = 1$–3 MK, and hence the mere existence of an Fe XII spectral line will tell you immediately that the plasma temperature is around 2 MK. Figure 8.8 shows spectra from different stars that each display different intensities in the emission lines. The existence and strength of the emission lines give us clues as to how hot each star is.

As mentioned in 6.1, the intensity of a coronal line is proportional to the emission measure ($\int N_e^2 dV$) of plasma in the temperature range where the line is emitted (also known as the differential emission measure, or DEM). Each observed line will provide a measure of DEM at a particular temperature, and can be brought together to provide information on different structures. A solar flare measurement will show much more material at higher temperatures than a region of quiet Sun. An example of a DEM curve for a region of the quiet Sun, measured using the Coronal Diagnostic Spectrometer on board *SOHO*, is shown in Figure 8.9. As mentioned earlier, the so called 'satellite lines' formed through dielectronic recombination are important in measuring electron temperature. The intensity of the dielectronic satellites lines is proportional to density in the same way as the standard 'resonance' transitions are, but has a different relationship with temperature. This means that if we measure the ratio between the resonance line and the satellite lines, then we can determine the electron temperature. Figure 8.10 shows a model for the He-like sulphur ion as a function of temperature. It can be seen that when the temperature is lower, the satellite lines become more enhanced, and as the temperature increases, the satellite line emission starts to decrease.

8.6.3   *How fast is it?*

The speed of line-emitting material along the line of sight is straightforwardly determined by measuring the Doppler shifts of spectral lines. The wavelength of a spectral line will be shifted to longer wavelengths (redshift) if the material emitting it is moving away from the observer, or to shorter wavelengths (blueshift) if the

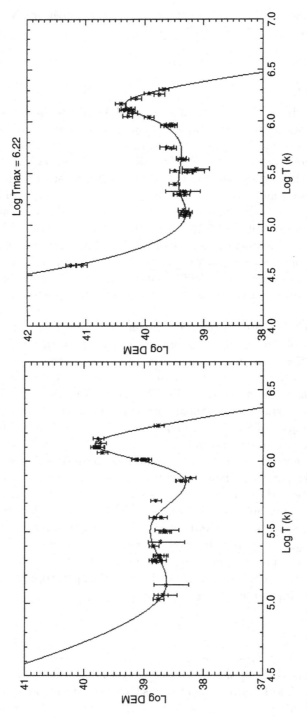

Figure 8.9 The differential emission measure (DEM) determined using data from the Coronal Diagnostic Spectrometer on board the *SOHO* spacecraft of a region of the quiet Sun and an active region. This gives an overall view of the temperature profile of the different regions. It can clearly be seen that the active region is significantly hotter than the region of the quiet Sun (Landi and Landini, 1998, *A&A* **340**, 265–276).

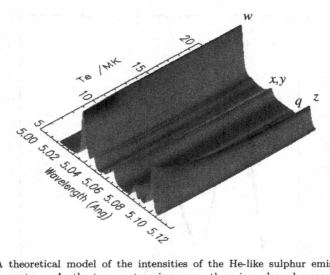

Figure 8.10 A theoretical model of the intensities of the He-like sulphur emission lines as a function of temperature. As the temperature increases, there is a clear decrease in the satellite line q, and the forbidden line z.

material is moving towards the observer. The amount of shift is proportional to the velocity.

An example of a large blueshift is shown in Figure 8.11. This observation was made with the CDS instrument on board *SOHO* during the ejection of a prominence. The prominence was located at the disc centre and hence a strong blueshift was observed in many emission lines covering a range of temperatures. In the case illustrated, there was a 'splitting' of the line, with the blueshifted component coming from material that was travelling towards us, and a stationary component coming from material that was static in the solar atmosphere.

The atmospheres of certain hot stars afford another dramatic illustration of the effect that the dynamical environment can have on the appearance of spectral lines. These atmospheres are being accelerated by intense radiation pressure from the central star, to form a stellar wind whose velocity increases with radius from the star (Figure 8.12). The atoms in the wind are excited by the stellar photons. In certain (resonance) transitions there is a high probability that the electron will decay back to its original energy level, emitting a photon with exactly the same energy as was absorbed. Whereas the exciting photons all come from one direction (the star), they can be re-emitted from the excited atom in any direction. The effect is to 'scatter' photons out of the light 'beam', and the process is known as resonance scattering. In the geometry illustrated in Figure 8.12, the observer sees a deficit of light coming from the region of the wind that is projected against the star, due to the resonance scattering. Since this is the part of the wind that is moving directly towards the observer, it is blueshifted, resulting in an absorption feature to the blue of the line rest wavelength. On the other hand, there is an

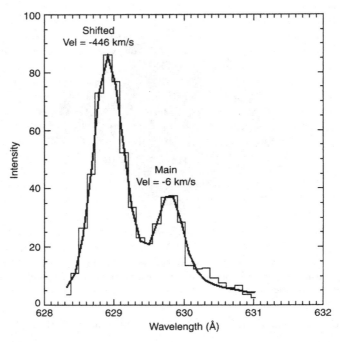

Figure 8.11   The behaviour of a spectral emission line under the influence of plasma moving towards the observer at a speed of 446 km/s. A smaller intensity stationary component remains.

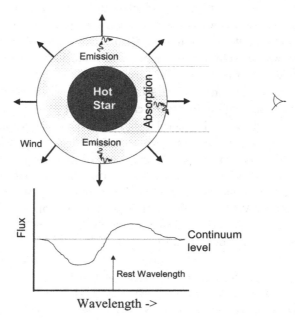

Figure 8.12   The formation of a 'P-Cygni' line profile in the radiation pressure accelerated steller wind of a hot, massive early-type star.

enhancement in light from that part of the wind moving away from the observer, due to scattering of photons *into* the line of sight. This causes excess emission long ward of the rest wavelength. This characteristic line shape is known as a 'P-Cygni' profile, after the star in which it was first observed. The lines are broad both because of the spherical geometry, and because the wind is accelerating outwards, so that different radii scatter different wavelengths in the continuum of the star, due to the Doppler effect.

8.6.4 *How turbulent is it?*

An optically thin coronal emission line has a spectral profile that follows a Gaussian form given by

$$I(\Delta\lambda) = I_0 \exp[-(\Delta\lambda/\Delta\lambda_D{}^2)^2]\,,$$

where

$$\Delta\lambda_D = \lambda_0(2kT_{\text{ion}}/m_{\text{ion}})^{1/2}/c\,,$$

T_{ion} being the ion kinetic energy. This broadening essentially represents the Doppler shift caused by the thermal motion of the ions. Hence, by measuring the line profile width, a measurement of the ion kinetic temperature can be made. Observations of this quantity in the Sun normally indicate that the spectral lines are broadened above the level that is indicated by the electron kinetic temperature. This excess broadening is hence non-thermal and is sometimes known as 'turbulence'. If you can work out the electron temperature, then a measure of the non-thermal velocity can be made through the equation

$$v^2 = \frac{2kT_e}{m_{\text{ion}}} + v_{\text{nt}}^2\,,$$

where v_{nt} is the non-thermal velocity and v is the combination of the thermal and non-thermal terms.

Non-thermal velocity is most prominent on the Sun during flares. An example is shown in Figure 8.13, with data from the Bragg Crystal Spectrometer on board the *Yohkoh* spacecraft. This shows a helium-like calcium spectrum during the impulsive phase of a flare. The electron temperature is shown to be 10 MK, whereas, the Doppler (ion) temperature reaches 63 MK in the early phase of this flare. This implies a non-thermal velocity of 150 km/s. Non-thermal velocities are evident even in the quiet Sun.

8.6.5 *What is it made of?*

The intensity of a spectral line depends directly on the abundance of the element that emits it. Normally abundances are referenced to that of hydrogen. However, in the solar corona, for example, hydrogen is completely ionised, and hence there are no hydrogen emission lines to measure. Thus abundances of elements are generally

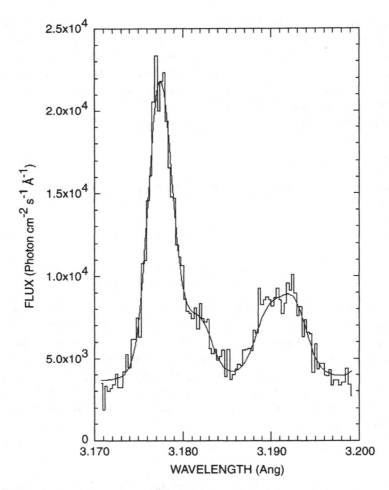

Figure 8.13   Data from the helium-like calcium spectra from the Bragg Crystal Spectrometer on board *Yohkoh* (histogram). The solid line shows the fit to the profiles. The line at this stage of the flare (during the early impulsive phase) are board, with the Doppler width being a factor of 6 larger than the width it should be if the electron temperature and ion temperature were equal. Emission from satellite lines can be seen on the long-wavelength wing of the resonance line at ~ 3.177 Å.

specified relative to other heavier elements instead. If the temperature behaviour of two different ions (the emissivity) is similar, then the ratio of the intensity of the two ions will give a relatively accurate determination of the relative abundance.

As mentioned earlier, inner-shell electron excitation is caused by the excitation of the inner level of — in the example in Section 4.1.4 — a lithium-like ion. The emission line produced is a 'satellite' line of the resonance line of a He-like ion. The ratio of these two emission lines will provide an accurate determination of the ion abundance ratio.

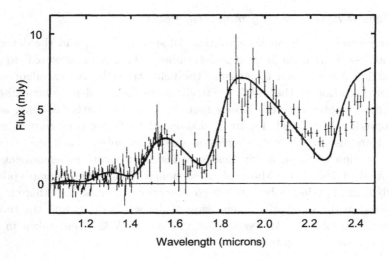

Figure 8.14   Infrared spectrum of the magnetic cataclysmic variable star AM Herculis, showing broad humps due to cyclotron emission. The wavelength of the humps at 1.55 and 1.90 μm is consistent with the fifth and fourth (respectively) harmonics of cyclotron emission from a 1400-tesla magnetic field. (After Bailey, Ferrario and Wickramasinghe, 1991.)

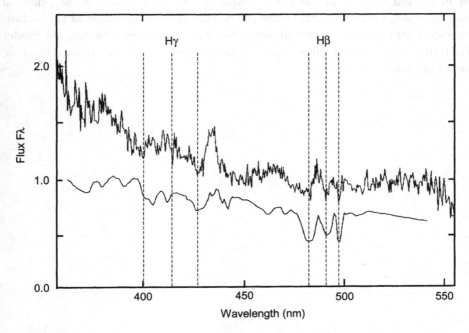

Figure 8.15   Optical spectrum of the magnetic cataclysmic variable (polar) V834 Cen, showing Zeeman absorption components due to the Balmer lines Hβ and Hγ. The best-fit dipole model is shown below, displaced downwards for clarity. This has a polar field strength of 2800 tesla, and is viewed at an angle of 45°. (Adapted from Puchnarewicz *et al.*, 1990.)

8.6.6   *Does it have a strong magnetic field?*

The Zeeman effect, which causes a splitting of spectral lines, and the detection of cyclotron emission can both be used to determine if there is a magnetic field present in the emitting plasma, and to measure the field strength. An excellent example of cyclotron radiation in the mildly relativistic case is found in polar cataclysmic variables (see Chapter 7), close binary systems in which matter is being accreted onto a magnetic white dwarf. Figure 8.14 shows the infrared spectrum of one such star, AM Herculis, in which the emission due to cyclotron harmonics is clearly visible. Detection of these harmonics provides an accurate measurement of the magnetic field of the star. Many of these polars also show Zeeman split lines, particularly during periods when the rate of accretion is low, so that there is greater contrast between the photospheric emission of the white dwarf and the (normally bright) accretion column. An example of the system V834 Cen, taken in such a 'low state', is shown in Figure 8.15.

8.7   Conclusions

Spectroscopy is a powerful tool for determining the physical characteristics of objects in the universe. It pervades every field of astrophysics, and is the principal means by which we can determine the physics of processes going on in the distant Universe. This chapter has introduced some of the basic concepts, but the reader is referred to the many detailed texts on the subject to fully appreciate the true wonders of quantum physics.

Chapter 9

An Introduction to Magnetohydrodynamics

Lidia van Driel-Gesztelyi

9.1 Introduction

Since magnetic fields are ubiquitous in the universe, one cannot properly understand most of the physical processes in stars, planetary atmospheres, interplanetary space and galaxies without taking the effects of magnetic fields into account.

Magnetohydrodynamics (MHD) is a branch of continuum mechanics, which studies the flow of electrically conducting fluid in the presence of an electromagnetic field. It is the marriage of fluid mechanics and electromagnetism, which created new physical concepts, like the Alfvén wave. Other names for MHD are hydromagnetics and magnetofluid dynamics.

The root of MHD lies in the mutual interaction of fluid flow (v) and magnetic field (B): the Lorenz force, $\mathbf{j} \times \mathbf{B}$, accelerates the fluid, while the electromotive force, $\mathbf{v} \times \mathbf{B}$, creates a current which modifies the electromagnetic field. It has important applications in astrophysics.

The behaviour of the continuous plasma is governed by the following equations:

- Maxwell's equations (simplified)
- Ohm's law
- gas law (equation of state)
- mass continuity equation
- motion equation
- energy equation

A combination of Maxwell's equations and Ohm's law leads to the induction equation which eliminates the electric field and relates $\mathbf{v}$ to $\mathbf{B}$.

In the MHD equations, vector combinations and operators are used. Though, presumably, the reader is already familiar with these mathematical tools, it may still be useful to summarise the main rules:

9.1.1   *Vector combinations*

The *dot product* produces a scalar from a pair of vectors:

$$\mathbf{a} \cdot \mathbf{b} = a_x b_x + a_y b_y + a_z b_z \,,$$

which is equal to the product of the length of the vectors times the cosine of the angle between them.

The *cross product* of two vectors produces a new vector, perpendicular to both:

$$\mathbf{a} \times \mathbf{b} = (a_y b_z - a_z b_y, a_z b_x - a_x b_z, a_x b_y - a_y b_x) \,.$$

The magnitude of the cross product is the product of the length of the vectors times the sine of the angle between them.

9.1.2   *Vector operators*

Gradient of f (vector):

$$\nabla f(x,y,z,t) = \left(\frac{\partial f(x,y,z,t)}{\partial x}, \frac{\partial f(x,y,z,t)}{\partial y}, \frac{\partial f(x,y,z,t)}{\partial z} \right).$$

Divergence of A (a dot product of two vectors, a scalar):

$$\nabla \cdot \mathbf{A} = \frac{\partial A_x(x,y,z,t)}{\partial x} + \frac{\partial A_y(x,y,z,t)}{\partial y} + \frac{\partial A_z(x,y,z,t)}{\partial z} \,.$$

Curl of A (a cross product of two vectors; produces a vector):

$$\nabla \times \mathbf{A}(x,y,z,t) = \left(\frac{\partial A_z(x,y,z,t)}{\partial y} - \frac{\partial A_y(x,y,z,t)}{\partial z}, \frac{\partial A_x(x,y,z,t)}{\partial z} \right.$$
$$\left. - \frac{\partial A_z(x,y,z,t)}{\partial x}, \frac{\partial A_y(x,y,z,t)}{\partial x} - \frac{\partial A_x(x,y,z,t)}{\partial y} \right),$$

which can be treated almost like any other vector, provided that the order of the operators and operands in the expression is retained.

9.2   Conservation Principles

The MHD equations incorporate the following conservation principles, which have been derived from the equations of fluid mechanics and electromagnetism.

9.2.1   *Mass conservation (mass continuity equation)*

$$\frac{d\rho}{dt} \equiv \frac{\partial \rho}{\partial t} + \nabla \cdot (\rho \mathbf{v}) = 0 \,,$$

where ρ is the mass density, $\mathbf{v}$ is the flow velocity, t is the time and d/dt is the convective derivative, which represents the time rate of change following the plasma element as it moves.

The mass continuity equation expresses the fact that the density at a point increases if mass flows into the surrounding region, whereas it decreases when there is a divergence rather than a convergence of the mass flux.

9.2.2 *Momentum conservation (motion equation)*

$$\rho \frac{d\mathbf{v}}{dt} = -\nabla p + \mathbf{j} \times \mathbf{B} + \nabla \cdot \underline{S} + \mathbf{F_g} ,$$

where ∇p is the plasma pressure gradient, $\mathbf{j}$ is the current density, $\mathbf{B}$ is the magnetic induction, $\mathbf{j} \times \mathbf{B}$ is the Lorenz force per unit volume, $\underline{S}$ is the viscous stress tensor, and F_g is an external force (e.g. gravity).

$\mathbf{B}$ is usually referred to as the 'magnetic field', though in fact the magnetic field is $\mathbf{H} = \mathbf{B}/\mu_0$, where μ_0 is the magnetic permeability of free space, $\mu_0 = 4\pi \, 10^{-7}$ Hm^{-1}. Ignoring the gravity and the magnetic field in the motion equation gives the standard form for the Navier–Stokes equations for a viscous fluid.

9.2.3 *Internal energy conservation (energy or heat equation)*

$$\rho T \frac{ds}{dt} = \rho \frac{de}{dt} + p\nabla \cdot \mathbf{v} = \nabla \cdot (\underline{\kappa} \cdot \nabla \mathbf{T}) + (\underline{\eta_e} \cdot \mathbf{j}) \cdot \mathbf{j} + Q_v - Q_r ,$$

where s is the entropy, $e = p/((\gamma - 1)\rho)$ is the internal energy per unit mass, $\underline{\kappa}$ is the thermal conductivity tensor, T is the temperature, $\underline{\eta_e}$ is the electrical resistivity tensor, Q_v is the heating by viscous dissipation, Q_r is the radiative energy loss, and $\gamma = c_p/c_v$ is the ratio of specific heats. The first term on the right hand side of the equation describes particle conduction, while second term ohmic dissipation.

The energy or heat equation expresses that the heat increases or decreases as the net effect of energy sinks and sources as the plasma moves in space.

The **equation of state** is given by

$$p = \mathcal{R}\rho T = nk_B T ,$$

where $\mathcal{R}$ is the universal gas constant (8.3×10^3 m^2 s^{-2}deg^{-1}), n is the total number of particles per unit volume, and k_B is Boltzmann's constant (1.381×10^{-23} J deg^{-1}).

The density can be written in terms of n (total number of particles per unit volume), where $\bar{m}$ is the mean particle mass, so that

$$k_B/\mathcal{R} = \bar{m} .$$

For a fully ionised hydrogen plasma with electron number density n_e, the pressure becomes $p = 2n_e k_B T$, and the plasma density is $\rho = n_p m_p + n_e m_e \approx n_e m_p$, where m_p is the proton mass.

9.3   Maxwell's Equations and Ohm's Law

9.3.1   *Poisson's equation, or Gauss's law for* E

$$\nabla \cdot \mathbf{E} = \frac{\rho_q}{\varepsilon_0},$$

where $\mathbf{E}$ is the electric field, ρ_q is the net charge density and ε_0 ($= 8.854.10^{-12}$ F m^{-1}) is the permittivity of free space.

9.3.2   *Faraday's law*

$$\nabla \times \mathbf{E} = -\frac{\partial \mathbf{B}}{\partial t}.$$

Poisson's equation and Faraday's law mean that either electric charges or time-varying magnetic fields may give rise to electric fields.

9.3.3   *Ampère's law*

$$\nabla \times \mathbf{B} = \mu_0 \left(\mathbf{j} + \varepsilon_0 \frac{\partial \mathbf{E}}{\partial t} \right),$$

where $\mathbf{j}$ is the current density and μ_0 ($= 4\pi\ 10^{-7}$ H m^{-1}) is the magnetic permeability.

Ampère's law means that either currents or time-varying electric fields may produce magnetic fields.

9.3.4   *Gauss's law for* B

$$\nabla \cdot \mathbf{B} = 0,$$

which can be understood in different ways:

(a) The magnetic flux Φ (measured in webers (Wb), with 1 Wb = 1 tesla (T) per square metre), defined by

$$\Phi = \int \mathbf{B} dS,$$

where S is a closed surface. No *net* magnetic flux crosses the surface.
(b) There are no magnetic sources or monopoles.

Maxwell's equations apply in any inertial reference frame; they are invariant by frame transformation.

It is noteworthy that Gauss's law for $\mathbf{B}$, for a divergence-free initial state, follows from Faraday's equation (taking the divergence of Faraday's equation and remembering that the divergence of the curl is always zero, one finds that $\partial(\nabla \cdot B)/\partial t = 0$, meaning that if $\nabla \cdot \mathbf{B} = 0$ initially, it will remain zero all the time).

9.3.5 *Ohm's law*

$$\mathbf{E}' = \mathbf{E} + \mathbf{v} \times \mathbf{B} = \underline{\eta_e} \cdot \mathbf{j},$$

where $\underline{\eta_e}$ is the electrical resistivity tensor. In most of the applications it is sufficient to use $\underline{\eta_e} = \eta_e \cdot \delta_{ij}$, where η_e (the scalar electrical resistivity) is the inverse of the electrical conductivity(σ). So Ohm's law can be written as

$$\mathbf{E}' = \mathbf{E} + \mathbf{v} \times \mathbf{B} = \frac{\mathbf{j}}{\sigma},$$

where $\mathbf{E}' = \mathbf{E} + \mathbf{v} \times \mathbf{B}$ gives the Lorenz transformation from the electric field ($\mathbf{E}$) in a laboratory frame of reference to the electric field ($\mathbf{E}'$) in a frame moving with the plasma.

Ohm's law expresses that the electric field (E') in the frame moving with the plasma is proportional to the current, or, in other words, *that the moving plasma in the presence of magnetic field B is subject to an electric field* $\mathbf{v} \times \mathbf{B}$, *in addition to* $\mathbf{E}$.

It is noteworthy that Ohm's law is an approximative law, which depends on the constitution of the medium.

9.3.6 *Maxwell's equations in MHD*

approximation Since in MHD the typical plasma velocities are low ($\mathbf{v} \ll \mathbf{c}$), second-order terms in v/c are negligible (c is the speed of light). Furthermore, the effects of local electric charge densities and displacement currents, of viscosity and of radiation pressure are neglected. Thus, Maxwell's equations will take the form

$$\text{Faraday's law: } \nabla \times \mathbf{E} = -\frac{\partial \mathbf{B}}{\partial t}$$

$$\text{Poisson's equation, or Gauss's law for } E: \ \nabla \cdot \mathbf{E} = 0$$

$$\text{Ampère's law: } \nabla \times \mathbf{B} = \mu_0 \mathbf{j}$$

$$\text{Gauss's law for } B: \ \nabla \cdot \mathbf{B} = 0$$

9.4 The MHD Induction Equation

The equations of the conservation of mass, momentum and energy, the Maxwell equations and Ohm's law are the basic MHD equations for the determination of 15 unknowns: $\mathbf{v}$, $\mathbf{B}$, $\mathbf{j}$, $\mathbf{E}$, ρ, p and T. Combining the basic MHD equations can lead to useful new equations and we can eliminate $\mathbf{E}$, $\mathbf{j}$ and T through substitution to remain with the unknowns $\mathbf{B}$, $\mathbf{v}$, p and ρ, which are considered to be the primary variables of MHD.

Combining Ampère's law $\nabla \times \mathbf{B} = \mu_0 \mathbf{j}$ and Ohm's law in a frame moving with the plasma, $\mathbf{j} = \sigma(\mathbf{E} + \mathbf{v} \times \mathbf{B})$, where σ is the electric conductivity and $\mathbf{v}$ is the

Table 9.1   Maxwell's equations in different systems of units.

	S1 units	Gaussian CGS units
Poisson equation:	$\nabla \cdot E = \dfrac{\rho_q}{\varepsilon_0}$	$\nabla \cdot E = 4\pi\rho_q$
Faraday's law:	$\dfrac{\delta B}{\delta t} = -\nabla \times E$	$\dfrac{1}{c}\dfrac{\partial B}{\partial t} = -\nabla \times E$
Ampère's law:	$\nabla \times B = \mu_0 \left(j + \varepsilon_0 \dfrac{\partial E}{\partial t} \right)$	$\nabla \times B = \dfrac{4\pi}{c} j + \dfrac{1}{c}\dfrac{\delta E}{\delta t}$
Gauss's law:	$\nabla \cdot B = 0$	$\nabla \cdot B = 0$
Lorenz-force law:	$F = q[E + v \times B]$	$F = q\left[E + \dfrac{1}{c} v \times B \right]$
Displacement field:	$D = \varepsilon_0 E$	$D = E$
Magnetic intensity:	$H = \dfrac{B}{\mu_0}$	$H = B$

Units:

	SI (*Système International*)	Gaussian CGS
E	Volt per metre ($\mathrm{Vm^{-1}}$)	Statvolt per centimetre
D	Coulomb per square metre ($\mathrm{Cm^{-2}}$)	Statcoulomb per square centimetre
ρ	Coulomb per square metre ($\mathrm{Cm^{-3}}$)	Statcoulomb per cubic centimetre
B	Tesla (T)	Gauss
H	Ampère per metre ($\mathrm{Am^{-1}}$)	Oersted
j	Ampère per square metre ($\mathrm{Am^{-2}}$)	Statampère per square centimetre

plasma velocity, we get

$$\nabla \times \mathbf{B} = \mu_0 \sigma (\mathbf{E} + \mathbf{v} \times \mathbf{B}).$$

Taking the curl of this, using Faraday's law, and supposing that $\sigma = $ const, we get

$$\nabla \times (\nabla \times \mathbf{B}) = \mu_0 \sigma \left(-\frac{\partial \mathbf{B}}{\partial t} + \nabla \times (\mathbf{v} \times \mathbf{B}) \right).$$

Introducing the ohmic magnetic diffusivity,

$$\eta = \frac{1}{\mu_0 \sigma},$$

and noticing that

$$\nabla \times (\nabla \times \mathbf{B}) = \nabla(\nabla \cdot \mathbf{B}) - \nabla^2 \mathbf{B} = -\nabla^2 \mathbf{B},$$

we then get

$$\frac{\partial \mathbf{B}}{\partial t} = \nabla \times (\mathbf{v} \times \mathbf{B}) + \eta \nabla^2 \mathbf{B}.$$

This is the *MHD induction equation,* which is the basic equation for magnetic behaviour in MHD. It expresses that *local change* can be due to *advection* and *diffusion.*

9.4.1 The magnetic Reynolds number

In the induction equation the magnitude of the advection (convective) term divided by that of the diffusive term is the magnetic Reynolds number:

$$R_{\mathrm{m}} = \frac{\nabla \times (\mathbf{v} \times \mathbf{B})}{\eta \nabla^2 \mathbf{B}} = \frac{v/l}{\eta/l^2} = \frac{vl}{\eta} = \mu_0 \sigma v l \,,$$

where v and l are characteristic values for velocity and length, μ_0 is the magnetic permeability and σ is the electric conductivity.

For example, in the solar corona above an active region, where $T \approx 10^6$ K, $\eta \approx 1$ m^2 s^1, $l \approx 10^5$ m, $v \approx 10^4$ m s^{-1}, the Reynolds number $R_{\mathrm{m}} \approx 10^9$, meaning that the diffusive term is negligible compared to the advective term in the induction equation.

The case of $R_{\mathrm{m}} \ll 1$ (found mainly in laboratory plasmas):
If $R_{\mathrm{m}} \ll 1$, the temporal evolution of the field is dominated by the diffusion term

$$\partial \mathbf{B}/\partial t = \eta \nabla^2 \mathbf{B}$$

and the plasma motions are not important. Magnetic flux leaks out of any concentration so that the gradients are reduced. The characteristic time scale for such ohmic diffusion is, with the fully ionised value of η,

$$t_d \approx \frac{l^2}{\eta} = \frac{l}{v} R_{\mathrm{m}} = 1.9 \cdot 10^{-8} l^2 T^{3/2} / \ln \Lambda \,,$$

where $\ln \Lambda$ is the Coulomb logarithm (Holt and Haskell, 1965):

$$\ln \Lambda \approx 16.3 + 3/2 \ln T - 1/2 \ln n \,, \quad \text{where } T < 4.2 \times 10^5 \text{ K} \,,$$
$$\ln \Lambda \approx 22.8 + \ln T - 1/2 \ln n \,, \qquad \text{where } T > 4.2 \times 10^5 \text{ K} \,,$$

where n is the total number of particles per unit volume.

For a copper sphere of $r = 1$ m, $t_d \approx 10$ s; for a sunspot (see Figure 9.1) it is about $30,000$ years ($\eta \approx 1$ m^2 s^{-1}, $l = 10^6$ m); for the total solar magnetic field t_d is of the order of 10^{10} years. For the terrestrial core $l = 3470$ km, with $\sigma = 5.10^5$ (ohm m^{-1}), $t_d \approx 24,000$ years, so if there were no fluid motions in the Earth's core, the terrestrial magnetic field would have disappeared long ago.

The case of $R_{\mathrm{m}} \gg 1$ (e.g. in the solar atmosphere):
If $R_{\mathrm{m}} \gg 1$, the induction equation becomes

$$\frac{\partial \mathbf{B}}{\partial t} = \nabla \times (\mathbf{v} \times \mathbf{B}) + \eta \nabla^2 \mathbf{B} \Rightarrow \frac{\partial \mathbf{B}}{\partial t} = \nabla \times (\mathbf{v} \times \mathbf{B}) \,.$$

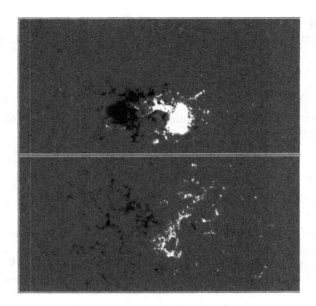

Figure 9.1 Bipolar sunspot groups are formed by *buoyant magnetic flux tubes* which emerge to the solar photosphere from the bottom of the convection zone. The upper image shows the magnetic field map of such a bipolar region a few days after its emergence, when it is fully formed. The lower image shows the same bipolar region one solar rotation, i.e. 28 days, later. By that time turbulent plasma motions dispersed the magnetic field, which outlines the boundary of supergranular cells. *SOHO/MDI* high resolution images taken on 26 November 1996 (upper panel) and on 23 December 1996 (lower panel) near to the centre of the Sun. White indicates positive polarity and black indicates negative polarity, respectively.

For such an ideal fluid Ohm's law is

$$\mathbf{E} + \mathbf{v} \times \mathbf{B} = 0$$

$(= \mathbf{j}/\sigma$, where $\sigma = \infty)$.

Hannes Alfvén (1943) showed that in such cases field lines move as though they were 'frozen' into the fluid (magnetic field line conservation), and the magnetic flux is constant through any closed contour moving with the plasma (magnetic flux conservation). There is virtually no relative motion between the plasma and the magnetic field perpendicular to the field. However, plasma can flow along the field without changing it.

In most astrophysical conditions (stellar atmospheres and convective envelopes) the magnetic Reynolds number is large, except for very small length scales which lie well under the observable scales. Thus, *observable changes in the magnetic field must involve transport of the field by plasma motions* through $\partial \mathbf{B}/\partial t = \nabla \times (\mathbf{v} \times \mathbf{B})$, and the electromagnetic state of the medium can be described by the magnetic field $\mathbf{B}$ alone. The magnetic field lines and flux tubes may be pictured as strings and rubber tubes that are transported and may be deformed by plasma flows.

This long-lasting field line identity breaks down at small R_m, i.e. at very small length scales l. In the photosphere $\eta \approx 1$ m^2 s^{-1} and the flows are as slow as $v = 10$ m/s; then R_m drops below 1 for scales $l < 1$ km. Then ohmic dissipation becomes important and magnetic field lines may reconnect, or simply 'field lines slip across the plasma'.

9.4.2 *An example: interaction between convective flows and magnetic fields*

Galloway and Weiss (1981) showed that for large magnetic Reynolds numbers (e.g. in the solar photosphere) the magnetic field ends up in the regions between persistent velocity cells or convective rolls (Figure 9.2(a)). After a few turnover times the initially weak, vertical, homogeneous magnetic field gets strongly sheared in the convective cell interiors. Then, the diffusive term $\eta \nabla^2 \mathbf{B}$ becomes dominant, and reconnection and field annihilation occur. The field disappears from the cell interior and becomes concentrated at the boundaries. By then, the Lorentz force

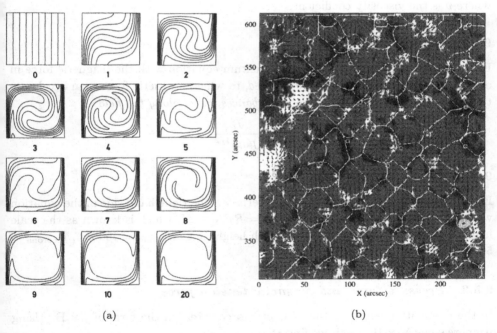

(a)   (b)

Figure 9.2   (a) Interaction between a velocity cell and weak vertical magnetic fields (Galloway and Weiss, 1981) — at large Reynolds numbers ($R_m = 250$) the magnetic field disappears from the interior of such velocity (convective) cells and becomes concentrated at their boundaries. (By kind permission of the American Astronomical Society.)   (b) *SOHO/MDI* magnetogram with superimposed (black) arrows indicating photospheric indeed flows, and white lines outline the convective cells. Note that the magnetic field concentrations are located along the intergranular lanes between convective cells. (Courtesy of the SOHO MDI consortium SOHO is a project of international cooperation between ESA and NASA.)

$\mathbf{j} \times \mathbf{B} = 1/\mu_0 \, (\nabla \times \mathbf{B}) \times \mathbf{B}$ gets sufficiently intensified to hamper the flow locally. Magnetic field strengths locally produced by this process can be estimated equating the *magnetic energy density $B^2/2\mu_0$ to the kinetic energy density $\rho v^2/2$ to yield the equipartition field strength*:

$$B_{\text{eq}} \equiv (\mu_0 \rho v^2)^{1/2}.$$

Indeed, in the solar photosphere magnetic field is mainly concentrated along the intergranular lanes between convective cells (Figures 9.1 and 9.2(b)).

9.5   Examining the Momentum Equation

In the MHD approximation of the second main equation of MHD, the momentum equation is

$$\rho \frac{d\mathbf{v}}{dt} = -\nabla p + \mathbf{j} \times \mathbf{B} + \rho \mathbf{g} + v \nabla^2 \mathbf{v},$$

where v is the viscosity coefficient.

9.5.1   *Alfvén speed*

Equating the left-hand side of the momentum equation with the magnetic force in order of magnitude gives the *Alfvén speed*, to which magnetic forces can accelerate the plasma, or at which *magnetic disturbances travel along the field*:

$$v_{\text{A}} \equiv \frac{B}{(\mu_0 \rho)^{1/2}}.$$

9.5.2   *Pressure scale height*

Equating the sizes of the first and third terms on the right-hand side of the equation with $p = R\rho T$ gives a length scale of $L = RT/g \equiv H$, which is known as the scale height for the fall-off of the pressure with height, e.g. $H_{\text{chrom}} = 500$ km, $H_{\text{corona}} = 50{,}000$ km.

9.5.3   *Pressure force and magnetic tension force*

In the momentum equation the magnetic force enters in the form of $\mathbf{j} \times \mathbf{B}$. Using Ampère's law $\nabla \times \mathbf{B} = \mu_0 \mathbf{j}$, we find that

$$\mathbf{j} \times \mathbf{B} = \frac{1}{\mu_0}(\nabla \times \mathbf{B}) \times \mathbf{B} = -\nabla \left(\frac{B^2}{2\mu_0} \right) + \frac{(\mathbf{B} \cdot \nabla)\mathbf{B}}{\mu_0},$$

where on the right-hand side the first term is the gradient of the isotropic magnetic pressure (*pressure force*), and the second term is the force arising from the magnetic tension (*magnetic tension force*).

If the field lines are not straight, but have a radius of curvature R, their tension B^2/μ_0 exerts a transverse force $B^2/(\mu_0 R)$ per unit volume. A net force is exerted on the plasma, if this stress arising from the curvature is not balanced by the magnetic pressure:

$$\nabla\left(\frac{B^2}{2\mu_0}\right) \neq \frac{(\mathbf{B}\cdot\nabla)\mathbf{B}}{\mu_0}\,.$$

It is noteworthy, however, that there is no net force along the magnetic field $\mathbf{B}$.

9.5.4 *Plasma-β parameter*

The magnitude of the plasma pressure p and magnetic pressure $\mathbf{B}^2/2\mu_0$ are compared in the plasma-β parameter, which is defined as

$$\beta \equiv \frac{2\mu_0 p}{B^2}\,.$$

9.5.5 *Force-free magnetic field conditions*

If $\beta \gg 1$, the pressure balance is determined by the plasma pressure distribution, and the influence of the magnetic field is negligible. However, for $\beta \ll 1$ the plasma pressure has virtually no influence and the *slowly evolving* magnetic field is essentially *force-free*:

$$\mathbf{j}\times\mathbf{B} = \frac{1}{\mu_0}(\nabla\times\mathbf{B})\times\mathbf{B} = \mathbf{0}\,,$$

where $\mathbf{j} = \nabla\times\mathbf{B}/\mu_0$ and $\nabla\cdot\mathbf{B} = 0$. *In force-free fields the current $\mathbf{j}$ runs parallel to $\mathbf{B}$*, so where α is a scalar function of position. Taking the divergence of the equation for force-free field, $\nabla\times\mathbf{B} = \alpha\mathbf{B}$, gives

$$\nabla\cdot(\nabla\times\mathbf{B}) = \nabla\cdot(\alpha\mathbf{B}) = \alpha\nabla\cdot\mathbf{B} + \mathbf{B}\cdot\nabla\alpha = 0 \Rightarrow \mathbf{B}\cdot\nabla\alpha = 0\,,$$

so that α *is constant along a field line.*

In a force-free field the gradient of the magnetic pressure $\nabla(\mathbf{B}^2/2\mu_0)$ is exactly balanced by the curvature force $(\mathbf{B}\cdot\nabla)\mathbf{B}/\mu_0$; hence there the *hydrostatic equilibrium* $\nabla p = \rho\mathbf{g}$ applies along each field line when the speed is sub-sonic $(\mathbf{v} \ll \mathbf{v}_{\text{sound}}\ (= c_s))$.

In the particular case where α is independent of the field line, i.e. it is uniform in space, taking the curl of the force-free equation gives

$$\nabla\times(\nabla\times\mathbf{B}) = \nabla\times(\alpha\mathbf{B})$$

$$\Rightarrow \nabla\times(\nabla\cdot\mathbf{B}) - \nabla^2\mathbf{B} = \alpha(\nabla\times\mathbf{B}) = \alpha^2\mathbf{B}$$

$$\Rightarrow (\nabla^2 + \alpha^2)\mathbf{B} = \mathbf{0}\,.$$

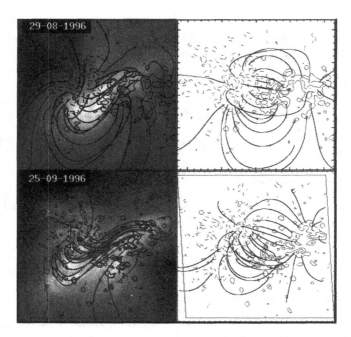

Figure 9.3 Linear force-free (left panels) and potential (right panels) extrapolations of a solar active region (NOAA 7986) during two consecutive solar rotations. *SOHO/MDI* magnetograms were used as photospheric boundary conditions. In the left-hand panels the model field lines are overlaid with *Yohkoh/SXT* soft X-ray images. The observed coronal loops show a good global match with the sheared linear force-free models (the α parameter is 1.2 and 1.4×10^{-2} Mm^{-1}, in the upper and the lower panel, respectively), while they do not match the potential model at all, implying that the coronal magnetic field is not current-free. Note that not only the shape of the field lines but also their magnetic connectivities are quite different in the current-carrying and current-free case. (Courtesy G. Aulanier.)

Solutions to that equation are called *linear* or *constant-α force-free fields*. They are well understood and widely used (see Figure 9.3 for an example, and for the method of magnetic extrapolations see e.g. Gary, 1989). The particular case $\alpha = 0$ gives potential field which is *current-free* (Figure 9.3). The potential field has the smallest energy of all the fields in a finite volume with a given $\mathbf{B_n}$ (the normal component of $\mathbf{B}$) on the boundary.

9.5.6 *Magnetostatic equilibrium*

Equilibrium means that there is a force balance in the magnetic structure:

$$-\nabla p + \rho \mathbf{g} - \nabla \left(\frac{B^2}{2\mu_0} \right) + \frac{1}{\mu_0} (\mathbf{B} \cdot \nabla) \mathbf{B} = 0 \,.$$

For example, in an isolated thin flux tube in the convective zone ($d < R; d < H_{\mathrm{p}}$) the magnetic (Lorentz) force has no component along the field lines, so this equation in the direction of the unit vector tangent to the field $\hat{\mathbf{l}}$ becomes

$$-\frac{\partial p}{\partial l} + \rho(\mathbf{g} \cdot \hat{\mathbf{l}}) = 0\,,$$

while in a direction $\check{\mathbf{n}}$ perpendicular to the tube it is

$$\frac{\partial}{\partial n}\left(p + \frac{B^2}{2\mu_0}\right) = 0\,.$$

In this example the magnetic field is considered to be confined to a small region, called the interior. There is a gas pressure jump $(p_e - p_i)$ between the internal (p_i) and external (p_e) pressures. This is compensated by the magnetic pressure inside the flux tube, so that

$$p_i + \frac{B^2}{2\mu_0} = p_e\,.$$

9.5.7 *Magnetic buoyancy*

A flux tube that is in thermal balance with its surroundings $(T_i = T_e)$ is buoyant, since the lower internal gas pressure $(p_i < p_e)$ implies lower mass density, $\rho_i < \rho_e$. The arising magnetic buoyancy force (see below) will lead to the rise of flux tubes.

The path of the flux tube in the convection zone is determined by the velocity shear in the ambient medium that couples to the tube through drag, the buoyancy of the flux tube, and the magnetic tension along the flux tube. For example, shear flows stretch the flux tube, and consequently they enhance B due to mass and flux conservation. The drag force F_d per unit length of the tube is

$$F_d = \frac{1}{2}\rho v_\perp^2 a C_d\,,$$

where $v_\perp$ is the relative velocity perpendicular to the flux tube, a is the flux tube radius and C_d is the drag coefficient (≈ 2 in convective envelopes). For a horizontal flux tube the buoyancy force per unit length is

$$F_b = \pi a^2 g(\rho_e - \rho_i)\,.$$

By balancing buoyancy against drag, the rate of rise $v_\uparrow$ for flux tubes

$$v_\uparrow = v_A \left(\frac{\pi}{C_d}\right)^{1/2}\left(\frac{a}{H_p}\right)^{1/2}\,,$$

where v_A is the Alfvén speed and H_p is the pressure scale height. With $B = 10^4$–10^5 G, $\alpha \approx 0.1\, H_p$, the rise of a flux tube from the bottom to the top of the convection zone takes weeks to months.

9.6 Magnetic Reconnection

The concept of reconnection is tied to the concept of a field line. A *field line* can be defined as a path tangent to the local direction of the magnetic field.

In a *perfectly conducting plasma* the field lines are 'frozen in' the plasma, i.e. two plasma elements located anywhere along the field line always remain on the same field line, and thus *no reconnection occurs*.

More realistically, reconnection occurs if the length scales are sufficiently small (R_m is locally of the order of unity). *Reconnection requires the dissipation of electric currents*, and thus the redirection of the field is associated with plasma no longer being tied to the field lines.

9.6.1 *Current sheet*

Cowling (1953) pointed out that, if a solar flare is due to ohmic dissipation, a current sheet only a few metres thick is needed to power it. Such a current sheet can indeed form by the collapse of the magnetic field near an X-type neutral point (Dungey, 1953).

A current sheet is a thin current-carrying layer across which the magnetic field changes either in direction or magnitude or both.

- It can only exist in a conducting medium such as a plasma, since by definition it contains a current;
- In plasmas, magnetic null points typically give rise to current sheets.

9.6.2 *Force balance in current sheets*

In equilibrium there is a total pressure balance between the current sheet and its surroundings at both sides:

$$p_2 + \frac{B_2^2}{2\mu_0} = p_0 + \frac{B_0^2}{2\mu_0} = p_1 + \frac{B_1^2}{2\mu_0} .$$

If $B_0 = 0$, it is a *neutral sheet*; if ambient plasma pressures $p_1 = p_2 = 0$, then

$$\frac{B_2^2}{2\mu_0} = p_0 = \frac{B_1^2}{2\mu_0} .$$

In the simplest case, when B changes sign and $B_1 = -B_2$, i.e. a complete field reversal occurs along the sheet, Ampère's law implies that

$$j_z = \frac{1}{\mu_0} \frac{dB_y}{dx} ,$$

meaning that this steep gradient in B_z produces a strong current along the sheet and perpendicular to the field lines.

9.6.3 *Kelvin–Helmholz instability in a simple current sheet*

Current sheets with tangential discontinuity are prone to instabilities such as the Kelvin–Helmholz instability. Consider a current sheet with uniform magnetic fields $\mathbf{B}_1$ and $\mathbf{B}_2$ and with uniform flows $\mathbf{v}_1$ and $\mathbf{v}_2$ on either side. In an incompressible plasma ($\rho = $ const, or $\nabla \cdot \mathbf{v} = 0$) an instability occurs when the kinetic energy of the tangential velocity difference exceeds the total magnetic energy density (magnetic field stabilises the current sheet, since the fluid must do work to distort the field lines)

$$\frac{1}{2}\mu_0\rho(\mathbf{v}_1 - \mathbf{v}_2)^2 > B_1^2 + B_2^2,$$

and

$$\frac{1}{2}\mu_0\rho\{[\mathbf{B}_1 \times (\mathbf{v}_2 - \mathbf{v}_1)]^2\} + \{[\mathbf{B}_2 \times (\mathbf{v}_2 - \mathbf{v}_1)]^2\} > (\mathbf{B}_1 + \mathbf{B}_2)^2.$$

The latter condition involves the relative orientations of flow and field. Flows perpendicular to the field are more susceptible to instability. The Kelvin–Helmholz instability has been studied in simple one-dimensional current sheets, surrounded by uniform fields and flows. Real current sheets at neutral points have highly non-uniform flows and fields on either side, and they are of finite length — little is known about the stability of such sheets.

9.6.4 *Tearing mode instability*

Furth, Killeen and Rosenbluth (1963) analysed the stability of a simple, static current sheet and found that a sheet whose length is at least 2π times greater than its width will spontaneously reconnect to form magnetic islands. This is called the *tearing mode instability* of the current sheet, and implies that reconnection is inherently a time-dependent process with episodes of impulsive and bursty releases of energy. This instability may be important for magnetospheric substorms, coronal heating, solar and stellar flares (e.g. Shivamoggi, 1985).

9.6.5 *The Sweet–Parker reconnection model*

Sweet (1958) and Parker (1957) considered the following process:

- Oppositely directed field lines are carried into a thin current sheet of length L and width l by a flow v_i.
- The plasma outside the current sheet is effectively frozen to the field.
- In the sheet the field is no longer frozen to the plasma and may slip through and reconnect, and eventually it is expelled from the ends of the sheet at the Alfvén speed (v_A).

The continuity of mass flux gives $L \cdot v_i = l \cdot v_A$, while the condition for steady state (speed of inflow = diffusion speed) gives $v_i = \eta/l$. Eliminating l we get

$$v_i^2 = \eta \frac{v_A}{L}$$

or, in dimensionless form,

$$M = \frac{1}{R_m^{1/2}},$$

where $M = v/v_A$ is the Alfvén Mach number and R_m is the magnetic Reynolds number in terms of the Alfvén speed. The Sweet–Parker mechanism gives reconnection rates of between 10^{-3} and 10^{-6} of the Alfvén speed. This reconnection rate is faster than simple diffusion (by a factor of L/l), but it is still much too slow for a flare (see below), while the length of the current sheet in this model is approximately the same as the global scale length of the flaring region.

9.6.6   *Conditions for fast reconnection*

Solar flares release stored energy in the corona within a period of 100 s, while the time scale for magnetic dissipation for a scale length of 10^5 km is about 10^6 years!

Flare-like phenomena require a conversion (part) of the stored magnetic energy within a few Alfvén time scales ($= L/v_A$, the time it takes to traverse the sheet at the Alfvén speed).

There are two main approaches applied in the theory to achieve fast reconnection. The first is to find a mechanism to create an *anomalous (high) resistivity*, which allows rapid dissipation. The other approach is to *reduce the dissipation length scale*, i.e. to find such a geometrical configuration, which greatly reduces the effective dissipation scale length.

9.6.7   *The Petschek reconnection model*

Petschek suggested that the *current sheet* or diffusion region could be *much smaller* and therefore the reconnection much faster if *slow-mode shock waves* stand in the flow and propagate from the ends of the current sheet. The shocks play an important role, because they accelerate the plasma to the Alfvén speed and convert magnetic energy into kinetic energy and heat. Thus, the shocks can change v and B, and at large distances they would be quite different from their values at the input to the reconnection region. Around the diffusion region the field lines curve (to allow the slow shocks to exist) and the field strength decreases from B_e to B_i:

$$B_i = B_e - \frac{4B_N}{\pi} \log_e \frac{L_e}{L},$$

where $B_N = v_e(\mu\rho)^{1/2}$. Considering $B_i = \frac{1}{2}B_e$, the reconnection rate is

$$M_e^* = \frac{\pi}{8 \log R_{me}} \approx 0.01 \,,$$

which is the maximal reconnection rate possible with the Petschek mechanism, and it is much larger than the Sweet–Parker rate. Petschek's model was the first of the *fast reconnection* models.

Since then, a new generation of more general almost-uniform and non-uniform models have been developed, which include the Petschek mechanism as a special case (see Priest and Forbes, 2000, for details).

9.6.8 *A key issue: the rate of reconnection*

In 2-D, the rate of reconnection is the rate at which field lines move through the X-type neutral point.

The following characteristic speeds play important roles:

(i) The external flow speed (v_e) at a fixed global distance (L_e) from the neutral point;
(ii) The Alfvén speed [$v_{Ae} = B_e/(\mu\rho)^{1/2}$] in terms of the external magnetic field B_e;
(iii) The global magnetic diffusion speed ($vd_e = \eta/L_e$).

From these, two independent dimensionless parameters are constructed:

- The external Alfvén Mach number,

$$M_e = \frac{v_e}{v_{Ae}} \,,$$

  which is a measure of the rate of reconnection at the boundary of the system;
- The global magnetic Reynolds number,

$$R_{me} = \frac{v_{Ae}}{v_{de}} \,,$$

based on the Alfvén speed (also called the Lundquist number), which measures the ability of the field lines to diffuse through the plasma.

The relative scaling of the Alfvén Mach number M_e with the magnetic Reynolds number R_{me} defines whether the reconnection is

- Super-slow:

$$M_e \leqslant R_{me}^{-1} \,.$$

The magnetic field diffuses like it does in a solid conductor, since the advective term in Ohm's law is unimportant, and such ordinary magnetic diffusion is very slow in astrophysical plasmas. For example, linear reconnection belongs to this category (Priest *et al.*, 1994).

• Slow:

$$R_{\mathrm{me}}^{-1} < M_{\mathrm{e}} \leqslant R_{\mathrm{me}}^{-1/2} \, .$$

When the reconnection speed exceeds $1/R_{\mathrm{me}}$, the $\mathbf{v} \times \mathbf{B}$ advection term in Ohm's law becomes important. The first model with such reconnection characteristics was the Sweet–Parker model (Sweet, 1958; Parker, 1957, 1963; see above).

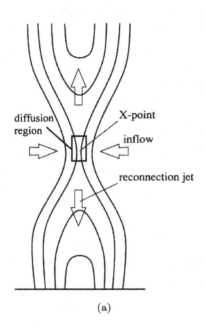

(a)

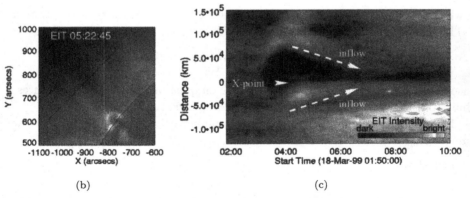

(b)   (c)

Figure 9.4   Cartoon of the reconnection configuration; (b) *SOHO/EIT* EUV (Fe 195 A) image of the cusp-shaped flare on 18 March 1999; (c) time evolution of the one-dimensional distribution of EUV intensity along the thick solid line in (b). The inflow speed is up to 5 km s^{-1}, which is an upper limit of the theoretically expected reconnection inflow speed and the calculated reconnection rate $M_{\mathrm{e}} = 0.001$–0.03. (Yokoyama *et al.*, 2001; with permission of the AAS.)

- Fast:

$$M_e > R_{me}^{-1/2} .$$

Petschek's model belongs to this category ($M_e \sim (\log R_{me})^{-1}$). Even for $R_{me} > 10^8$, which is common in astrophysical plasmas, the reconnection rate is in the range of 0.01–0.1. By now, the Petschek mechanism is regarded as a special case in the new generation of fast reconnection models (Priest and Forbes, 2000).

In many astrophysical applications the Reynolds numbers are huge (10^6–10^{12}), so the fast reconnection mechanisms are the most relevant. However, even the slow reconnection can become applicable when R_{me} is reduced by kinetic effects or turbulence.

9.6.9 *An application: reconnection in solar flares*

Solar flares are thought to be caused by reconnection, during which, in the simplest case, antiparallel magnetic field lines change connectivity and dissipate at and around a point (neutral or X-point) in the corona. The tension force of the re-connected field lines then accelerates the plasma out of the dissipation region. The inflowing plasma carries the ambient magnetic field lines into the dissipation region. These field lines continue the reconnection cycle. Through this process, the magnetic energy stored near the neutral point is released to become the thermal and bulk-flow energy of the plasma. The reconnection phase is followed by the formation of hot (newly reconnected) loops. The ones which are below the reconnection region frequently have a cusp shape and after cooling down appear as cool (e.g. Hα) loops. The reconnected loops, which are above the reconnection region, may form a plasmoid, which is progressively ejected. However, the formation of such plasmoids occurs only in certain adequate magnetic topologies.

- *Pre-flare phase*: an active region filament (and its overlying arcade) starts to rise due to weak eruptive instability or loss of equilibrium.
- *Flare onset*: under the rising filament the stretched-out field lines start to recon-nect; the impulsive energy release occurs, and the cutting of the field lines leads to the acceleration in the filament eruption.
- *Main phase*: continuing reconnection forms hot X-ray loops and Hα ribbons at their footpoints, where the particles accelerated during the reconnection process impact the chromosphere; the reconnection region rises as field lines initially more and more distant from the magnetic inversion line reconnect, so the resulting X-ray loops are larger and larger; Hα ribbons move apart.

The initial instability may be triggered by:

- A twisted flux tube, of which the twist or height is too great; or
- A sheared magnetic arcade, in which the magnetic energy is being built up by shearing motions, reaching a critical value.

The above 2-D model best describes the scenario of the so-called long-duration eruptive flares, which occur in a dominantly bipolar magnetic configuration. A typical example is shown in Figure 9.5. However, in more complicated, and thus more realistic, magnetic topologies 2-D and half and 3-D reconnection models have to be applied.

The main changes when moving from 2-D to 3-D are that

- The reconnecting magnetic fields are not necessarily antiparallel;
- The emphasis is put on the field line connectivities — separatrices or quasi-separatrix layers separate domains with drastically different magnetic connectivities;
- The magnetic reconnection occurs along these separatrices, which have a complex 3-D shape.

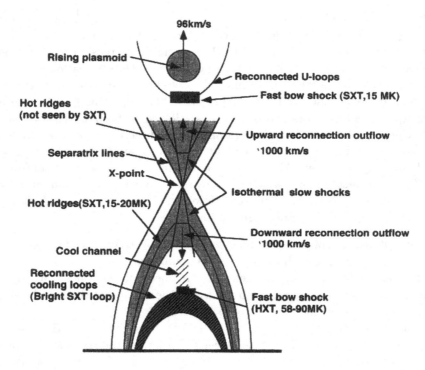

Figure 9.5 Schematic magnetic and temperature structure of a two-ribbon flare. The hot ridges located below the X-point are heated by the pair of slow shocks and are clearly seen in the *Yohkoh/SXT* temperature maps. There must be (less dense) symmetrical hot ridges heated by another pair of the slow shocks located above the X-point. This is not seen by *SXT*, probably owing to the contamination of the line-of-sight corona with lower temperature. The downward and upward reconnection outflows collide with the bright soft X-ray loop below the X-point and with the rising plasmoid above the X-point, and form fast (perpendicular) shocks. The hot source heated by the fast shock below the X-point is seen by *Yohkoh/HXT* (50–90 MK), and that above the X-point is seen by *SXT* (15 MK). (Tsuneta, 1997; with permission of the AAS.)

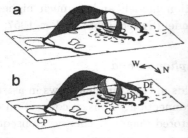 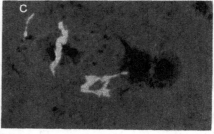

Figure 9.6 Example of 3-D reconnection in a quadrupolar magnetic configuration. (a) Pre-reconnection (i.e. pre-flare) magnetic connectivities. (b) Post-reconnection magnetic connectivities — note the change in the dark grey (light grey) shaded field lines domains. (c) Hα image shows the located of the bright flare ribbons, which appear at the footpoints of the reconnected magnetic field lines, along the intersection of the 3-D separatrix layer with the chromosphere (Ottava River Solar Observatory image taken on 27 May 1978). (Adaptation of figures by Gaizauskas *et al.* (1998); courtesy P. Démoulin.)

A typical example for 3-D reconnection in a quadrupolar magnetic configuration is shown in Figure 9.6, after Gaizauskas *et al.* (1998). For theoretical details on 3-D reconnection, refer to Priest and Forbes (2000).

9.6.10 *The first observation of the reconnection inflow and outflow*

One of the most important applications of reconnection is in solar flares, in which many of the morphological and physical characteristics of the reconnection scenario had been confirmed by observations. However, one of the criticisms has always been towards the reconnection theory of flares that the reconnection inflow and outflow had not been observed. First, Forbes and Acton (1996), through a careful analysis of soft X-ray loops, showed how the cusped loops shrink and relax into a roughly semi-circular shape as a consequence of the retraction of the reconnected field lines due to the magnetic tension force. This is an indirect evidence for the existence of an *outflow from the magnetic reconnection region*. More recently, McKenzie and Hudson (1999) found downward plasma motion above an arcade formed in a long-duration flare and coronal mass ejection, confirming the existence of such outflow.

Furthermore, very recently, using *SOHO/EIT* and *Yohkoh/SXT* data, a clear *evidence for the existence of* such *inflow* was presented by Yokoyama *et al.* (2001). They observed a flare on the solar limb, which displayed a geometry and scenario (e.g. plasmoid ejection, cusp, X-point) highly resembling the 2-D reconnection simulations. Following the ejection of a plasmoid, a clear motion with $v = 1.0$–4.7 km s^{-1} was observed towards the reconnection region (X-point; see Figure 9.3). Based on X-ray data, a magnetic field strength of $B \approx (16\pi n k_B T)^{1/2} = 12$–$40$ G was estimated and an Alfvén speed was computed to be in the range of $v_A = 160$–970 km s^{-1}. The reconnection rate, which is defined as the ratio of the inflow speed to the Alfvén velocity, derived from this observation was $M_A = 0.001$–0.03, which is roughly consistent with Petschek's (1964) reconnection model, and is much higher than would be expected from the Sweet–Parker model (Sweet, 1958; Parker, 1957, 1963).

9.6.11   *Reconnection in astrophysical phenomena*

Due to the ubiquitous nature of magnetic fields in the Universe, flows in astrophysical plasmas lead inevitably to magnetic reconnection. Thus, reconnection occurs everywhere, leading to dynamic releases of stored magnetic energy. Consequently, magnetic reconnection has a very wide range of applications in astrophysics.

The most traditional applications of the reconnection theory are to the solar corona and the terrestrial magnetosphere. Observing magnetic fields and magnetic phenomena outside the solar system is quite a challenging task, but new observing and modelling techniques like Doppler imaging and recent space-borne observations led to considerable advances in this field (for more details see e.g. Mestel, 1999).

Reconnection is thought to be an important mechanism for the heating of the solar and stellar coronae to million-degree temperatures, which has a huge literature. In the generation of solar, stellar, planetary and galactic magnetic fields by self-excited magnetic dynamo, reconnection is supposed to play an important role. Stellar flares are natural analogies to the processes happening in solar flares and the galactic magnetotail is paralleled with the Earth's magnetosphere. For more details on these processes, refer to Chapters 4 and 6.

9.7   Magnetic Helicity

Magnetic helicity plays a key role in MHD, because it is one of the few global quantities, which is conserved even in resistive MHD on a timescale less than the global diffusion timescale (Berger, 1984). Magnetic helicity is a measure of both the twist and the linkage of magnetic flux tubes. Recent observations (e.g. twisted magnetic flux emergence, helical structures in CMEs; see Figure 9.7) and theoretical development in solar physics highlighted the importance of magnetic helicity.

Magnetic helicity is defined by an integral in a magnetically closed volume V (i.e. a volume with no outgoing magnetic field line; the magnetic field component normal to the boundary vanishes, $B_n = 0$):

$$H_m = \int_V \mathbf{A} \cdot \mathbf{B} dV ,$$

where $\mathbf{A}$ is the vector potential, expressed in Tesla-metres $\mathbf{B} = \nabla \times \mathbf{A}$ and $\mathbf{B}$ is the magnetic field. For a given $\mathbf{B}$ there exist an infinite number of possible $\mathbf{A}$'s within an additive term of which the curl is zero, but when $B_n = 0$ it gives only one helicity value.

When $B_n \neq 0$ along the boundary S of volume V, a relative magnetic helicity can be computed by subtracting the helicity of a reference field $\mathbf{B}_0$, which has the

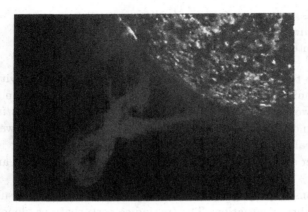

Figure 9.7   An example of prominence eruption. Note the flux rope structure indicating the significantly twisted magnetic field erupting prominences/CMEs carry away from the Sun (*SOHO/EIT* 304 Å image taken on 18 January 2000). (Courtesy of the SOHO EIT consortium. SOHO is a project of international cooperation between ESA and NASA.)

same B_n distribution on S as $\mathbf{B}$ (Berger and Field, 1984):

$$H_r = \int_V \mathbf{A} \cdot \mathbf{B} dV - \int_V \mathbf{A_0} \cdot \mathbf{B_0} dV \,.$$

Since magnetic helicity is a well-preserved quantity, it can only be modified in volume V by helicity flux crossing S (Berger and Field, 1984):

$$\frac{dH_r}{dt} = -2 \int_S [(\mathbf{A_0} \cdot \mathbf{v})\mathbf{B}) - (\mathbf{A_0} \cdot \mathbf{B})\mathbf{v}] d\mathbf{S} \,,$$

where the first term is helicity generation by plasma motions parallel to S like e.g. differential rotation in the photosphere (DeVore, 2000; Démoulin *et al.*, 2002a), while the second term is 'inflow' and 'outflow' of helicity through the boundary S increasing the helicity through twisted flux emergence (Leka *et al.*, 1996), or torsional Alfvén waves (Longcope and Welsch, 2000), or depleting the helicity via coronal mass ejections (Low, 1996; Démoulin *et al.*, 2002b).

It is noteworthy that the *current helicity* density $H_c = \mathbf{B} \cdot \nabla \times \mathbf{B}$, widely used in solar helicity studies, *is not a conserved MHD quantity*.

9.8   Waves

A common response of physical systems to (periodic) perturbations is that they emit small-amplitude oscillations, or waves. In hot astrophysical plasmas the basic modes of wave motion can be classified by the restoring forces they are driven by:

- magnetic tension $\rightarrow$ *Alfvén wave*
- Coriolis force $\rightarrow$ inertial wave

- magnetic pressure → *compressional Alfvén wave*
- plasma pressure → sound waves
- gravity → (internal) gravity waves

However, when the magnetic pressure, plasma pressure and gravity act together, they produce only two magneto-acoustic-gravity wave modes. In the absence of gravity these two modes (slow and fast) are called *magneto-acoustic waves*. When there is no magnetic field, they are referred to as acoustic gravity waves. The scope of this chapter allows only a brief description of these waves. For a more detailed treatment, refer to Priest (1982), Kivelson (1995), and Goertz and Strangeway (1995).

Wave motions or propagating disturbances are described as results of a slightly disturbed equilibrium situation. The (periodic) perturbations, denoted by suffix 1, can be written as

$$\mathbf{v} = \mathbf{v}_1, \rho = \rho_0 + \rho_1, p = p_0 + p_1, \mathbf{B} = \mathbf{B}_0 + \mathbf{B}_1 \, .$$

The basic equations are deduced from the basic MHD equations decribed at the beginning of this chapter, i.e. the conservation principles, the Maxwell equations and the induction equation. The governing equations are written in a reference frame rotating with the Sun at an angular velocity $\mathbf{\Omega}$ relative to the inertial frame, which gives rise to the Coriolis force $(-2\rho\mathbf{\Omega} \times \mathbf{v})$ and the centrifugal force $(\frac{1}{2}\rho\Delta|\mathbf{\Omega} \times \mathbf{v}|^2)$. The latter is normally combined with the gravitational term. The gravitational force is $-\rho g\hat{\mathbf{z}}$, where g is considered to be constant, and the z-axis is directed radially outwards, normal to the solar surface.

The equations are used in a linearised form, neglecting squares and products of the small physical quantities (v_1, ρ_1, p_1, and $\mathbf{B}_1$ — the small departures from the equilibrium value). This set of linearised equations is then reduced to a single equation for a vertically stratified stationary plasma with a uniform magnetic field ($\mathbf{B}_0$), uniform temperature (T_0), and with density and pressure which are proportional to $\exp(-z/\Lambda)$, where $\Lambda = p_0/(\rho_0 g)$ is the scale height. After differentiating the momentum equation with respect to time and substituting the time derivatives of ρ_1, p_1, and $\mathbf{B}_1$ from the other MHD equations, a generalised wave equation is derived for the disturbance velocity ($\mathbf{v}_1$):

$$\frac{\partial^2 \mathbf{v}_1}{\partial t^2} = c_s^2\nabla(\nabla{\cdot}\mathbf{v}_1) - (\gamma-1)g\hat{\mathbf{z}}(\nabla{\cdot}\mathbf{v}_1) - g\nabla v_{1z} - 2\mathbf{\Omega}\times\frac{\partial \mathbf{v}_1}{\partial t} + [\nabla\times(\nabla\times(\mathbf{v}_1\times\mathbf{B}_0))]\times\frac{\mathbf{B}_0}{\mu\rho_0} \, ,$$

where c_s is the sound speed, $\hat{\mathbf{z}}$ is a unit vector in the z-direction and γ is the polytropic index, which is the ratio of the specific heat at constant pressure to the specific heat at constant volume. The *sound speed* can be written as

$$c_s = \frac{\gamma p_0}{\rho_0} = \frac{\gamma k_B T_0}{m} \, ,$$

where γ is the politropic index, p_0 is the gas pressure, ρ_0 is the gas density and k_B is Boltzmann's constant (1.381×10^{-23} J deg^{-1}).

For a plane wave with wavelength λ and frequency f the oscillating quantities can be taken proportional to $e^{i(\mathbf{k}\cdot\mathbf{r}-\omega t)}$, so e.g. $\mathbf{v}_1$ can be written as

$$\mathbf{v}_1(\mathbf{r}, t) = \mathbf{v}_1 e^{i(\mathbf{k}\cdot\mathbf{r}-\omega t)} ,$$

where $\mathbf{k}$ is the *wave number vector* and $\omega = 2\pi f$. The period of the wave is $2\pi/\omega$, while its wavelength is $\lambda = 2\pi/k$, and $\hat{k} = \mathbf{k}/k$ is a unit vector showing the propagation direction of the wave.

The assumed exponential dependence on $\mathbf{r}$ and t of the wave properties implies that the time derivatives $\partial/\partial t$ can be replaced by $-i\omega$ and the $\mathbf{r}$ derivatives ∇ by $i\mathbf{k}$ in the generalised wave equation.

In the presence of magnetic field ($\mathbf{B}_0 \neq 0$), assuming that the wavelength λ of the perturbations is much smaller than the scale height $\Lambda = c_s^2/\gamma g$, the plasma density ρ_0 may be regarded as locally constant, and the equation can be written as

$$\omega^2\mathbf{v}_1 = c_s^2\mathbf{k}(\mathbf{k}\cdot\mathbf{v}_1)+i(\gamma-1)g\hat{\mathbf{z}}(k\cdot\mathbf{v}_1)+igkv_{1z}-2i\boldsymbol{\Omega}\times\mathbf{v}_1+[\mathbf{k}\times(\mathbf{k}\times(\mathbf{v}_1\times\mathbf{B}_0))]\times\frac{\mathbf{B}_0}{\mu\rho_0} .$$

This equation forms a basis for the discussion of the fundamental wave modes. It defines the *dispersion relations*, which impose relations between the frequency and the vector wave number that must be satisfied in order for the wave to exist in the plasma. Such well-known dispersion relations express, for example, that electromagnetic waves in free space propagate with the speed of light ($\omega/k = c$), and sound waves in a neutral gas propagate with the speed of sound ($\omega/k = c_s$). The roots of the dispersion relations give the *phase velocity* $\mathbf{v}_p = (\omega/k)\hat{\mathbf{k}}$ for MHD waves, where ω/k gives the magnitude of the speed of propagation in the direction $\hat{\mathbf{k}}$ for a wave specified by a single wave number. When the phase velocity varies with wavelength, the wave is *dispersive*. A packet or group of waves has a range of wave numbers and travels with a *group velocity*, which gives the velocity at which the energy is transmitted and, in general, differs both in magnitude and direction from the phase velocity, unless ω is linearly proportional to k, i.e. the wave is *non-dispersive*. In general, the wave propagation is *anisotropic*, and the phase speed varies with the direction of propagation. The three preferred directions are of the magnetic field, gravity and rotation, which introduce great complexity.

9.8.1 *Acoustic waves*

A common response of a physical system to pressure perturbations when $g = \mathbf{B}_0 = \boldsymbol{\Omega} = 0$ is the emission of acoustic (sound) waves. In this case the wave equation is reduced to

$$\omega^2\mathbf{v}_1 = c_s^2 k(k\cdot v_1) .$$

Taking the scalar product of the above equation with $\mathbf{k}$ and assuming that $\mathbf{k} \cdot \mathbf{v}_1 \neq 0$, i.e. $\nabla \cdot \mathbf{v}_1 \neq 0$, and thus the plasma is compressible, we get

$$\omega^2 = k^2 c_s^2 \,.$$

The square root of this equation gives the *dispersion relation* for acoustic or sound waves:

$$\omega = k c_s \,,$$

which means that acoustic waves propagate in all directions at a phase speed $v_p \equiv \omega/k = c_s$, and a group velocity $v_g = d\omega/dk = c_s$ in the direction $\mathbf{k}$. In the solar atmosphere the sound speed varies from about 10 km s^{-1} (photosphere) to 200 km s^{-1} (corona).

Sound waves are polarised along the direction of propagation; the polarisation direction is that of the gradient of the fluctuating pressure. For a closed system, the oscillations are the combinations of standing waves, whose frequencies are governed by (i) the size of the system and (ii) its material properties.

Acoustic waves have very important applications in studying the internal structure of the Sun, since due to the sharp density gradient under its surface the Sun acts as a resonant cavity, in which acoustic waves are trapped. Helioseismology, due to the input of a wealth of measurements obtained from the ground (GONG network) and with space instruments on the *SOHO* spacecraft (SOI/MDI and GOLF), is one of the most dynamically evolving branches of solar research. For more details see the chapter on the physics of the Sun in this book (Harra, 2003).

9.8.2   *Alfvén waves or magnetohydrodynamic waves*

As mentioned in the introduction of this chapter, Alfvén waves are physical phenomena, which exist neither in fluid mechanics nor in electromagnetism, but can be described only in their combined form, magnetohydrodynamics (MHD). The Lorenz force, $\mathbf{j} \times \mathbf{B}$, drives these waves, either along or across the field.

Alfvén waves are pictured in an incompressible ($\nabla \cdot \mathbf{v}_1 = 0$), inviscid, perfectly conducting fluid, pervaded by a uniform field $\mathbf{B}_0$, with $g = \mathbf{\Omega} = p_0 = (c_s) = 0$. In this case, in the basic wave equation only the last term remains on the right-hand side. Introducing $\hat{\mathbf{B}}_0$ as a unit vector in the direction of equilibrium magnetic field $\mathbf{B}_0$, and Θ_B as the angle between the direction of the wave propagation $\hat{\mathbf{k}}$ and the $\mathbf{B}_0$, the wave equation can be written as

$$\omega^2 \mathbf{v}_1/v_A^2 = k^2 \cos^2 \Theta_B \mathbf{v}_1 - (\mathbf{k} \cdot \mathbf{v}_1) k \cos \Theta_B \hat{\mathbf{B}}_0 + [(\mathbf{k} \cdot \mathbf{v}_1) - k \cos \Theta_B (\hat{\mathbf{B}}_0 \cdot \mathbf{v}_1)] \mathbf{k} \,.$$

The scalar product of this with $\mathbf{k}$ gives

$$(\omega^2 - k^2 v_A^2)(\mathbf{k} \cdot \mathbf{v}_1) = 0 \,,$$

which has two solutions, leading to (i) the shear and (ii) the compressional Alfvén waves. In the above equations v_A is the Alfvén speed:

$$v_A = \frac{\mathbf{B}_0}{(\mu\rho_0)^{1/2}}.$$

9.8.3 Alfvén waves

In the incompressible case ($\nabla \cdot \mathbf{v}_1 = 0$, i.e. $\mathbf{k} \cdot \mathbf{v}_1 = 0$) the positive square root of the dispersion relation gives

$$\omega = k v_A \cos \Theta_B,$$

which shows that the wave has a *phase speed* of $v_A \cos \Theta_B$, which is just the Alfvén speed along the magnetic field, and gives zero speed in a direction normal to $\mathbf{B}_0$. The *group velocity* is $\mathbf{v}_g = v_A \hat{\mathbf{B}}_0$, which implies that the energy propagates with the Alfvén speed along the direction of the magnetic field.

The total pressure is not affected by the wave; only the magnetic stress (or tension) $\mathbf{B}_0^2/2\mu$ stretching of the field lines needs to be considered. The field lines are 'frozen into' the plasma, so a unit cross-sectional area behaves like a string of mass ρ_0 per unit length stretched under the tension $\mathbf{B}_0^2/2\mu$. When a field line is 'plucked' it transmits a transverse, nondispersive wave, called an (shear) Alfvén wave, moving with the Alfvén velocity. In the solar photosphere $v_A = 10$ km s^{-1}, while it is about 1000 km s^{-1} in the corona.

The condition $\mathbf{k} \cdot \mathbf{v}_1 = 0$ implies that the velocity perturbation is perpendicular to the direction of propagation, so the waves are transverse. It can also be shown that the magnetic field perturbation $\mathbf{B}_1$ is normal to the equilibrium magnetic field $\mathbf{B}_0$. Shear Alfvén waves are polarised perpendicular to $\mathbf{B}_0$ and to the electric field of the wave. The mean energy density of the wave is equipartitioned between the motion and the magnetic field:

$$\frac{\mathbf{B}_1^2/2\mu_0}{\frac{1}{2}\rho_0\mathbf{v}_1^2} = 1.$$

The wave energy is completely reflected from perfectly conducting or insulating walls. When dissipative effects are considered in the case of small magnetic Prandtl number $P_m = v/\eta$ (v is the kinematic viscosity), the wave is attenuated over a distance $v_A\tau_\eta$, where $\tau_\eta = \lambda^2/\eta$ and λ is its wavelength. Waves on the same scale as the system ($\lambda = L$) can cross it without serious attenuation only if the Lundquist number is large, i.e. the Alfvénic time scale $\tau_A = L/v_A$ is small compared to τ_η.

9.8.4 Compressional Alfvén waves

The other solution to the magnetic wave equation gives a dispersion relation of

$$\omega = k v_A,$$

which describes the compressional Alfvén waves. Their *phase speed* is v_A, independent of the direction of propagation, while their *group velocity* is $\mathbf{v}_g = v_A \mathbf{k}$, meaning that the energy propagates isotropically.

In general, these waves lead to both density and pressure changes. In the special case where the wave propagates normal to $\mathbf{B}_0$, and thus $\mathbf{v}_1$ is parallel to $\mathbf{k}$, the wave is longitudinal and it is driven by the magnetic pressure alone. However, when the wave propagates along the magnetic field, it becomes the shear Alfvén wave, which is transverse and driven by the magnetic tension alone.

Centre-to-limb variation in the upper chromospheric, transition region and coronal lines can be used to detect Alfvén waves and distinguish them from other types of waves, like magneto-acoustic waves (see below). Erdélyi *et al.* (1998) detected line broadening in upper chromospheric and transition region lines (*SOHO*/SUMER data), what they interpreted as the effect of Alfvén waves.

9.8.5 *Magnetoacoustic waves*

When both the magnetic force and pressure gradient are important, but $g = \Omega = 0$, there are two distinct solutions of the dispersion relation for outward-propagating disturbances:

$$\frac{\omega}{k} = \left[\frac{1}{2}(c_s^2 + v_A^2) \pm \frac{1}{2}\sqrt{c_s^4 + v_A^4 - 2c_s^2 v_A^2 \cos 2\Theta_B} \right]^{1/2} .$$

The higher frequency solution corresponds to a fast magnetoacoustic wave, and that at lower frequency to slower magnetoacoustic waves. For propagation along the field ($\Theta_B = 0$), $\omega/k = c_s$ or v_A, and propagation across the field ($\Theta_B = 1/2\pi$) leads to $\omega/k = (c_s^2 + v_A^2)^{1/2}$ or 0.

The two magnetoacoustic modes can be regarded as:

- A sound wave modified by the magnetic field;
- A compressional Alfvén wave modified by the plasma pressure.

With no magnetic field, $v_A = 0$ ($B = 0$), the slow wave vanishes, and the fast wave becomes a sound wave. In vacuum, where $c_s = 0$ ($p = 0$), the slow wave vanishes, and the fast wave becomes a compressional Alfvén wave.

Observational evidence for magnetoacoustic waves propagating along coronal structures was obtained with *SOHO*/EIT, and *TRACE* (e.g. DeForest and Gurman, 1998) has renewed interest in whether MHD waves can heat coronal loops. For more discussion of their role see the chapter in this book on the physics of the Sun (Harra, 2003).

9.8.6 *Internal gravity waves*

When a plasma blob, which is in pressure equilibrium with the surrounding plasma and has adiabatic internal pressure, changes, is moved away from its equilibrium situation vertically by a distance of ∂z, where both the pressure and the density are slightly different from those of its original height ($p_0 + \partial p_0, \rho_0 + \partial \rho_0$ at height $z + \partial z$), it is subjected to a buoyancy force which leads to a harmonic motion.

Considering $\boldsymbol{\Omega} = 0$ in the basic wave equation, then taking the scalar product with $\mathbf{k}$ and $\hat{\mathbf{z}}$, supposing that N is much smaller than that of sound waves, implying that the wavelength λ is much smaller than the scale height, we get

$$\omega^2 c_s^2 \approx (\gamma - 1)g^2(1 - k_z^2/k).$$

Introducing $\Theta_g(= \cos^{-1}(k_z/k))$ as the angle between the direction of propagation and the z-axis, this can be written as

$$\omega = N \sin \Theta_g,$$

where N is the Brunt–Väisälä frequency,

$$N^2 = -g \left(\frac{1}{\rho_0} \frac{d\rho_0}{dz} + \frac{g}{c_s^2} \right).$$

Gravity waves do not propagate in the vertical direction, which would not allow interaction with the plasma at the same height, but along cones centred along the z-axis. The *z-component of the group velocity* for an upward-propagating wave is

$$v_{gz} = -\frac{\omega k_z}{k^2},$$

which is negative and implies that upward-propagating internal gravity waves carry energy downwards and vice versa. The group velocity vector is perpendicular to the surface of propagation cone of the wave.

9.8.7 *Inertial waves*

When only the Coriolis force is taken into account, an inertial wave is created with the dispersion relation

$$\omega = \pm \frac{2(\mathbf{k} \cdot \boldsymbol{\Omega})}{k},$$

which, introducing the angle between the rotation axis and propagation direction Θ_Ω becomes

$$\omega = \pm 2\Omega \cos \Theta_\Omega.$$

These waves are transverse and circularly polarised, since the velocity vector rotates about the direction of propagation as the wave moves.

9.9 The Future

This chapter has concentrated on basic MHD theory and how it can be applied to different circumstances. The linkage between theory and observation in the Sun and magnetosphere has become more realistic as fantastic new datasets are becoming available. In the future it is expected that we will be able to obtain much more information on the reconnection process through missions such as *Solar-B* that will provide high spatial observations of the magnetic field, along with accurate measurements of the corresponding plasma flows in the corona. Astrophysics missions such as *XMM-Newton* are now providing high-resolution spectral datasets from e.g. stellar flares and AGN that will be able to be compared to MHD theory.

Chapter 10

'Minimal' Relativity

Kinwah Wu

10.1 Prelude

There are many excellent textbooks on the subject of relativity, and some very good material also on the World-Wide Web. It is difficult for me to do a better job than the masters. I therefore use this opportunity to simply explore some of the wonders that I have encountered since my student years, and give a taste of the power of Einstein's theory of relativity.

I shall consider a minimal approach and start with some simple mathematics and classical physics. I shall then discuss some geometrical aspects related to relativity and end the chapter with several examples that are of astrophysical interest.

I will avoid using tensor calculus whenever possible. The special and the general theory will not be discussed separately and I will use the approach of raising questions rather than providing answers. I hope this will encourage the reader to explore the subject further.

I will adopt the following notation:

(a) Einstein's summation, i.e.

$$a_{ij}dx^i dx^j = \sum_{i=1}^{3}\sum_{j=1}^{3} a_{ij}dx^i dx^j, \quad i,j = 1,2,3, \quad \text{3-dimensional space};$$

$$g_{\mu\nu}dx^\mu dx^\nu = \sum_{\mu=0}^{3}\sum_{\nu=0}^{3} g_{\mu\nu}dx^\mu dx^\nu, \quad \mu,\nu = 0,1,2,3, \quad \text{4-dimensional space–time};$$

(b) The Landau and Liftshitz (1971) notation for the metric, Reimann and Einstein tensors, i.e.

$$g_{\mu\nu} = \begin{bmatrix} + & & & \\ & - & & \\ & & - & \\ & & & - \end{bmatrix} ;$$

Reimann: +ve ;

Einstein: +ve ;

(c) The convention '$c = G = 1$', unless otherwise stated, where c is the speed of light and G is the gravitational constant.

10.2 Some Mathematics

10.2.1 *Algebra*

Consider two positive real numbers β_1 and β_2, and β_1 and β_2 are an element of the set $[0, 1]$, i.e. β_1 and $\beta_2 \in [0, 1]$. Their algebraic sum, $\beta = \beta_1 + \beta_2$, is also a positive real number, and it can be larger than 1. Therefore, $\beta \in [0, 2]$.

Here we see that the number resulting from addition makes a new set of numbers. When we continue the operation n times, we obtain a successive 'mapping':

$$[0, 1] \rightarrow [0, 2] \rightarrow [0, 4] \rightarrow \cdots \rightarrow [0, 2^n] ,$$

with new sets being generated recursively. Consider the same numbers β_1 and β_2 again. Now, instead of using the conventional addition, we apply a different addition rule. We define an operator $\oplus$ such that

$$\beta_1 \oplus \beta_2 \equiv \frac{\beta_1 + \beta_2}{1 + \beta_1 \beta_2} ,$$

where $+$ is the conventional addition. This new operation gives a new number $\beta = \beta_1 \oplus \beta_2$ in the set $[0, 1]$. This operation maps the set $[0, 1]$ into itself.

10.2.2 *Geometry*

Consider a real variable x and a continuous smooth function $y = f(x)$. If the derivative of y,

$$y' = \frac{d}{dx} f(x) ,$$

exists for all $x \in \Re$, where $\Re$ is the set of real numbers, then for any x in $\Re$, we can find y and y', and y' is tangential to y (Figure 10.1).

We can extend this to higher dimensions. Take a curve Γ on a smooth surface $S(x_1, x_2, \ldots, x_n) = 0$ in an n-dimensional space spanned by n orthogonal vectors $\mathbf{x}_1, \mathbf{x}_2, \ldots, \mathbf{x}_n$. If at each point $\mathbf{x}_0$ along the curve Γ on S, a vector $\mathbf{t} \equiv \nabla S|_\Gamma$ exists, where ∇ is the gradient operator, then we have a set of vectors, which contains

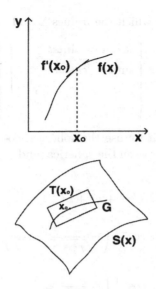

Figure 10.1   *Top*: The tangent of a curve $f(x)$ at x_0. *Bottom*: The tangential plane $T(\mathbf{x})$ of a surface $S(\mathbf{x})$ at $\mathbf{x}_0$.

all tangent vectors $\mathbf{t}$ corresponding to the curves Γ passing through the point $\mathbf{x}_0$. This set of vectors generate a plane $T(\mathbf{x}_0)$, which is tangential to the surface S (Figure 10.1).

10.2.3   *Co-ordinate transformation*

Suppose that we have a physical quantity, specified as $A(\mathbf{x})$ in a co-ordinate frame Σ and as $A'(\mathbf{x}')$ in another co-ordinate frame Σ'. How are $A(\mathbf{x})$ and $A'(\mathbf{x}')$ related?

Let us use the following as an illustration. Consider two co-ordinate frames, Σ and Σ', related by the transformation

$$\begin{bmatrix} x' \\ y' \\ z' \\ t' \end{bmatrix} = \begin{bmatrix} 1 & & & -v \\ & 1 & & \\ & & 1 & \\ & & & 1 \end{bmatrix} \begin{bmatrix} x \\ y \\ z \\ t \end{bmatrix},$$

where (x, y, z) is the space vector and t is time. The positions of an object are $\mathbf{x}$ and $\mathbf{x}'$ in the Σ and Σ' frames respectively. The velocity is the time derivative of the position, i.e. the velocities of the objects are $\mathbf{u} = \frac{d\mathbf{x}}{dt}$ and $\mathbf{u}' = \frac{d\mathbf{x}'}{dt'}$ in the two frames. It follows that

$$\begin{bmatrix} u'_x \\ u'_y \\ u'_z \end{bmatrix} = \begin{bmatrix} u_x - v \\ u_y \\ u_x \end{bmatrix}.$$

Consider another case, in which the frames Σ and Σ' are related by

$$\begin{bmatrix} x' \\ y' \\ z' \\ t' \end{bmatrix} = \begin{bmatrix} \cos\omega t & -\sin\omega t & & \\ \sin\omega t & \cos\omega t & & \\ & & 1 & \\ & & & 1 \end{bmatrix} \begin{bmatrix} x \\ y \\ z \\ t \end{bmatrix}.$$

Here, the algebra is simpler if we use the spherical co-ordinate system. For the Σ frame, the transformations between the spherical and Cartesian co-ordinate systems are

$$r = \sqrt{x^2 + y^2 + z^2}\,,$$

$$\theta = \cos^{-1}\left(\frac{z}{\sqrt{x^2 + y^2 + z^2}}\right),$$

$$\varphi = \cos^{-1}\left(\frac{x}{\sqrt{x^2 + y^2}}\right).$$

For the Σ' frame, the expression is the same. We can easily show that $\mathbf{u}'$ and $\mathbf{u}$ are related by

$$\begin{bmatrix} \dot{r}' \\ \dot{\theta}' \\ \dot{\varphi}' \end{bmatrix} = \begin{bmatrix} \dot{r} \\ \dot{\theta} \\ \dot{\varphi} - \omega \end{bmatrix}.$$

The two examples above demonstrate that transformations of the physical quantities represented in two different frames can be obtained when we know the co-ordinate transformations between the frames.

10.3   From Classical Physics to Relativity

10.3.1   *Maxwell equations*

The basis of classical electrodynamics can be summarised in the four Maxwell equations, which read

$$\nabla \cdot \mathbf{E} = 4\pi\rho\,,$$

$$\nabla \times \mathbf{E} = -\frac{1}{c}\frac{\partial}{\partial t}\mathbf{B}\,,$$

$$\nabla \cdot \mathbf{B} = 0\,,$$

$$\nabla \times \mathbf{B} = \frac{1}{c}\frac{\partial}{\partial t}\mathbf{E} + \frac{4\pi}{c}\mathbf{j}\,,$$

where $\mathbf{E}$, $\mathbf{B}$ are the electric and magnetic field strengths. The Maxwell equations can be combined to yield the wave equations.

For charge-free and current-free conditions, i.e. $\rho = 0$ and $\mathbf{j} = 0$, we have the homogeneous wave equations

$$\left[\nabla^2 - \frac{1}{c^2}\frac{\partial^2}{\partial t^2}\right](\mathbf{E},\mathbf{B}) = 0\,,$$

and the general solutions are

$$\mathbf{E} = \mathbf{f}(\mathbf{x} \pm \mathbf{c}t)\,,$$

$$\mathbf{B} = \mathbf{g}(\mathbf{x} \pm \mathbf{c}t)\,.$$

A simple form of the solution is

$$(\mathbf{E},\mathbf{B}) \sim \sin(\mathbf{k}\cdot\mathbf{x} \pm \omega t)\,,$$

which represents plane waves propagating with a speed c. The speed is related to the angular frequency ω by the wave number k; thus $ck = \omega$. The Michelson–Morley experiment showed that this speed (of light) is universal; independent of the relative motion of the observers. Thus, a co-ordinate transformation must satisfy the condition that the speed of propagation of an electromagnetic wave is invariant.

As a digression, let us examine the Maxwell Equation more carefully. Notice that even in the charge-free and current-free case the four equations are not completely symmetric; in particular, the right sides of the two inhomogeneous equations have different signs.

Imagine that the Maxwell equations take the forms

$$\nabla \cdot \mathbf{E} = 0\,,$$

$$\nabla \times \mathbf{E} = -\frac{1}{c}\frac{\partial}{\partial t}\mathbf{B}\,,$$

$$\nabla \cdot \mathbf{B} = 0\,,$$

$$\nabla \times \mathbf{B} = -\frac{1}{c}\frac{\partial}{\partial t}\mathbf{E}\,.$$

If we follow the same procedure as we use to derive the wave equations, we obtain

$$\left[\nabla^2 + \frac{1}{c^2}\frac{\partial^2}{\partial t^2}\right](\mathbf{E},\mathbf{B}) = 0\,,$$

which is a Laplace equation. Now, c is not the speed of propagating electromagnetic waves. What is the physical meaning of c in this new set of equations? Also, what would happen to the world that we live in if the Maxwell equations were to take this form?

10.3.2   *Newton's second law of motion*

In classical mechanics Newton's second law of motion states that

$$\mathbf{F} = m\mathbf{a} \,,$$

where $\mathbf{F}$ is the action force, m is the mass of the object and $\mathbf{a}$ is the acceleration. We can see that the force $\mathbf{F}$ may be an interaction between the object and another object. It can be described properly without the need to invoke a coordinate frame. The acceleration $\mathbf{a}$ is a dynamical quantity, however, and not very meaningful without a reference coordinate frame.

Suppose that we have two reference frames Σ and Σ', and the frame Σ' has an acceleration $\boldsymbol{\alpha}$ relative to the frame Σ. Let the acceleration of the object be $\mathbf{a}$ when observed in frame Σ and $\mathbf{a}'$ when observed in frame Σ'. Then we have $\mathbf{a}' = \mathbf{a} - \boldsymbol{\alpha}$, and

$$\mathbf{F} = m(\mathbf{a}' + \boldsymbol{\alpha}) \,.$$

Only when $\boldsymbol{\alpha} = 0$ is $\mathbf{F} = m\mathbf{a}'$. In order to preserve Newton's second law, the reference frames are forbidden to have relative acceleration. These 'non-accelerating' reference frames are called inertial frames.

This is not at all a satisfactory situation. Why does nature pick these reference frames preferentially? More seriously, with respect to what frame is the acceleration of a reference frame measured? The need for an inertial reference frame led Newton to introduce the concept of 'absolute space', so that he could define the inertial frames as the reference frames that have no acceleration with respect to the absolute space.

10.3.3   *Rotating-bucket experiment*

Newton proposed an experiment to demonstrate the existence of the absolute space. The apparatus in the experiment is a bucket half-filled with water. The water surface is flat when the bucket is at rest, which is obvious. When we rotate the bucket, the water is dragged to rotate and its surface becomes curved (Figure 10.2). Newton

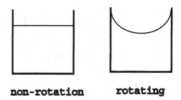

non-rotation   **rotating**

Figure 10.2   *Left*: The water has a flat surface when the bucket is at rest with respect to the absolute space. *Right*: The water surface is curved when the water in the bucket is rotating, i.e. the water has an acceleration relative to the absolute space.

proposed that we could distinguish whether or not we have acceleration with respect to the absolute space simply by observing the surface of the water.

The rotating-bucket experiment seems to provide a means of determining the acceleration relative to the absolute space, but it has not defined the absolute velocity. There is also a philosophical difficulty with the absolute space. If absolute space is an object of reality then we agree that its existence affects the dynamics of objects in the universe. However, this definition also requires that the dynamics of the objects have no effects on it at all. It is therefore a one-way relation. Although such a relation is not forbidden mathematically, it is unsatisfactory in the perspective of physics.

Mach criticised Newton's concept of absolute space, and argued that the presence of inertia is a consequence of the body interacting with the rest of the Universe (Mach's principle). In Mach's opinion, there is no absolute space, no absolute acceleration and no absolute speed.

10.3.4 *Newton's law of gravity*

In Newton's law of gravity, the force is related to the object mass m (see Figure 10.3):

$$\mathbf{F} = -G \sum_{i=1}^{n} \frac{m m_i (\mathbf{r} - \mathbf{r}_i)}{|\mathbf{r} - \mathbf{r}_i|^3}.$$

This is different to electromagnetic and nuclear forces, which are not proportional to the object mass. Newton's second law of motion states that $\mathbf{F} = m\mathbf{a}$. The second law also implies that the force $\mathbf{F}$ is related to the mass of the object. Combining Newton's law of gravity and his second law of motion yields

$$\mathbf{a} = -G \sum_{i=1}^{n} \frac{m_i (\mathbf{r} - \mathbf{r}_i)}{|\mathbf{r} - \mathbf{r}_i|^3}.$$

The expression above indicates that acceleration due to gravity is independent of the mass of the object. As a result, objects with different masses will have the same trajectory when the initial velocity is specified.

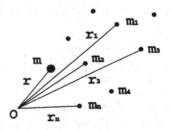

Figure 10.3 The gravitational interaction between an object of mass m located at $\mathbf{r}$ and other objects of masses $m_1, m_2, m_3, \ldots$ located at $\mathbf{r}_1, \mathbf{r}_2, \mathbf{r}_3, \ldots$ respectively.

Let us take a step backwards and consider the following question: Are the masses m in $\mathbf{F} = m\mathbf{a}$ and in $\mathbf{F} = -G \sum m m_i(\mathbf{r} - \mathbf{r}_i)/|\mathbf{r} - \mathbf{r}_i|^3$ the same? If they are not, we have difficulty justifying that acceleration due to gravity is independent of the mass of the object.

Suppose that the masses are not the same. Now Newton's second law states that

$$\mathbf{F} = m_{\mathrm{I}}\mathbf{a},$$

and the law of gravity states that

$$\mathbf{F} = -G \sum_{i=1}^{n} \frac{m_{\mathrm{G}}^{\mathrm{P}} m_{\mathrm{G}i}^{\mathrm{A}}(\mathbf{r} - \mathbf{r}_i)}{|\mathbf{r} - \mathbf{r}_i|^3}.$$

We call m_{I} inertial mass, and $m_{\mathrm{G}}^{\mathrm{P}}$ and $m_{\mathrm{G}}^{\mathrm{A}}$ the passive and active gravitational masses respectively. Can we show that the three masses m_{I}, $m_{\mathrm{G}}^{\mathrm{P}}$ and $m_{\mathrm{G}}^{\mathrm{A}}$ of an object are equivalent? In other words, are the ratios $m_{\mathrm{I}}/m_{\mathrm{G}}^{\mathrm{P}}$ and $m_{\mathrm{G}}^{\mathrm{A}}/m_{\mathrm{G}}^{\mathrm{P}}$ constant?

10.3.5   *Galileo's experiment and the principle of equivalence*

To investigate the different masses, let us consider an experiment by Galileo. He released objects of different kinds from the same height and measured the time the object took to reach the ground (Figure 10.4). The gravitational force of the Earth accelerates an object released from a height downwards. From Newton's laws, we have

$$m_{\mathrm{I}}a = m_{\mathrm{G}}^{\mathrm{P}} \left[\frac{G m_{\mathrm{G}\oplus}^{\mathrm{A}}}{R_{\oplus}^2} \right],$$

where $m_{\mathrm{G}\oplus}^{\mathrm{A}}$ and $R_{\oplus}$ are the gravitational mass and radius of the Earth respectively. Hence, the acceleration is

$$a = \left(\frac{m_{\mathrm{G}}^{\mathrm{P}}}{m_{\mathrm{I}}} \right) \left[\frac{G m_{\mathrm{G}\oplus}^{\mathrm{A}}}{R_{\oplus}^2} \right].$$

For an object dropped from a height h (Figure 10.4), the free-fall time t to reach the ground is given by

$$t = \sqrt{2h \left(\frac{m_{\mathrm{I}}}{m_{\mathrm{G}}^{\mathrm{P}}} \right) \left[\frac{G m_{\mathrm{G}\oplus}^{\mathrm{A}}}{R_{\oplus}^2} \right]^{-1}},$$

Objects with the same ratio $m_{\mathrm{I}}/m_{\mathrm{G}}^{\mathrm{P}}$ therefore take the same time to fall to the ground. Galileo demonstrated that different objects do indeed take the same time to fall, implying that the inertial mass and gravitational mass of an object are equivalent, that is $m_{\mathrm{I}} \propto m_{\mathrm{G}}^{\mathrm{P}}$. How about $m_{\mathrm{G}}^{\mathrm{A}}$ and $m_{\mathrm{G}}^{\mathrm{P}}$? For two objects (of gravitational

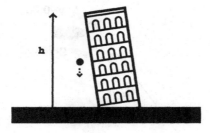

Figure 10.4   An illustration of Galileo's experiment. Objects of different mass were released at the same height. The time that it took for them to reach the ground was measured and shown to be the same.

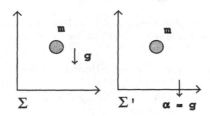

Figure 10.5   *Left*: An object of mass m is in free fall under gravity when observed in the frame Σ. *Right*: The same object is observed in a reference Σ', which has acceleration $\boldsymbol{\alpha}$ equal to the gravitational acceleration with respect to the frame Σ.

masses m_{G1} and m_{G2} respectively) the gravitational force on object 1 due to the presence of object 2 is

$$\mathbf{F}_{2 \to 1} = -G m_{G1}^{P} m_{G2}^{A} \frac{\mathbf{r}_1 - \mathbf{r}_2}{|\mathbf{r}_1 - \mathbf{r}_2|^3}.$$

The gravitational force on object 2 due to the presence of object 1 is

$$\mathbf{F}_{1 \to 2} = -G m_{G2}^{P} m_{G1}^{A} \frac{\mathbf{r}_2 - \mathbf{r}_1}{|\mathbf{r}_2 - \mathbf{r}_1|^3}.$$

Recall Newton's third law, $\mathbf{F}_{2 \to 1} = -\mathbf{F}_{1 \to 2}$. Therefore we have

$$\frac{m_{G1}^{P}}{m_{G1}^{A}} = \frac{m_{G2}^{P}}{m_{G2}^{A}} = \kappa.$$

Setting the proportionality constant $\kappa = 1$ yields $m_G^A = m_G^P$.

An important consequence of $m_I = m_G$ is that choosing a reference frame properly can eliminate the gravitational force. The acceleration $\mathbf{a}$ of an object falling under gravity $\mathbf{g}$, in the reference frame Σ (Figure 10.5), is given by

$$m_I \mathbf{a} = m_G \mathbf{g} + \mathbf{f}_{ext},$$

where $\mathbf{f}_{ext}$ is an external force. When we observe the object in another reference frame Σ', which has an acceleration $\boldsymbol{\alpha} = \mathbf{g}$ relative to the frame Σ (Figure 10.5),

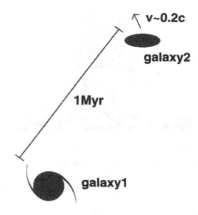

Figure 10.6   Two galaxies, 1 and 2, are separated by a distance of a million light years. Galaxy 2 is moving at a speed of 20% of the speed of light relative to galaxy 1.

the acceleration $\mathbf{a}'$ is given by $\mathbf{a}' = \mathbf{a} - \mathbf{g}$. Hence, we have

$$m_{\text{I}}(\mathbf{a} - \mathbf{g}) = m_{\text{G}}\mathbf{g} + \mathbf{f}_{\text{ext}},$$

Because of the equivalence of m_{I} and m_{G} we can eliminate the gravity terms to obtain

$$m_{\text{I}}\mathbf{a}' = \mathbf{f}_{\text{ext}}.$$

This shows that in a gravitational field, we can always find a local inertial reference frame such that all interactions are the same as in the absence of gravity. As a result, we cannot distinguish an inertial frame and a frame in free fall (principle of equivalence).

10.3.6   *Some thoughts*

10.3.6.1   *Newton's third law of motion*

Consider two objects separated by a large distance — say, two galaxies separated by one million light years. Suppose also that galaxy 2 is moving with a large speed, about 20% the speed of light, with respect to galaxy 1 (Figure 10.6). Imagine that we are residents of galaxy 1. We can only know the position of galaxy 2 a million years ago but not its present location. Are we still sure that $\mathbf{F}_{1\to 2}$ is along $\mathbf{r}_1 - \mathbf{r}_2$ and $\mathbf{F}_{2\to 1}$ is along the direction $\mathbf{r}_2 - \mathbf{r}_1$? If we cannot ensure that $\mathbf{F}_{1\to 2} = -\mathbf{F}_{2\to 1}$, can we still have $m_{\text{G}}^{\text{A}}/m_{\text{G}}^{\text{P}}$ be a constant?

10.3.6.2   *Negative masses?*

The conclusion that $m_{\text{I}} = m_{\text{G}}^{\text{P}} = m_{\text{G}}^{\text{A}}$ requires that the three masses must be proportional to each other, but it does not forbid the masses to take negative

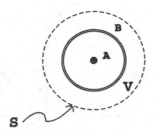

Figure 10.7   A spherical shell B made of material that has negative mass. A spherical body A made of material that has positive mass is placed at the centre of the shell. S is an imaginary surface enclosing the shell, and V is the volume bounded by the surface.

values. Newton's law of gravity tells us that two objects, with masses m_1 and m_2, are attracted to each other and that their accelerations are in the directions of the gravitational forces acting upon them. Imagine that the mass of object 2 is negative, i.e. $m_2 < 0$, but the mass of object 1 is positive. The forces on the objects become repulsive. Object 1 is accelerated along the same direction as the repulsive force. Because object 2 has a negative mass, its acceleration is in the opposite direction to the force on it. We end up with an interesting scene in which object 2 is chasing wildly after object 1!

What else do we deduce if the mass of an object can be negative? Construct a device (Figure 10.7) consisting of a spherical hollow shell B made of material with negative mass. Inside the shell there is a spherical object A made of material with positive mass. Now consider a fictitious spherical surface S, which encloses the device and the body, and let the volume of the region bounded by the surface S be V.

From Newton's law of gravity, we have

$$\frac{1}{4\pi} \int_v d^3x \ \nabla \cdot \mathbf{g} = -G \sum_i m_i \,,$$

where m_i are all the masses within the volume V. Using the divergence theorem, we obtain

$$\oint_s d\mathbf{a} \cdot \mathbf{g} = -G \sum_i m_i \,.$$

If we choose the mass of the shell such that it equals the mass of the body inside, then $\mathbf{g} = 0$. Hence, we shield the gravitational force, in the same way as we can shield the electric field of charge particles.

Does this mean that the inertial frame is no longer confined to be local and that we can construct a more global inertial frame? Also, what happens to Mach's principle now? Can we still define the inertia in terms of the interaction with the universe?

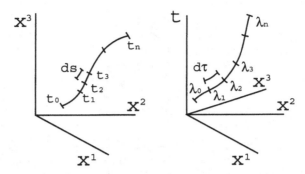

Figure 10.8   *Left*: The motion of a particle in a three-dimensional space. The position of the particle is specified by the time parameter t. The separation of two spatial points along the particle trajectory is ds. *Right*: The motion of a particle in a four-dimensional space–time as a consequence of events. The events are specified by a parameter λ. The interval between two events is $d\tau$.

10.4   Geometrical Aspect

10.4.1   *Metrics*

Consider a co-ordinate system $x^\mu = \{x^0, x^1, x^2, \ldots, x^n\}$, and a parameter λ. We can 'label' the motion of a particle by a sequence of events $x^\mu(\lambda_1)$, $x^\mu(\lambda_2)$, $x^\mu(\lambda_3)$, $\ldots$. In the Newtonian world, we need only a co-ordinate system in three dimensions (Figure 10.8). As space and time are not related, it is convenient to choose time t as the labelling parameter. The spatial separation of the places where two events occur is

$$ds^2 = g_{ij} dx^i dx^1 \,,$$

where g_{ij} is called the metric. When the Cartesian co-ordinate (x, y, z) system is used, the metric is

$$g_{ij} = \begin{bmatrix} 1 & & \\ & 1 & \\ & & 1 \end{bmatrix} .$$

If the spherical co-ordinate (r, θ, φ) system is used, the metric is

$$g_{ij} = \begin{bmatrix} 1 & & \\ & r^2 & \\ & & r^2 \sin^2 \theta \end{bmatrix} .$$

We can also construct a four-dimensional Euclidean space when we include time in the co-ordinate system (Figure 10.8). The Minkowski space is an example. In the

Cartesian co-ordinates an event is specified by $x^\mu = (t, x, y, z)$, and the metric is

$$g_{\mu\nu} = \begin{bmatrix} 1 & & & \\ & -1 & & \\ & & -1 & \\ & & & -1 \end{bmatrix}.$$

The interval $d\tau$ between two events is given by

$$d\tau^2 = dt^2 - dx^2 - dy^2 - dz^2.$$

In the spherical co-ordinate system the metric is

$$g_{\mu\nu} = \begin{bmatrix} 1 & & & \\ & -1 & & \\ & & -r^2 & \\ & & & -r^2 \sin^2\theta \end{bmatrix},$$

and the interval is

$$d\tau^2 = dt^2 - [dr^2 + r^2(d\theta^2 + \sin^2 -d\varphi^2)].$$

Unlike the distance ds^2 in the three-dimensional space of the Newtonian world, $d\tau^2$ in the Minkowski space can take either positive or negative values. For $d\tau^2 > 0$, the interval separating the two events is time-like, and the two events may have a causal relation. For $d\tau^2 < 0$, it is space-like, and the two events do not have causal consequences on each other. The paths along which light propagates are those with $d\tau^2 = 0$.

10.4.2 *Extremum*

The separation between two points in a three-dimensional flat space is a minimum. It can be shown as follows. Consider two points A and B in a reference frame Σ. The separation is an invariant, and we can rotate the co-ordinates in such a way that in a new reference frame Σ' both points are on the y'-axis (Figure 10.9). We can see that for any point P in Σ',

$$S_{AB} \leq S_{AP} + S_{PB},$$

and S_{AB} is the minimum route from A to B.

Is an interval in the Minkowski space also a minimum? Select a co-ordinate frame in which the points A and B are on the time axis (Figure 10.10). The interval between A and B is then given by $\tau_{AB}^2 = t_B^2$. For any point P, we have

$$\tau_{AP}^2 = t_P^2 - x_P^2,$$

$$\tau_{PB}^2 = (t_B - t_P)^2 - x_P^2.$$

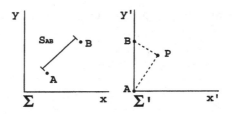

Figure 10.9   *Left*: The separation of two spatial points A and B in a reference Σ. *Right*: In a new reference frame, Σ', the two points are both on the y-axis. P is an arbitrary point in the space.

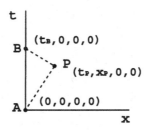

Figure 10.10   The interval between two points, A and B, in a four-dimensional Minkowski space. The reference frame is chosen such that both points are on the t-axis.

It follows that

$$\tau_{AP} + \tau_{PB} = \sqrt{t_P^2 - x_P^2} + \sqrt{(t_B - t_P)^2 - x_P^2} \le \tau_{AB} .$$

In the Minkowski space, the 'straight line' between two events is therefore a maximum, not a minimum.

10.4.3   *Geodesics and world lines*

A world line, $x^\mu(\tau)$, can be regarded as a continuous sequence of events, and the interval between two infinitesimally close events is given by

$$d\tau^2 = g_{\mu\nu} dx^\mu dx^\nu .$$

The interval between two specific events, A and B, is

$$\tau_{AB} = \int_A^B \sqrt{g_{\mu\nu} dx^\mu dx^\nu}$$

$$= \int_A^B d\tau \sqrt{g_{\mu\nu}[x^\sigma(\tau)] \frac{dx^\mu(\tau)}{d\tau} \frac{dx^\nu(\tau)}{d\tau}} .$$

The geodesics are extrema, and can be obtained by considering the variation $\delta\tau_{AB} = 0$.

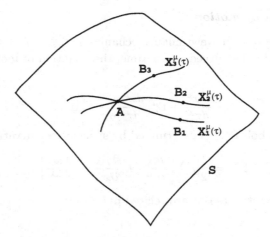

Figure 10.11   There can be many geodesics, $x_i^\mu(\tau)$, passing through a point A on a surface S.

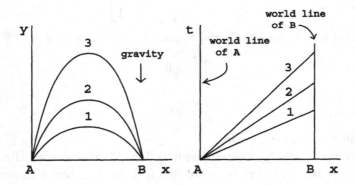

Figure 10.12   *Left*: Three particles are projected from point A to point B with different initial velocities. The parabolic curves 1, 2 and 3 represents trajectories of the particles. *Right*: The straight lines 1, 2 and 3 are the world lines corresponding to the trajectories 1, 2 and 3 of the particles. The world lines of point A and B are also shown.

There can be many geodesics, $x_1^\mu(\tau)$, $x_2^\mu(\tau)$, $x_3^\mu(\tau)$, ..., passing through a point A (an event) on the surface S (Figure 10.11), provided that the geodesics satisfy the variation condition $\delta\tau_{AB_i} = 0$ $(i = 1, 2, 3, ...)$. The particular geodesics with τ_{AB_i} are called the null geodesics. In Einstein's theory of relativity, the world lines of particles in free fall are time-like geodesics. The world lines of massless particles, e.g. photons, are null geodesics.

We must not confuse the world lines in the space–time with the trajectories of particle in a three-dimensional space. Keep in mind that there can be many free-fall trajectories between two points in a three-dimensional space for a particle, but there is only one world line representing the motion in gravity for two specific events (see the illustration in Figure 10.12).

10.4.4 *Equation of motion*

The equation of motion in Newtonian mechanics is $\frac{d^2x}{dt^2} - f = 0$, where f is the acceleration force. In the curved space–time, the equation of motion derived from the geodesics is

$$\frac{d^2x^\mu}{d\tau^2} + \Gamma^\mu_{\nu\sigma}\frac{dx^\nu}{d\tau}\frac{dx^\sigma}{d\tau} = 0\,.$$

The Christoffel symbol, $\Gamma^\mu_{\nu\sigma}$, is determined by space–time curvature as

$$\Gamma^\mu_{\nu\sigma} = \frac{1}{2}g^{\mu\lambda}\left[\frac{\partial g_{\lambda\sigma}}{\partial x^\nu} + \frac{\partial g_{\lambda\nu}}{\partial x^\sigma} - \frac{\partial g_{\nu\sigma}}{\partial x^\lambda}\right]\,,$$

where $g^{\mu\nu}$ and the metric tensor are related by

$$g^{\mu\lambda}g_{\lambda\nu} = \delta^\mu_\nu = \begin{bmatrix} 1 & & & \\ & 1 & & \\ & & 1 & \\ & & & 1 \end{bmatrix}\,.$$

It can be shown that in a local inertial frame

$$\frac{d^2t}{d\tau^2} = 0\,,$$

$$\frac{d^2\mathbf{r}}{d\tau^2} = 0\,,$$

the same as in the Newtonian world.

Does the term $\Gamma^\mu_{\nu\sigma}\frac{dx^\nu}{d\tau}\frac{dx^\sigma}{d\tau}$ correspond to a 'force'? Also, can we eliminate this force when the space–time is globally not flat, and the $\frac{\partial g_{\lambda\sigma}}{\partial x^\nu}$ terms do not vanish?

10.4.5 *Curved space–time*

To illustrate the concept of curvature, we can use a plane, a spherical surface and a cylindrical surface, which are two-dimensional space. Most of us will correctly conclude that the plane is flat and that the spherical surface is curved. We may perhaps have difficulty in telling whether the cylindrical surface is curved or flat. Although the cylindrical surface looks curved to us, it is actually flat. We therefore need a method to distinguish curved and flat surfaces.

Gauss proposed a scheme as follows. First, construct a circle C with a radius a on the surface. Measure its circumference c. Then define a (Gaussian) curvature

$$k \equiv \lim_{a\to0}\frac{3}{\pi}\left(\frac{2\pi a - c}{a^3}\right)\,.$$

If k is non-zero, the surface is curved. Now we easily find that $k = 0$ for both the plane and the cylindrical surface. The spherical surface is intrinsically curved since

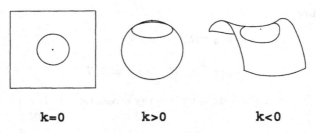

$$k=0 \qquad\qquad k>0 \qquad\qquad k<0$$

Figure 10.13 Examples of two-dimensional surfaces (a plane, a spherical surface and a saddle surface) with different Gaussian curvatures.

the circumferences of circles on it are smaller than those we would expect for circles in a plane, i.e. $k > 0$. We can also show that a saddle surface (Figure 10.13) has a negative curvature.

The general form of the Gaussian curvature is

$$k = \frac{1}{2g_{11}g_{22}} \left\{ \frac{\partial^2 g_{11}}{\partial (x^2)^2} - \frac{\partial^2 g_{22}}{\partial (x^2)^2} + \frac{A_1}{2g_{11}} + \frac{A_2}{2g_{22}} \right\},$$

where

$$A_1 = \left[\frac{\partial g_{11}}{\partial x^1} \frac{\partial g_{22}}{\partial x^1} + \left(\frac{\partial g_{11}}{\partial x^2} \right)^2 \right],$$

$$A_2 = \left[\frac{\partial g_{11}}{\partial x^2} \frac{\partial g_{22}}{\partial x^2} + \left(\frac{\partial g_{22}}{\partial x^1} \right)^2 \right].$$

For multi-dimensional spaces, curvature is specified by a set of tensors, known as the Reimann tensors. The Reimann tensors are defined as

$$R_{\mu\nu\alpha\beta} \equiv g_{\mu\lambda} R^{\lambda}_{\nu\alpha\beta},$$

$$R^{\mu}_{\nu\alpha\beta} \equiv \frac{\partial}{\partial x^{\alpha}} \Gamma^{\mu}_{\nu\beta} - \frac{\partial}{\partial x^{\beta}} \Gamma^{\mu}_{\nu\alpha} + \Gamma^{\mu}_{\lambda\alpha} \Gamma^{\lambda}_{\nu\beta} - \Gamma^{\mu}_{\lambda\beta} \Gamma^{\lambda}_{\nu\alpha}.$$

For a space with n dimensions, there are $n^2(n^2 - 1)/12$ independent terms in the Reimann tensor. Thus, for a surface, which is a two-dimensional space, we need only one quantity, the Gaussian curvature, to describe the space curvature.

10.4.6   Einstein's field equation

From the Reimann tensor, we can construct the Ricci tensor

$$R_{\mu\nu} = g^{\alpha\beta} R_{\mu\alpha\nu\beta}$$

and the curvature scalar

$$R = g^{\mu\nu} R_{\mu\nu},$$

$$R_{\mu\nu} - \frac{1}{2} g_{\mu\nu} R - \lambda g_{\mu\nu} = 8\pi T_{\mu\nu}.$$

The tensor $T_{\mu\nu}$ on the right side is the energy stress tensor. For a perfect gas, it is given by

$$T^{\mu\nu} = pg^{\mu\nu} + (p + \rho) \frac{dx^\mu}{d\tau} \frac{dx^\nu}{d\tau},$$

where p is the pressure and ρ is the matter density.

Einstein's field equation is more general than the Poisson equation

$$\nabla^2 \Phi = 4\pi\rho$$

in the Newtonian formulation. The field equation describes how the presence of matter warps the space–time, which is clearly seen by inspecting the terms on the two sides of the equation.

10.5   Astrophysical Examples

10.5.1   *Schwarzschild metric and non-rotating black holes*

To understand quantitatively how matter curves the space–time, we must solve Einstein's field equation. For space–time that is spherical, symmetric and non-rotating, the interval is given by

$$d\tau^2 = f(r)dt^2 - [g(r)dr^2 + r^2 d\Omega],$$

where $d\Omega = d\theta^2 + \sin^2\theta d\varphi^2$. For flat space–time, we have $f(r) = g(r)^{-1} = 1$. If there is a point mass m at the origin ($\mathbf{r} = 0$), then

$$d\tau^2 = \left(1 - \frac{2m}{r}\right) dt^2 - \left[\left(1 - \frac{2m}{r}\right)^{-1} dr^2 + r^2 d\Omega\right],$$

and the metric is

$$g_{\mu\nu} = \begin{bmatrix} \left(1 - \dfrac{2m}{r}\right) & & & \\ & -\left(1 - \dfrac{2m}{r}\right)^{-1} & & \\ & & -r^2 & \\ & & & -r^2 \sin^2\theta \end{bmatrix}.$$

This is the well-known Schwarzschild metric.

There is a singularity at $r_s \equiv 2m$, the Schwarzschild radius. When $r > 2m$, $d\tau^2 > 0$, and the interval is time-like. When $r < 2m$, $d\tau^2 < 0$, and the interval is space-like. At $r = 2m$, there is a boundary called the event horizon, which separates the two regions where $d\tau^2$ has different properties. It is also where radiation suffers infinite redshift. Radiation emitted from inside the event horizon will not reach an observer outside the horizon.

Note that the singularity at the Schwarzschild radius is in fact fictitious. It is a consequence of the co-ordinate system that we use in the representation. This is analogous to the north and south poles in the spherical co-ordinate system. The two poles do not correspond to any true physical singularities at all. This singularity at the Schwarzschild radius can be eliminated by using an appropriate co-ordinate system, e.g. the Kruskal co-ordinates.

The Schwarzschild metric is useful in describing phenomena near compact astrophysical objects with strong gravity. It is particularly essential in the study of radiation from matter in the vicinity of black holes, where the space–time is severely curved. A black hole is formed when the forces of nature cannot provide enough support to maintain the structure of the object. The object thus implodes and is buried below the surface defined by the Schwarzschild radius. After the implosion, the object is no longer visible to a distant observer.

10.5.2   Kerr metric and rotating black holes

The Schwarzschild metric describes a type of black hole that does not rotate (i.e. with zero angular momentum). Black holes in the universe are more likely to rotate. The metric for rotating black holes is the Kerr metric,

$$d\tau^2 = \left(1 - \frac{2mr}{\Sigma}\right)dt^2 + 4\left(\frac{amr}{\Sigma}\right)\sin^2\theta dt d\varphi$$

$$- \frac{\Sigma}{\Delta}dr^2 - \Sigma d\theta^2 - \left[r^2 + a^2 + 2\left(\frac{a^2 mr}{\Sigma}\right)\sin^2\theta\right]\sin^2\theta d\varphi^2$$

(in Boyer–Linquist co-ordinates), where

$$\Delta = r^2 - 2mr + a^2 \,,$$

$$\Sigma = r^2 + a^2\cos^2\theta \,,$$

a being the angular momentum.

Notice that the t and φ terms are fixed together in the Kerr metric, and that the space and time are 'mixed'. The rotation of the black hole drags the space–time in its vicinity, causing it to rotate. This is known as the inertial-frame dragging. We illustrate schematically the frame dragging using the trajectories of photons around a rotating and a non-rotating black hole (Figure 10.14).

Figure 10.14   A schematic illustration of inertial-frame dragging. The black hole on the left is non-rotating and the one on the right is rotating. As viewed by a distant observer, the trajectory of a photon changes direction before falling into a rotating black hole if the photon and the black hole have angular momenta with opposite directions.

The general metric for black holes is the Reissner–Nordstrom metric, which takes account of the three essential parameters of black holes (the mass, the angular momentum and the charge) allowed by the 'no-hair theorem'. The expression of the Reissner–Nordstrom metric can be found in textbooks and is not given here. Charged black holes should be rare in astrophysical environments, as they can be neutralised quickly by the opposite-charged particles attracted to the hole.

10.5.3   *How black is a black hole?*

Recall that light is deflected by the presence of a gravitating body. When we shine a torch beam towards a Schwarzschild black hole, will the light disappear in the hole? In fact light can circulate around a black hole without falling into it. For a Schwarzschild black hole, the radius at which photons have a circular orbit is $r = 3M$. Thus, if we aim at the black hole properly, the beam will return to us, and we shall see a circular ring around the hole. In comparison, if we were to shine the beam onto a perfectly reflecting sphere, we would see only a bright point. In this sense, therefore, a black hole is the better 'reflector'.

One of the characteristics of black holes is their event horizon. The event horizon of a charge-free rotating black hole is located at

$$r_+ = M + \sqrt{M^2 - a^2}\,.$$

When $a = 0$, we obtain the Schwarzschild radius $r = 2m$. The area of the event horizon is given by

$$A = 4\pi(r_+^2 + a^2)\,.$$

Hawking showed that the area of the event horizon of a black hole never decreases and can only increase. The area of the event horizon is in fact related to the entropy of the black hole by

$$S = \frac{1}{4}kA\,,$$

where k is the Boltzmann constant. Black holes turn out to be not as black as we think, but they are a perfect blackbody radiator with a temperature

$$T = \frac{1}{2\pi k}\left[\frac{\sqrt{M^2 - a^2}}{2M(M + \sqrt{M^2 - a^2})}\right].$$

10.5.4   Least-action principle

The least-action principle states that 'of all the possible paths along which a dynamical system may move from one point to another in configuration space within a specified time interval, the path is that which minimises the time integral of the Lagrangian function of the system'. Let $f(x, \dot{x}; t)$ be the Lagrangian function and

$$J = \int_{t_1}^{t_2} dt \quad f(x, \dot{x}; t).$$

The requirement of $\delta J = 0$ for the extremum implies that

$$\frac{\partial f}{\partial x} - \frac{d}{dt}\frac{\partial f}{\partial \dot{x}} = 0,$$

which is the Euler–Lagrange equation.

A central theme in general relativity is that the trajectories of free particles in space–time are geodesics. To determine the equation of motion is therefore a matter of finding an extremum as in the least-action principle.

For the principle of extreme length, we have $\delta\tau = 0$, where

$$\tau = \int_a^b d\tau \sqrt{g_{\mu\nu}\frac{dx^\mu}{d\tau}\frac{dx^\nu}{d\tau}}.$$

If we replace the geometric extremal principle by the dynamic extremal principle, then we have

$$I = \frac{1}{2}\int_A^B d\lambda \left(g_{\mu\nu}\frac{dx^\mu}{d\lambda}\frac{dx^\nu}{d\lambda}\right)$$

and

$$\frac{\partial \mathcal{L}}{\partial x^\mu} - \frac{d}{d\lambda}\left(\frac{\partial \mathcal{L}}{\partial \dot{x}^\mu}\right) = 0,$$

where the Lagrangian is

$$\mathcal{L} = \frac{1}{2}g_{\mu\nu}\dot{x}^\mu\dot{x}^\nu,$$

$\dot{x}^\mu$ being $\frac{dx^\mu}{d\lambda}$, and λ are affine parameters (see also Figure 10.8). It follows that

$$g_{\sigma\nu}\frac{d^2 x^\nu}{d\lambda^2} + \frac{1}{2}\left[\frac{\partial g_{\sigma\nu}}{\partial x^\mu} + \frac{\partial g_{\sigma\mu}}{\partial x^\nu} - \frac{\partial g_{\mu\nu}}{\partial x^\sigma}\right]\frac{dx^\mu}{d\lambda}\frac{dx^\nu}{d\lambda} = 0,$$

and hence the equation of motion

$$\frac{d^2 x^\sigma}{d\lambda^2} + \Gamma^\sigma_{\mu\nu} \frac{dx^\mu}{d\lambda} \frac{dx^\nu}{d\lambda} = 0 \,.$$

10.5.5   *Conservation laws and equations of motion*

Using the Lagrangian formulation we can derive the equation of motion of particles near a black hole. The Lagrangian for the Schwarzschild metric is

$$\mathcal{L} = \frac{1}{2} \left[\left(1 - \frac{2m}{r}\right) \dot{t}^2 - \left(1 - \frac{2m}{r}\right)^{-1} \dot{r}^2 - r^2 \dot{\theta}^2 - r^2 \sin^2 \theta \dot{\varphi}^2 \right] \,.$$

It follows that in the t co-ordinate,

$$\frac{d}{d\lambda} \left[\dot{t} \left(1 - \frac{2m}{r}\right) \right] = 0 \,,$$

which yields

$$E = \dot{t} \left(1 - \frac{2m}{r}\right) \,,$$

where the constant E is the particle energy. For the azimuth co-ordinate φ, we have

$$\frac{d}{d\lambda} [\dot{\varphi} r^2 \sin^2 \theta] = 0 \,,$$

and hence

$$L = r^2 \dot{\varphi} \sin^2 \theta \,,$$

where L is the angular momentum of the particle. We can derive the equation of motions for r and θ using similar variation procedures.

The Lagrangian of the Kerr metric is

$$\mathcal{L} = \frac{1}{2} \left[\left(1 - \frac{2mr}{\Sigma}\right) \dot{t}^2 + \frac{4amr \sin^2 \theta}{\Sigma} \dot{t} \dot{\varphi} - \frac{\Sigma}{\Delta} \dot{r}^2 - \Sigma \dot{\theta}^2 \right.$$
$$\left. - \left(r^2 + a^2 + \frac{2a^2 mr \sin^2 \theta}{\Sigma}\right) \sin^2 \theta \dot{\varphi}^2 \right] \,.$$

The corresponding equations of motion are

$$\dot{t} = \frac{[\Sigma(r^2 + a^2) + 2a^2 mr \sin^2 \theta] E - 2amrL}{(\Sigma - 2mr)(r^2 + a^2) + 2a^2 mr \sin^2 \theta} \,,$$

$$\dot{r}^2 = \frac{\Delta}{\Sigma} [E\dot{t} - L\dot{\varphi} - \Sigma \dot{\theta}^2) \,,$$

$$\dot{\theta}^2 = \frac{1}{\Sigma^2}[Q + a^2\cos^2\theta E - \cot^2\theta L^2),$$

$$\dot{\varphi} = \frac{2amr\sin^2\theta E + (\Sigma - 2mr)L}{(\Sigma - 2mr)(r^2 + a^2)\sin^2\theta + 2a^2mr\sin^4\theta},$$

where

$$Q = \Sigma^2\dot{\theta}^2 - a^2\cos^2\theta E + \cot^2\theta L^2$$

is the Carter constant.

We can use it to derive the equations of motion for particles in a Keplerian (geometrically thin, pressure-free) disc around a rotating black hole. What we need is to set $\theta = \pi/2$ in the Lagrangian for the Kerr metric, i.e.

$$\mathcal{L} = \frac{1}{2}\left[\left(1 - \frac{2mr}{r}\right)\dot{t}^2 + \frac{4amr}{r}\dot{t}\dot{\varphi} - \frac{r^2}{\Delta}\dot{r}^2 - \left(r^2 + a^2 + \frac{2a^2m}{r}\right)\dot{\varphi}^2\right]$$

and consider the variation. The two resulting equations are

$$\dot{t} = \frac{[r^3 + a^2r + 2a^2m]E - 2amL}{r\Delta},$$

$$\dot{\varphi} = \frac{2amE + (r - 2m)L}{r\Delta}.$$

For further information on the subject matter of this chapter, see Fang and Ruffini (1983), Landau and Lifshitz (1971), Luminet (1992), and Misner, Wheeler and Thorne (1973).

Chapter 11

Cosmology

Mat Page

11.1 Introduction

Cosmology is the science of the Universe as a whole: its contents, its shape, its history and its fate. As a subject, cosmology has a huge scope, and draws on virtually every branch of physics — from quantum mechanics, which describes the Universe at the sub-atomic level, to general relativity, which describes the behaviour of the Universe on the largest imaginable scales.

In this chapter I will begin with a brief history of cosmology as a science, followed by the axioms that are applied to investigate cosmology, and the limitation this imposes on scientific investigations. I will provide a brief history of the Universe according to current wisdom, from the initial singularity, through the radiation-dominated inferno to the present-day, matter-dominated era. I will then review the contents of the Universe, both known and suspected. I will discuss the formation of structures, the crucial role played by gravity, and the models which seek to explain this process. I will discuss the cosmic background radiation, and the observational data that are now being accumulated to understand the history of the Universe. The importance of recent data at X-ray and far infrared wavelengths will be emphasised for its ability to tell us about the role of massive black holes in the formation of galaxies. Finally, I will describe some of the exciting new results we can expect from programmes taking place in this first decade of the new millennium.

11.2 A Brief History of Cosmology

Cosmology is a subject that has interested mankind for as long as we have been able to think. Virtually every culture or religion has its own creation story; while these stories are generally neither accurate nor scientific, they demonstrate mankind's universal curiosity about our Universe.

- 580–322 B.C.: Greek philosophers, including Pythagoras, Plato and Aristotle, proposed various systems for the motions of the planets. Pythagoras' system had all the bodies (including the Sun) revolving around a central fire, while the others were geocentric. The spherical nature of the Earth was realised in these theories.
- Aristarchus (~ 280 B.C.) proposed a heliocentric cosmology, but neither his contemporaries nor successors favoured it.
- Copernicus (1473–1543 A.D.) proposed a heliocentric cosmology.
- Newton (1643–1726) formulated the law of gravitation.
- Herschel (1738–1822) proposed that the nebulae are 'island universes'.
- Einstein formulated the theory of general relativity, which was published in 1916.
- Friedmann and Lemaitre found expanding universe models that satisfy Einstein's equations of gravity, in 1922 and 1925 respectively.
- Hubble announced his observations of galaxy recession in 1929.
- Bondi, Gold and Hoyle postulated a 'steady state' cosmology in 1948.
- Penzias and Wilson discovered the cosmic microwave background in 1965. The Big Bang cosmology achieved widespread acceptance after this discovery.

11.3   Axioms, Principles and Limitations

In order to pursue the science of cosmology we need to make at least one very significant assumption about the Universe. This axiom of cosmology is called the 'cosmological principle', and states that the Universe is spatially homogeneous and isotropic on a large scale. This may sound like a rather strange axiom — after all, we can test whether the Universe is isotropic from observations — but it does have some important consequences: it implies that we do not have a special vantage point, thereby ruling out the conclusion that we are at the centre of the Universe, from the observation of isotropy. It also implies that those parts of the Universe we cannot observe are similar to the parts that we can observe, and hence that the conclusions we make about cosmology are valid everywhere in the Universe.

Juxtaposed with this is the 'anthropic principle', which states that any Universe that can be observed must be capable of supporting life. One important consequence of this is that it provides an explanation as to why we might observe the Universe at a 'special' time when life can exist.

These two principles somewhat limit the scope of cosmology as a science: we are handicapped by the fact that we can only observe a finite portion of the Universe's space–time. What if the cosmological principle is not correct? If some parts of the Universe are very different to the portion we have access to, they might be unable (or unlikely) to sustain life. Consequently, large expanses of inhospitable universe might be hidden from us by the anthropic principle.

11.4   A Brief History of the Universe

In big bang cosmology the Universe began with a singularity: extrapolating backwards in time, the size scale of the Universe tends to zero as time tends to zero, and consequently, temperature and density tend to infinity, and conditions become so extreme that classical physics is invalid. The earliest time at which classical physics can be valid is known as the Planck time, 10^{-43} s after the singularity. Prior to this moment, the wavelengths of typical particles would have been smaller than their Schwarzschild radii; we do not (yet) have physics that can describe the Universe under such extreme conditions.

Between 10^{-36} and 10^{-33} s it is thought that the Universe underwent a phase change, from a state in which the nuclear and electromagnetic forces were unified and symmetrical, to the present state, in which the nuclear force far overwhelms the electromagnetic force. This phase change resulted in a large-scale homogenising process of rapid expansion known as 'inflation', which smoothed out primordial irregularities and imperfections.

At this early stage of the Universe matter and radiation were in equilibrium: the energy density of photons was sufficiently high that pairs of hadrons and leptons were constantly being created, and annihilated. After about 10^{-4} s the temperature had dropped such that it was no longer possible to produce pairs of hadrons from the radiation field. The majority of hadrons annihilated, but leptons could still be created spontaneously from the radiation field. Of the remaining protons, many merged with the abundant electrons to form free neutrons. After about a second, the temperature had dropped to the point where leptons could no longer be produced. The majority of leptons annihilated, and the production of neutrons ceased.

After about 1 min the temperature had dropped sufficiently for neutrons and protons to combine, to form successively deuterium, tritium and finally the most stable of atomic nuclei, helium. Within a matter of minutes the majority of free neutrons were used up in the cosmological nucleosynthesis of helium.

After the production of particles had ceased, the majority of the energy density of the Universe was in radiation. The temperatures of radiation and matter were coupled through electron scattering and this meant that the Universe was optically thick. After about 10,000 years of expansion the temperature of the radiation field had declined enough that the energy density of matter exceeded that of radiation. After about 300,000 years the temperature of the radiation field had reached the point where the radiation could no longer ionise hydrogen. Protons and electrons combined to form hydrogen atoms. With the majority of free electrons taken up into atoms, electron scattering of the radiation field ceased and the Universe became optically thin. Matter and radiation had become decoupled, and the transition from an optically thick to an optically thin Universe marks the earliest time that can be observed by means of electromagnetic radiation. The microwave background dates from this time. It now has a temperature of 2.7 K, corresponding to a redshift of about 1000, and this epoch is sometimes referred to as the 'surface of last scattering'.

As time progressed, tiny fluctuations in density grew through gravitational instability. From these density enhancements emerged the structures that would eventually form galaxies and clusters of galaxies. Within these structures gas clouds collapsed to form the first stars, and black holes began to grow by accretion to form quasars. Quasars and hot young stars illuminated the Universe with fresh ionising radiation. The low-density gas filling the voids between collapsing structures was thus once again ionised. Elements heavier than lithium, exceedingly rare in the initial cosmological nucleosynthesis, were now synthesised by nuclear reactions in the cores of stars. Winds from stars and supernovae enriched the interstellar medium with metals, and successive generations of stars formed from this material. Gas was ejected from galaxies by winds from supernovae and quasars, and poured into the deep potential wells of clusters of galaxies to form the hot, intracluster medium.

The structures grew, galaxies condensed, stars formed and died, and gradually over billions of years the Universe developed to its present day appearance.

11.5   The Geometry and Fate of the Universe

Einstein's remarkable insight that gravity is equivalent to the curvature of four-dimensional space–time leads naturally to an understanding of the Universe's expansion and possible recollapse in terms of its geometry. If the Universe has sufficient density for gravity to eventually halt its expansion and eventually cause it to collapse back to a singularity (the 'big crunch'), the Universe is said to be 'closed', and to have positive curvature, a geometry somewhat analogous to that of a sphere. If, on the other hand, the Universe is not dense enough for gravity to halt its expansion, it will continue to expand forever and is called 'open'; in this case the Universe has negative curvature somewhat analogous to the geometry of a hyperboloid. The intermediate case, in which the expansion will halt asymptotically after infinite time, is termed 'flat', and is said to have zero curvature. The ultimate fate of the Universe is not quite as simple as this story would suggest, however, if there is enough energy density in the vacuum. This is related to the 'cosmological constant', Λ in Einstein's gravitational field equations. The presence of vacuum energy density implies positive Λ, and is analogous to a large-scale repulsive force. This could cause the Universe to expand forever even if its geometry is closed. The latest results from Hubble Space Telescope observations of supernovae in other galaxies imply that there is indeed a cosmological constant (Perlmutter *et al.*, 1999).

11.6   Age and Distance Scales

Even though it is now more than 70 years since Hubble announced his observation of the recession of galaxies, the value of the 'Hubble constant', the constant of proportionality between distance and recession velocity, is still subject to uncertainties of more than 10%.

For more distant objects, the linear relation between redshift and distance does not hold: the distance–redshift relation depends on the large-scale geometry of the Universe as well as the Hubble constant. High redshift objects are more distant if the Universe is open than if it is closed. Similarly, light from high redshift objects has taken longer to reach us if the Universe is open than if it is closed, and the age of the Universe itself will be larger in open models than in closed models. A recent estimate places the age of the Universe at approximately 12.5 billion years (Freedman *et al.*, 2001).

11.7   What Does the Universe Contain?

Not all of the Universe's contents are easily observed. We know that the Universe contains normal matter, distributed in gas, dust, stars, planets and other small bodies. The Universe also contains photons of electromagnetic radiation, neutrinos, and a number of other fundamental particles that have now been observed in accelerator experiments. There is also now strong observational evidence for the existence of black holes, which we observe indirectly by the radiation of material that is accreting onto them (see Chapter 7).

However we are also aware that a large fraction of the matter in the Universe is unaccounted for. This is called 'dark matter', because it does not appear to radiate electromagnetic waves. Its presence is inferred by a variety of means (see the section on cosmological probes below) but we do not know with any certainty what sorts of particles make up the dark matter. Some have proposed that the dark matter is normal baryonic matter, for example in compact or failed stars, in massive interstellar planets or perhaps even in black holes. However, recent observations of microlensing of the Large Magellanic Cloud (Alcock *et al.*, 2001) suggest that these types of objects cannot account for the total mass known to reside in dark matter. The alternative is that the dark matter is made of a fundamentally different, non-baryonic, weakly interacting type of matter. Such material is a vital ingredient in the leading models for structure formation. Obviously, an important goal of future particle physics experiments will be to detect a particle with the necessary properties to fit the profile of dark matter — a particle that is probably the dominant source of mass in the Universe.

The vacuum energy density corresponding to a positive cosmological constant is another possible (but poorly understood) constituent of the Universe, and is now often termed 'dark energy', in analogy to dark matter.

Finally, it has been proposed that phase transitions at very early times could have resulted in some rather strange relics know as 'topological defects', including 'magnetic monopoles' and 'cosmic strings'. Though the existence of such entities could have very important consequences for the development of structure, they have not so far been observed.

11.8 Key Cosmological Probes

The vast scope of cosmology and the huge range of different things we would like to understand about the Universe mean that very diverse phenomena are used as cosmological probes.

- The cosmic microwave background allows us to study conditions in the early Universe, before stars or galaxies had formed. The spectrum of fluctuations in the microwave background can tell us a great deal about the geometry of the Universe.
- Other background radiation, at a variety of wavelengths from metre wavelength radio to gamma rays, provides indicators as to the relative importance of different types of object and different radiation processes over the history of the Universe.
- Stars in our own galaxy and in external galaxies can be used as age indicators. Obviously, stars (or populations of stars) that appear to be older than the Universe at the epoch in which they are observed indicate a problem with the cosmological (or stellar evolution) model.
- Quasars have proven extremely useful cosmological probes because their high luminosities allow them to be observed at great distances and therefore early epochs of the Universe. As well as directly providing measurements of the evolution of massive black holes (and by inference their host galaxies), absorption lines in quasar spectra allow us to study the distribution, densities, ionisation states and elemental abundances of gas clouds along the line of sight.
- Galaxies tell us about the history of luminous matter while the clustering of galaxies is used to measure structure and its evolution. The velocity distributions of galaxies within clusters, gravitational lensing of galaxies, and the rotation curves of the galaxies themselves have provided us with measures of the importance of dark matter.
- Clusters of galaxies are the most massive bound systems, and their masses, space densities and evolution place important constraints on models for cosmological and structure formation. Because clusters are easily resolved objects at any distance, and often have relatively regular shapes, comparison of their line of sight and angular dimensions allows the geometry of cosmological models to be tested.
- Supernovae, as well as being important probes of star formation history, can be used as 'standard candles', thereby allowing us to measure the distance scale and the geometry of the Universe.
- Elemental abundances in stars, gas clouds or even in the materials that make up the solar system allow us to test models for nucleosynthesis. In regions that have undergone little or no star formation, elemental abundances tell us about primordial nucleosynthesis; in other regions we can use elemental abundances to learn about preceding generations of stars and supernovae.

11.9 Formation of Structure

Structure formation is the process through which material took up its present day distribution in galaxies and clusters of galaxies. Photographic galaxy surveys in the 1950s showed nicely that galaxies appear to be distributed on the sky in filaments and knots. Redshift surveys revealed this extraordinary structure in three dimensions, and the latest surveys (particularly the 2dF redshift survey conducted on the Anglo-Australian Telescope; Colless *et al.*, 2001) show this three-dimensional structure in extraordinary detail.

Gravitational instability must have been responsible for the collapse of the relatively homogeneous universe observed at the time of the microwave background, to the clumpy, structured universe observed today. Small inhomogeneities at the surface of last scattering must have been present to seed the collapse. The way in which these early fluctuations developed into the galaxies and clusters of galaxies observed at later times depends critically on the properties of dark matter, since this dominates the mass density of the Universe. Some might regard this as a rather embarrassing situation for cosmology: a form of matter that we have never directly observed controls the manner by which our Universe has attained its present day appearance. However, by observing the formation of structure and comparing it to models containing different types of dark matter, we are in fact learning about the dark matter itself. If the dark matter is hot (i.e. is made from particles with high random velocities), then perturbations on small scales will be damped and only large-scale perturbations will survive long enough to undergo gravitational collapse. In this case, the largest structures form first and the smaller structures must form from fragmentation. On the other hand, if the dark matter is cold, then small-scale perturbations survive long enough to collapse, small-scale structures form first and combine to form larger structures. This is called bottom-up, or hierarchical structure formation. As well as having simpler mathematics, cold dark matter (CDM) models appear to match observations better than hot dark matter.

11.9.1 *The complexity of galaxy formation*

Galaxies show a range of optical morphologies. Some appear to be flattened like a disc, often showing beautiful spiral structures that shine with the blue light of young stars. Elliptical galaxies have a spheroidal appearance and appear to contain only old stars. Many galaxies are a composite of these two shapes, with a spheroidal central bulge and a surrounding disc. Irregular galaxies are just that, with a knotty, asymmetric appearance. Galaxy discs are rich in gas, while galaxy bulges and elliptical galaxies (known collectively as 'spheroids') appear to contain relatively little gas. Ellipticals are found preferentially in clusters, while spiral galaxies are more evenly distributed through space.

Understanding how galaxies evolved to their present day appearances is difficult because of the number and complexity of processes responsible. Galaxies must involve the collapse or accretion of primordial gas, but they also collide with one another and must interact with the hot gas that forms the intracluster medium in galaxy clusters. One particularly important and complex process that has shaped the appearance of galaxies is star formation. Radiation pressure and winds from young stars might trigger the collapse of surrounding gas clouds, leading to increased star formation; but, conversely, if enough energy is given to the surrounding gas clouds, for example by supernovae, the gas can be blown right out of the plane of a galaxy, effectively halting star formation. This 'feedback' from star formation is a vital, but at present rather arbitrary, component of models that try to simulate the formation of realistic galaxies

There has been a very long debate as to whether elliptical galaxies and spiral bulges formed at early epochs or are the result of mergers of smaller galaxies. The hierarchical picture favours the latter origin, and observations of clusters of galaxies with the Hubble Space Telescope support this hypothesis: blue galaxies showing signs of interaction in high redshift clusters are replaced by red, spheroidal galaxies in low redshift clusters (Dressler *et al.*, 1998). However, the population of extremely luminous, high redshift galaxies recently detected at submillimetre wavelengths suggests that massive ellipticals formed in a high redshift epoch (Smail *et al.*, 2002).

11.9.2   *The history of star formation*

One subject that has advanced very dramatically in the last decade is the determination of the Universe's star formation rate as a function of cosmic epoch (Figure 11.1). While measurements using different techniques show some considerable scatter, we can be fairly certain that the stars that shine in the galaxies of today were formed over a relatively wide redshift range. However, much of the information we would like to know about the history of star formation cannot be contained in a simple plot such as Figure 11.1: did the stars form in large numbers of small galaxies or in a comparatively few massive ones? Did the stars form gradually, or were the stars in each individual galaxy formed in violent 'starburst' events? How did the environments of the galaxies affect their star formation rate? To answer such questions requires detailed studies of the luminosity function (the space density of objects as a function of luminosity) of galaxies and its evolution, at a variety of wavelengths and divided by morphological type and environmental conditions.

11.10   Cosmic Background Radiation

Figure 11.2 shows a schematic of the cosmic backgrounds, the integrated emission from all sources of radiation outside our galaxy. It can be used as a measure of the contributions of different processes to the present day radiation density. The largest

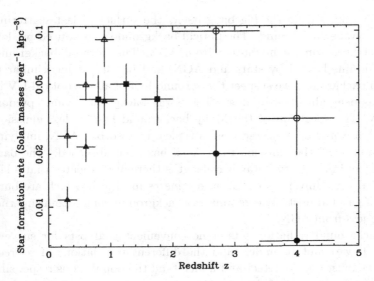

Figure 11.1 Star formation rate per Mpc^3 as a function of redshift, based on a compilation by Blain (2000). The closed triangles come from Lily *et al.* (1996), the filled squares are from Connolly *et al.* (1997), and the filled circles are from Madau *et al.* (1996). The open triangles and circles have been corrected for dust extinction and come from Flores *et al.* (1999) and Pettini *et al.* (1998) respectively. Although there is considerable scatter in the data, it is clear that the star formation rate has been in decline since a redshift of 1, and that the Universe's stars have formed over a wide range of redshift.

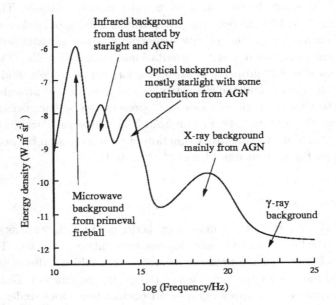

Figure 11.2 Schematic of the cosmic background radiation and its origin, based on Hasinger (2001).

contribution comes from the big bang itself: the surface of last scattering is seen as the microwave background. The optical background is produced predominantly by starlight with some contribution from AGN. The infrared background comes from dust grains heated by stars and AGN, and because it has similar power to the optical background, we suspect that around half of all the optical-UV light ever emitted has been absorbed by dust. The X-ray background comes predominantly from AGN. The spectrum of the X-ray background has far less emission at low energies than would be expected from a typical unabsorbed AGN, implying that a large fraction of all the emission from AGN has been absorbed. The background radiation from infra-red to X-ray wavelengths therefore indicates that a significant portion of all the Universe's radiation originates in objects which are buried deep in clouds of gas and dust. The gamma-ray background comes from supernovae and relativistic jets from AGN.

The background radiation sets some convenient goalposts for surveys of the Universe. If we want to understand the different populations of sources in the Universe, we must aim to detect and understand the populations responsible for all (or at least the majority) of the background emission.

11.11   Observations of Evolution

11.11.1   *Clusters of galaxies*

Clusters of galaxies are the largest, most massive bound systems. The diffuse X-ray-emitting gas that fills clusters, the gravitational lensing of background galaxies seen through clusters, and the velocity dispersions of the galaxies within clusters all provide good measures of cluster gravitational potential wells. The formation of clusters is an essential property of structure formation models, and the relative epochs at which clusters and their constituent galaxies formed provide an obvious observational test for hierarchical models. At present the best measurements of the evolution of the cluster population come from X-ray surveys. Perhaps surprisingly, very little evolution of the luminosity function of clusters is seen between a redshift of about 0.7 and the present day (Jones *et al.*, 2001).

11.11.2   *Galaxies*

For many decades, counting the number of galaxies per unit sky area as a function of brightness was the best means available for assessing galaxy evolution. This method, while capable of detecting the presence of evolution, could not distinguish between many potential models for the evolution of the galaxy population. Galaxy redshift surveys using multi-object spectrographs on optical telescopes undertaken in the last decade have revolutionised this field. In particular, two deep redshift surveys (Ellis *et al.*, 1996; Lilly *et al.*, 1995) demonstrated quite convincingly that while the luminosity function of red galaxies has not changed appreciably since a redshift of about 0.6, the luminosity function of blue galaxies has declined, particularly at low luminosity. This can be explained if the amount of star formation in relatively low luminosity galaxies has decreased over this redshift interval. HST studies of some of the galaxies in these surveys suggest that the changes in the luminosity function are accompanied by a change in the morphological mix of the galaxies: morphologically peculiar galaxies form a much larger fraction of the population at redshift 0.6 (Brinchmann *et al.*, 1998). One possibility is that merger-induced star formation has declined and post-merger galaxies have settled into more regular morphologies.

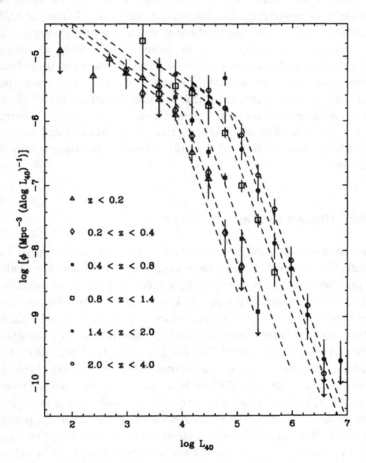

Figure 11.3 X-ray luminosity function (the number of objects per unit logarithmic interval of X-ray luminosity per unit volume of space) of AGN as a function of redshift z, from Page *et al.* (1997). In successively higher redshift shells the luminosity function is shifted towards higher luminosities; the dashed lines indicate model fits to this evolution. This shows that the typical luminosities of AGN at redshift 2 were more than a factor of 10 higher than they are in the present day.

11.11.3 *Active galactic nuclei and quasars*

Soon after their discovery, it was realised that quasars are a rapidly evolving population (Schmidt, 1968). However, it was the end of the 1980s when quasar surveys became large enough that the manner in which the luminosity function is evolving could be firmly determined (Boyle, Shanks and Peterson, 1988). In the following decade, studies at X-ray and other wavelengths achieved similar levels of statistical quality and led to similar conclusions: the shape of the luminosity function changes little, but appears to have shifted to lower luminosity between redshift 2 and the present day (see Figure 11.3). This implies that the massive black holes at the centres of galaxies were accreting material much more rapidly at earlier epochs than they do in the present day. The integrated luminosity density of AGN shows a similar evolution with cosmic time to the star formation rate shown in Figure 11.1. Star formation and accretion appear to be somehow intimately connected. Potentially, star formation and nuclear activity in galaxies trace the underlying pace of the formation of structure: as halos of dark matter merge, the galaxies they contain interact and coalesce, triggering star formation and funnelling material towards nuclear black holes. There is one large caveat to our current understanding of AGN evolution: the X-ray background tells us that most AGNs must be intrinsically absorbed, implying that we have so far only been able to study a minority of the AGN population.

11.12 Black Holes and Stars

The histories of star formation and accretion onto black holes appear to be qualitatively similar, but recently there have been some developments that imply an even closer link between the stellar and black hole components of galaxies. Imaging studies with HST combined with long slit spectroscopy of nearby galaxies have facilitated the determination of the central black hole mass in nearby galaxies. The masses of the black hole components appear to be approximately proportional to the masses of the surrounding spheroidal stellar components (Magorrian *et al.*, 1998). In individual galaxies, the processes that controlled the formation of the stars in the bulge and the growth of the central black hole must have been related to each other. One of the simplest, and therefore most appealing, possibilities is that the stars and central black holes grew from the same reservoirs of gas, over approximately the same time period. We know that individual AGNs cannot be long-lived phenomena, and hence for this scenario to be correct the majority of stars in individual galaxy spheroids must have formed over relatively short time spans ($< 10^9$ years). Evidence for this has recently been detected in the form of a population of extremely luminous submillimetre galaxies (Smail *et al.*, 2001). Observations of high redshift, X-ray-absorbed AGNs show that a significant fraction of these objects also have extremely high submillimetre luminosities (and by inference star formation rates), suggesting that the stellar and black hole components of galaxies could indeed grow at the same time and from the same reservoirs of gas (Page, Stevens, Mittaz and Carrera, 2001).

11.13 Prospects

Some extremely exciting advances in the field of cosmology can be anticipated in the next decade.

Large galaxy and quasar redshift surveys such as the 2dF survey, taking place on the Anglo-Australian Telescope, and the Sloan Digital Sky Survey, will soon be completed. Redshifts of hundreds of thousands of galaxies will allow extremely accurate measurements of clustering in the local Universe, and clustering of quasars will be measured to redshifts of greater than 2. These surveys will place strong constraints on models for the evolution of structure.

Important constraints on the cosmological world model, and in particular the Hubble constant, should also come from studies of galaxy clusters, using the latest generation of X-ray observatories (*Chandra* and *XMM-Newton*) to observe directly the intracluster gas, combined with observations of the Compton scattering effect of this gas on the microwave background (known as the Sunyaev–Zeldovich effect).

The latest X-ray observatories that are now in orbit will reveal a great deal about the population of obscured AGNs over the next decade. Until now these sources were hidden from view inside clouds of gas and dust, but the dramatically improved sensitivity of the latest observatories to higher-energy, more penetrating X-rays will allow us to see through these screens. The majority of accretion has taken place in obscured AGNs, so we should obtain a much clearer picture of the history of accretion and its relation to star formation and the growth of structure.

Near-future observatories at infrared/submillimetre wavelengths will revolutionise our understanding of the history of star formation and its influence on galaxy evolution by allowing us to measure the reprocessed emission from dust. Several satellite observatories are planned: The NASA mission *SIRTF* is due to fly within the next couple of years and the ESA mission *Herschel* is scheduled for launch in 2007. Surveys performed with these observatories will reach unprecedented depths and will detect significant numbers of high redshift dust enshrouded galaxies. In addition, the Atacama Large Millimetre Array (ALMA) is planned to start operation later this decade. ALMA will have excellent (a tenth arc second) spatial resolution and very impressive collecting area, and will be able to make deep millimetre-wavelength spectroscopic observations of distant star forming galaxies.

Large-scale, deep digital surveys of the sky at optical and infrared wavelengths planned in the next decade (for example using the United Kingdom Infrared Telescope and the planned UK/ESO VISTA telescope) will push back the frontiers of redshift to 7 or 8 in surveys or quasars — potentially allowing us to observe the epoch in which the intergalactic gas was re-ionised.

However, perhaps the most important near-future cosmological observations will be the space-, balloon- and ground-based experiments to measure fluctuations in the cosmic microwave background. In particular the ESA mission *Planck*, scheduled for launch in 2007, will map the whole microwave sky with an angular resolution of 10 arc minutes, improving on measurements currently being made by NASA's *MAP* spacecraft. By accurately measuring the spectrum of fluctuations in the background to this relatively fine scale, *Planck* should allow us to determine the main cosmological parameters such as the Hubble constant and the density of the Universe very accurately, and provide strong constraints on the nature of dark matter and dark energy. Observations of the polarisation of the microwave background have the potential to detect the effects of inflation on the surface of last scattering.

Chapter 12

Topics in Practical Statistics

Mark Cropper

12.1 Introduction

A practical knowledge of a core body of statistics is essential to any practising space scientist. Although the overall discipline is very extensive, it is surprising how a relatively small set of concepts can address the majority of situations.

This chapter sets out to address statistical topics that are relevant in day-to-day data reduction and analysis. It also sets out to clarify aspects of statistical analysis that often appear to be confusing, and provide a quick and simple reference. It makes no attempt to be rigorous or complete, and the topic choice is to a certain extent subjective.

The chapter is divided into four main sections. The first deals with probability distributions, the second with populations, the third with aspects of fitting data, and the final section considers error propagation.

12.2 Probability Distributions

12.2.1 *Central value: means, medians and modes*

Consider the example of the earnings data for a population. How do we characterise these earnings in a meaningful way? Of course the most complete description is a plot of the distribution, as in Figure 12.1. However, often, single numbers are sought, which distil the essence of the information in the distribution. Possibilities that come to mind include the average earnings, or alternatively, the value of the earnings that the most people enjoy — if the distribution of data is not symmetrical, these are not necessarily the same.

Distributions are usually characterised by the *mean*, the *mode* and the *median*. The mean is the average value of the distribution, the median the characteristic value and the mode the most frequently occurring value.

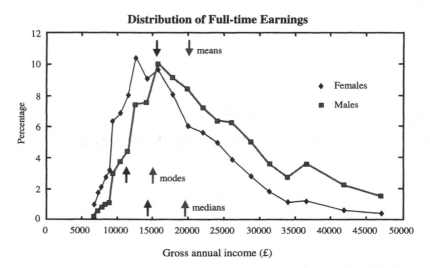

Distribution of Full-time Earnings

Figure 12.1   The distribution of female and male full-time earnings in the United Kingdom in 2000 (from New Incomes Survey 2000, UK National Statistics Office).

Before we go further we have to distinguish between, on one hand, lists of values (for example original data) and, on the other, a distribution function assembled from the list of values. In the case of a list of values, we can define the mean, median and mode as follows:

mean = average,

$$= \left(\sum_{i=1}^{n} x_i \right) / n \, ;$$

median = characteristic value, obtained by sorting the data from smallest to largest, then selecting the middle value (or the mean of the two middle values) ;

mode = most frequently occurring value obtained by sorting the data and finding that value which occurs most frequently. There may be more than one mode, or none at all, in which case the concept is not relevant .

For example, consider a set of raw data samples (in this case pollutant levels):

{18 18 21 21 21 21 21 21 21 21 25 25 25 25 25 25 35 35 50 70} parts per million (ppm)

The sample size $n = 20$. The mean pollutant level $= (\sum_{i=1}^{n} x_i)/n = 544/20 = 27.2$ ppm. The sample median pollutant level is the mean of the 10th and the 11th value in the sorted data set $= (21 + 25)/2 = 23$ ppm. The sample mode pollutant level is 21 ppm since this is the most frequently occurring value.

In the case of a distribution function $y(x)$:

$$\text{mean at} \quad \left(\sum_{i=1}^{n} x_i y_i\right) \Big/ \sum_{i=1}^{n} y_i \, ;$$

$$\text{median where} \quad \int_{-\infty}^{x_{\text{med}}} y(x)\,dx = \frac{1}{2} \, ;$$

$$\text{mode at} \quad \max(y(x)) \, .$$

For the salary sample, modes are arguably the most appropriate indicator, since this is the most frequently occurring value. In Figure 12.1, the mode for male full-time earnings is $\sim$ £16,000 per annum, while that for female full-time earnings is $\sim$ £12,500 per annum.

12.2.2 *Width: standard deviations, variances, standard errors*

Distributions can be analysed in terms of moments where the rth moment about any x value a is defined as

$$\mu_r(a) = \left(\frac{1}{n}\right) \sum_{i=1}^{n} (x_i - a)^r \qquad \text{(for a list)} \, ,$$

$$\mu_r(a) = \frac{1}{\sum_{i=1}^{n} y_i} \sum_{i=1}^{n} y_i (x_i - a)^r \qquad \text{(distribution function)} \, .$$

Note that $\mu_r(\text{mean})$ are the principle moments and $\mu_1(\text{mean}) = 0$.

Also, the second principle moment about the mean

$$\mu_2(\text{mean}) = \left(\frac{1}{n}\right) \sum_{i=1}^{n} (x_i - x_{\text{mean}})^2 \qquad \text{(list)} \, ,$$

$$\mu_2(\text{mean}) = \frac{1}{\sum_{i=1}^{n} y_i} \sum_{i=1}^{n} y_i (x_i - x_{\text{mean}})^2 \qquad \text{(distribution function)} \, ,$$

is the called the variance, which characterises the width of the distribution.

The standard deviation (denoted by σ) is an important quantity, and is the square root of the variance,

$$\sigma = \sqrt{\mu_2(\text{mean})} = \sqrt{\left(\frac{1}{n}\right) \sum_{i=1}^{n} (x_i - x_{\text{mean}})^2} \qquad \text{(list)} \, ,$$

$$\sigma = \sqrt{\mu_2(\text{mean})} = \sqrt{\left(\frac{1}{\sum_{i=1}^{n} y_i}\right) \sum_{i=1}^{n} y_i (x_i - x_{\text{mean}})^2} \qquad \text{(distribution function)}$$

(sometimes $1/(n-1)$ is used in the denominator). The third principle moment is the skewness, and the fourth the kurtosis. Note that for a list, the standard deviation σ is the square root of the mean of the squares of the data about the mean.

The variance and standard deviation characterise the distribution or list. How do we characterise the uncertainty of the mean? It can be shown that the uncertainty in the mean follows from the standard deviation and is

$$\sigma_m = \frac{\sigma}{\sqrt{n}} = \sqrt{\frac{1}{n(n-1)} \sum_{i=1}^{n} (x_i - x_{\text{mean}})^2} \, ,$$

called the standard deviation of the mean or the standard error.

Thus, to summarise, we have the variance $= \sigma^2$, the standard deviation $= \sigma$, and the uncertainty of the mean $= \sigma/\sqrt{n}$.

12.2.3   *Important theoretical probability distributions*

12.2.3.1   *Binomial*

The basic probability distribution is the binomial distribution. This describes the probability P that there will be a particular outcome x out of a total number of n trials each with a probability of p.

$$P(x : n, p) = \frac{n! p^x (1-p)^{(n-x)}}{x!(n-x)!} \, .$$

An example of the distribution is shown in Figure 12.2.

There are important special cases of the binomial distribution:

Poisson: limit when $p \ll 1$;

Gaussian: limit when $n \to \infty$.

We discuss these next.

12.2.3.2   *Poisson*

The Poisson distribution is particularly important for counting statistics when the count rate is low. It is given by

$$P(x : \mu) = \frac{\mu^x e^{-\mu}}{x!} \, , \quad \text{where } \mu = np \, .$$

This distribution is characterised only by μ. An example is given in Figure 12.3. In this case, given a total probability $\mu = 7$, the mode is $x \sim 6$. $P(x)$ has a characteristic width of $\sigma^2 = \mu$. The distribution tends to Gaussian for $\mu > 20$.

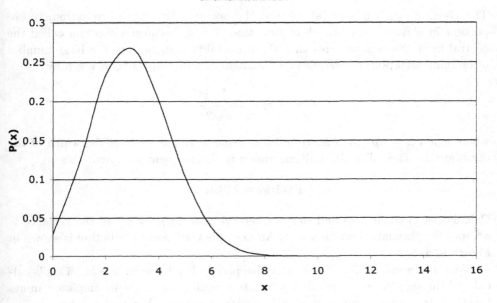

Figure 12.2   The binomial distribution for $n = 10$ and $p = 0.3$.

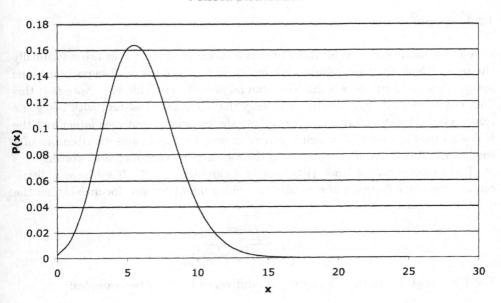

Figure 12.3   A Poisson distribution for $\mu = 7$.

12.2.3.3 *Gaussian*

The Gaussian distribution (also called the normal distribution) is extremely important in almost every branch of statistics. This is because a theorem called the central limit theorem requires that the probability distribution of a large number of random deviations converges to a Gaussian distribution. This is given by

$$P(x : \mu\sigma) = \frac{e^{-\frac{1}{2}\left(\frac{x-\mu}{\sigma}\right)^2}}{\sigma\sqrt{2\pi}} ,$$

where again $\mu = np$ (the mean) and σ is a parameter describing the width of the distribution. The full width half maximum is also frequently quoted: this is

$$\text{FWHM} = 2.354\sigma .$$

Calculating moments, we find that the second moment $\mu_2 = \sigma^2$, so the variance $= \sigma^2$, and the standard deviation $= \sigma$. An example Gaussian distribution is shown in Figure 12.4.

We often want to know the cumulative probability between $\mu \pm \delta x$. This is calculated through the error function, which is a special case of the incomplete gamma function. It can be shown that for a Gaussian, the region between $\mu \pm 1\sigma$ contains 68% of the cumulative probability, that between $\mu \pm 2\sigma$ contains 95% cumulative probability and that between $\mu \pm 3\sigma$ contains 99.7% cumulative probability.

12.3 Populations

Given two distributions, it is often of interest to know whether they are statistically the same. Generally this question is couched in terms of the 'null hypothesis': the approach is to determine whether they can be proved to be different. Note that this does not prove that they are the same, only that they are consistent with being the same. Generally there are two approaches: the chi-square test (see later) and the Kolmogorov–Smirnov test, which is generally used for continuous distributions of a single variable (for example comparing the incomes between men and women).

The Kolmogorov–Smirnov (KS) test is a cumulative test. If we have a distribution y as some function of a variable x (which might be, say, incomes), then the cumulative distribution is

$$S(x\text{cum}) = \frac{\int_{-\infty}^{x\text{cum}} y(x)dx}{\int_{-\infty}^{\infty} y(x)dx} .$$

The KS statistic is simply the maximum difference between two samples:

$$D = \max_{-\infty < x < \infty} |S_1(x) - S_2(x)| .$$

Gaussian Distribution

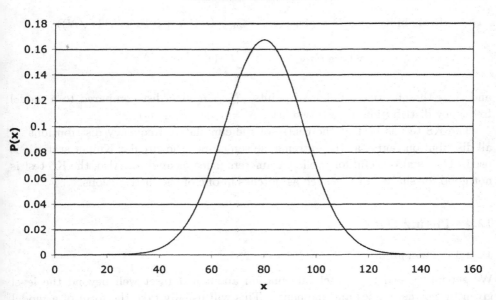

Figure 12.4   The Gaussian distribution for a mean $\mu = 80$ and $\sigma = 15$. This has FWHM $= 35$.

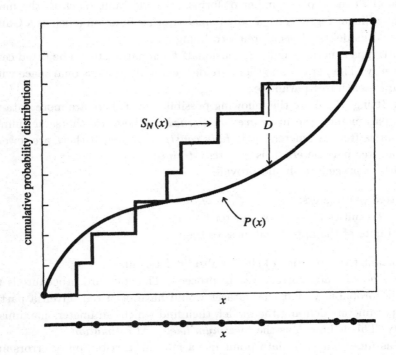

Figure 12.5   The cumulative probability distribution (from *Numerical Recipes*).

Then this can be checked against the formula

$$\text{Probability that } D > \text{observed} = Q_{KS}(D(\sqrt{N} + 0.12 + 0.11/\sqrt{N}))$$

$$\text{where } Q_{KS}(\lambda) = 2 \sum_{j=1}^{\infty} (-1)^{(j-1)} e^{-2j^2\lambda^2}$$

and N is the effective number of points. $N = n$ for a list compared to a model frequency distribution, or, for two lists, $N = n_1 n_2/(n_1 + n_2)$.

The KS test is best at the centre of the cumulative frequency distributions — all distributions agree at the extrema, so sometimes the circular Kuiper statistic is used. This is also useful for cyclic parameters (such as angles). Also, the KS test is not good for sharp features such as notches in one of the distributions.

12.4   Fitting Data

12.4.1   *Overview*

We generally need to extract information about a dataset well beyond the level given by means, modes and moments. This will usually take the form of a model fit to the data.

The model can be of a number of forms: a single value (such as the mean), a straight line (linear regression), a polynomial or other function (such as a Gaussian) or a physical model (such as a bremsstrahlung model).

The fitting is generally multi-dimensional. Each parameter to be fitted can vary independently. The parameters therefore define a multi-dimensional space with one dimension for each free parameter.

When fitting, there are the following possibilities: if there are more data points than free parameters, the fit is *overconstrained*; if there are the same number of data points as free parameters, it is *fully constrained*; and if there are fewer data points than free parameters, it is *underconstrained*.

The fitting procedure should provide

(a) the fitted parameters;
(b) the uncertainties for each parameter;
(c) an estimate of the appropriateness of the fit.

It is important that (b) and (c) should also be determined.

There is no *formally* correct way to proceed. The standard procedure is to ask what is the probability that the dataset would have occurred, given a particular choice of parameters defining the model, then find which parameters maximise this probability. This is known as the *maximum likelihood estimation*.

If we assume that each data point has a normal distribution of errors around the model, then maximising the probability is equivalent to minimising the least square deviations from the model (see e.g. *Numerical Recipes*, Press *et al.*, 1996).

12.4.2 *The chi-squared fit*

Minimising the least square deviations from the model, defined by

$$\chi^2 = \sum_{i=1}^{n} (y_i - y(c_1 \cdots c_m))^2 / \sigma_i^2 \,,$$

where y_i are the data points, $y(c_1 \cdots c_m)$ is the model with parameters $(c_1 \cdots c_m)$ and σ_i^2 is the variance of the data point, is called chi-squared fitting. Note that this is the same as the moments formula, but with normalisation by the variances. $\nu = (n - m)$ is known as the number of degrees of freedom. As an example: for the arithmetic mean, $y(c_1 \cdots c_m) = y_{mean}$, then finding the minimum

$$\frac{d[\sum_{i=1}^{n} (y_i - y_{mean})^2 / \sigma_i^2]}{dy_{mean}} = 0$$

$$\Rightarrow \sum_{i=1}^{n} \frac{(y_i - y_{mean})}{\sigma_i} = 0 \,,$$

which is the definition of the arithmetic mean.

For linear models the least squares fit can have an explicit expression (called the *normal equations*). What do we mean by linear models? Linear models are models that are linear combinations of functions (which themselves can be non-linear), i.e. models that can be written as

$$y(x) = \sum_{j=1}^{m} a_j F_j(x) \,.$$

Here there are m functions (called the basis functions) $F_j(x)$, $j = 1 \cdots m$, each with a coefficient a_j. Typical examples are polynomials or sinusoids (the latter are Fourier expansions). Thus

$$\chi^2 = \sum_{i=1}^{n} \left(y_i - \sum_{j=1}^{m} a_j F_j(x) \right)^2 \bigg/ \sigma_i^2 \,.$$

All we have to do is minimise χ^2, i.e. differentiate χ^2 with respect to the coefficients. Doing this,

$$0 = \sum_{i=1}^{n} \left(\left(y_i - \sum_{j=1}^{m} a_j F_j(x) \right) \bigg/ \sigma_i^2 \right) F_k(x) \,, \quad k = 1 \cdots m \,,$$

which is a matrix equation which can be solved by standard linear algebra techniques. Take the example of a set of sinusoids:

$$y = a_0 + \sum_{k=1}^{m} (a_k \cos kx + b_k \sin kx) \,.$$

Multiplying (orthogonal terms vanish), we are left with Fourier components

$$a_0 = \left(\frac{1}{n}\right) \sum_{i=1}^{n} y_i \,,$$

$$a_k = \left(\frac{2}{n}\right) \sum_{i=1}^{n} y_i \cos\left(\frac{2\pi k i}{n}\right) ,$$

$$b_k = \left(\frac{2}{n}\right) \sum_{i=1}^{n} y_i \sin\left(\frac{2\pi k i}{n}\right) .$$

Note that it is very useful for cross-terms to vanish, so orthogonal functions are particularly important.

In general, however, the normal equation approach is not useful for nonlinear fitting, i.e. where the model function is not a linear combination of other functions. This is the case for fitting even simple physical models, such as bremsstrahlung or power laws. The solution is to expand χ^2 into a polynomial form then discard terms higher than the quadratic and all second derivatives of the parameters. This is acceptable close to the minimum. Further from minimum, we can simply use the method of steepest descent down the χ^2 multi-dimensional surface to obtain the next guess for the parameters:

$$a_j(\text{iteration } i+1) = a_j(i) - c\nabla\chi^2$$

(where c is an appropriately small constant). This is the basis of the Levenberg–Marquardt method, which smoothly changes over from steepest descent to linearised χ^2 as the minimum is reached.

12.4.3   *The appropriateness of the fit*

The probability that χ^2 exceeds any particular value is given by the incomplete gamma function, which is tabulated in many statistical handbooks (see also *Numerical Recipes*). As a general rule, according to the gamma function, a good fit is given by

$$\chi^2 \approx \nu \quad \text{or } \chi_\nu^2 \equiv \frac{\chi^2}{\nu} \approx 1 \,,$$

where χ_ν^2 is the reduced chi-squared. Recall that $\nu = n - m$ is the number of degrees of freedom (n = number of data points; m = number of parameters to fit). Actually, we should use

$$\chi^2 = \nu \pm \sqrt{2\nu} \quad \text{or } \chi_\nu^2 = 1 \pm \sqrt{\frac{2}{\nu}} \,.$$

So, what happens if, when doing the fit, $\chi_\nu^2 \neq 1$? There are three main reasons for this:

(i) The model is inappropriate and can be rejected (at the level given by the incomplete gamma function). This results in $\chi_\nu^2 > 1$.

(ii) Estimates for the uncertainties σ_i^2 are wrong. If the uncertainties quoted are larger than they actually are, then $\chi_\nu^2 < 1$, while if the uncertainties quoted are smaller than they actually are, then $\chi_\nu^2 > 1$.

(iii) The uncertainties σ_i are not normally distributed: generally then $\chi_\nu^2 > 1$.

12.4.4   *Confidence intervals and error calculation*

We now have a prescription for the best estimate of the fitted parameters (a above) and an estimate of the appropriateness of the fit (c above). Now we deal with the uncertainties for each parameter (b above).

We had for the minimum in the linear case

$$0 = \sum_{i=1}^{n} \left(\left(y_i - \sum_{j=1}^{m} a_j F_j(x) \right) \Big/ \sigma_i^2 \right) F_k(x), \quad k = 1 \cdots m.$$

It can be shown fairly simply from the above (see for example *Numerical Recipes*, Press *et al.*, 1996) that the variances of the parameters $\sigma^2(a_j)$ are

$$\sigma^2(a_j) = C_{jj},$$

where

$$C_{jk} = \left(\sum_{i=1}^{n} F_j(x_i) F_k(x_i) \Big/ \sigma_i^2 \right)^{-1} \quad \text{(i.e. inverse of matrix)}.$$

C_{jk} are the *covariances* between parameters a_j and a_k. All this treatment is formally for the linear case, but the treatment is generally the same for the nonlinear case, since close to the optimal fit the behaviour can be considered approximately linear.

We need a relationship between the formal standard deviation $\sigma(a_j) = \sqrt{C_{jj}}$ and the confidence interval δa_j. When there is only one free parameter $\delta a_j = \delta a_1$ fitted, then it is simple:

$$\delta a_1 = 1\sigma(a_1) = \sqrt{C_{11}} \quad \text{is the 68\% confidence interval},$$

$$\delta a_1 = 2\sigma(a_1) = 2\sqrt{C_{11}} \quad \text{is the 95\% confidence interval},$$

$$\delta a_1 = 3\sigma(a_1) = 3\sqrt{C_{11}} \quad \text{is the 99.7\% confidence interval},$$

as in a standard normal (Gaussian) distribution.

When there is more than one free parameter δa_j, then the factors 1, 2 and 3 change (see Table 12.1 later).

So now we have an estimate in the uncertainties of the parameters and in the cross talk between the parameters — the answer to (b) above. However, the estimates from $\sigma(a_j) = \sqrt{C_{jj}}$ are *strongly dependent* on the normal distribution of uncertainties, and, in addition, the nonlinear case can sometimes be computationally difficult. Therefore, parameter uncertainties are generally calculated from the χ^2 distribution itself, or from simulated datasets using the fitted model (Monte Carlo simulations) or fits to subsets of the dataset selected at random (bootstrap).

We start with the χ^2 distribution itself. We can move in the m-dimensional parameter space (m parameters a_j, $j = 1 \cdots m$) and watch the χ^2 change. If we set a particular χ^2, then this will give a maximum range for any parameter a_j, so that

$$(a_j - \delta a_j) \leq a_j \leq (a_j + \delta a_j).$$

If we define $\sigma(a_j)$ to include, say, 68% of the likely values of a_j, or 90%, this translates to the following question: What is the χ^2 corresponding to the 68% or 90% limit?

Then, if we change a_j from its central (best-fitted) value until χ^2 reaches this level, we can say that the uncertainty on a_j is $\pm\delta(a_j)$ (68% or 90%).

So, what are the values for χ^2 appropriate for 68% or some other limit? This depends on the number of parameters we want to find the joint confidence interval of. We can call these the 'interesting' parameters. There may be other parameters that we allow to float free when we achieve the best fit as we step through a grid of values for the interesting parameters. It can be shown that if $\chi^2_{\min}$ is the minimum χ^2 when all parameters are free, and χ^2_k is the χ^2 when there are k interesting parameters, then

$$\Delta\chi^2_k = \chi^2_k - \chi^2_{\min}$$

is *also* distributed as a chi-squared distribution. So, again, we can use the gamma function to work out the probabilities.

Table 12.1 gives $\Delta\chi^2$ as a function of confidence level and number of interesting parameters.

Table 12.1   Change in $\Delta\chi^2$ for different confidence levels, as a function of the number of free parameters in the fit.

Confidence level	Number of interesting parameters					
	1	2	3	4	5	6
68.3%	1.00	2.30	3.53	4.72	5.89	7.04
90%	2.71	4.61	6.25	7.78	9.24	10.6
95.4%	4.00	6.17	8.02	9.70	11.3	12.8
99%	6.63	9.21	11.3	13.3	15.1	16.8
99.73%	9.00	11.8	14.2	16.3	18.2	20.1
99.99%	15.1	18.4	21.1	23.5	25.7	27.8

For linear fits this leads to confidence interval ellipses (Figure 12.6).

To clarify these points with an example, assume a number of data points $n = 8$ and a number of parameters to fit $m = 3$, with a minimum chi-squared that is found

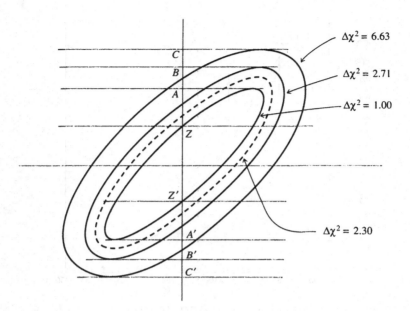

Figure 12.6 Elliptical confidence intervals. Here the solid lines define 68% (A–A′), 90% (B–B′) and 99% (C–C) intervals for one free parameter, while the dotted line is the 68% interval for two free parameters. (From *Numerical Recipes*, Press *et al.*, 1996.)

to be $\chi^2_{\min} = 5.1$. In this case, the number of degrees of freedom $\nu = n - m = 5$. The reduced chi-squared is $\chi^2_\nu = \chi^2/\nu = 5.1/5 = 1.02$. Now, if we are interested in confidence intervals of only two of the three parameters, then from Table 12.1, $\Delta\chi^2 = 6.17$ for the 95.4% confidence interval, so that the χ^2 defining the 95.4% confidence interval is $\chi^2 = 5.1 + 6.17 = 11.27$. Then simply finding the values of the interesting parameters that yield $\chi^2 = 11.27$ (by, say, calculating over a grid of parameter values) defines an ellipse (or ellipsoid in more than two dimensions).

Figure 12.7 shows a fit to the brightness of a star in different colours, with a blackbody model including reddening from interstellar dust (seven points, three free parameters, two of which are interesting in each of the plots).

Now we turn to bootstraps and Monte-Carlo techniques. Both are very valuable for getting the possible range in parameters, especially when errors are not normally distributed. This means that they are often used. Both are very simple.

The Monte-Carlo procedure starts with a least-squares fit to the data to get the best-fit parameters. A model is then generated with those parameters, and noise added to produce a synthetic dataset. A fit is made to this synthetic dataset, and the best-fit parameters recorded. A new set of noise is generated and applied to the model. Another fit is made and the parameters recorded. This process is repeated as many times as possible, after which the best fits are plotted in the parameter space and the contours drawn encompassing the desired confidence level (generally 68% and 90%). This is shown schematically in Figure 12.8.

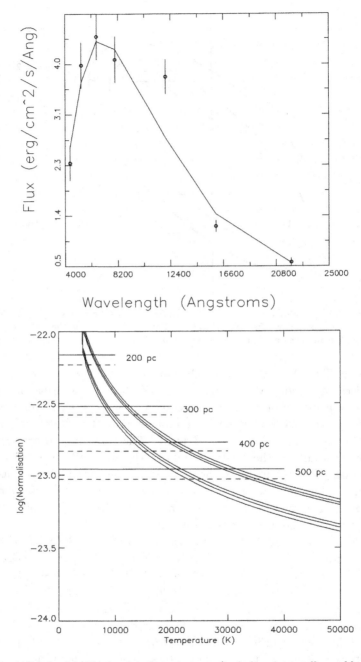

Figure 12.7 A fit of a blackbody emission spectrum (including interstellar reddening) to photometric flux measurements of a hot star. The fit is not good, with $\chi^2_\nu = 4.6$. The plot below shows the 68%, 90% and 94% confidence intervals in the temperature-normalisation plane. (From Ramsay *et al.*, 2002.) (Figure reproduced by permission of Blackwell Publishing.)

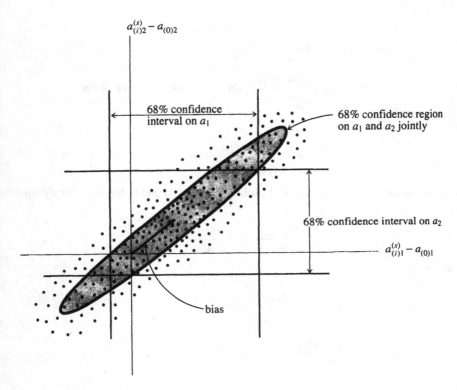

Figure 12.8 Plot of all the best fits in the two dimensions of the parameter space with contour encompassing 68% of them. (From *Numerical Recipes*, Press *et al.*, 1996.)

In the bootstrap procedure, a random number generator is used to select data points from the dataset, to make one of the same size (i.e. some points will not be selected, while others will be selected two or more times). A fit to this dataset is then made and the best fit recorded. This is repeated as many times as possible before finally plotting the 68% or other confidence contours, as in the Monte-Carlo technique.

12.5 Error Propagation

Now we have best-fit parameter and associated uncertainties. How do we combine, scale or operate on these parameters in some way? For example, if for parameters b and a, $b_j = Ca_j$ (as is often the case for a change of units), or, more generally, $b_j = f(a_j)$ (such as the magnitude scale in optical astronomy where $m_\nu = -2.5 \log_{10}(I)$), then how do we deal with the uncertainties? This is done by *propagation of errors*.

The basis for this procedure is as follows. If

$$y = f(a_k), \quad k = 1 \cdots m,$$

then

$$y + dy = f(a_k + \delta a_k)$$

$$= f(a_k) + \sum_{k=1}^{m} \frac{\partial f(a_k)}{\partial a_k} \partial a_k + \text{terms of higher order}.$$

Thus

$$dy \approx \sum_{k=1}^{m} \frac{\partial f(a_k)}{\partial a_k} \partial a_k.$$

As a simple example, put $m = 2$, and assuming equality near the limit for dy above, then

$$dy = \frac{\partial f(a_1)}{\partial a_1} \partial a_1 + \frac{\partial f(a_2)}{\partial a_2} \partial a_2.$$

Now we had

$$\sigma_y^2 = \sum_{i=1}^{n} \frac{(y_i - y_{\text{mean}})^2}{n-1}$$

$$= \sum_{i=1}^{n} \frac{(dy_i)^2}{n-1}$$

$$= \frac{1}{n-1} \sum_{i=1}^{n} \left(\frac{\partial f(a_1)}{\partial a_1} \partial a_1 + \frac{\partial f(a_2)}{\partial a_2} \partial a_2 \right)^2.$$

Since we can ignore crossed terms if the errors are uncorrelated,

$$\sigma_y^2 = \left(\frac{\partial f(a_1)}{\partial a_1} \right)^2 \frac{1}{n-1} \sum_{i=1}^{n} (\partial a_1)^2 + \left(\frac{\partial f(a_2)}{\partial a_2} \right)^2 \frac{1}{n-1} \sum_{i=1}^{n} (\partial a_2)^2$$

$$= \left(\frac{\partial f(a_1)}{\partial a_1} \right)^2 \sigma_{a_1}^2 + \left(\frac{\partial f(a_2)}{\partial a_2} \right)^2 \sigma_{a_2}^2.$$

This is the well-known RMS addition of errors.

As an example, assume that we are walking in a straight line in the countryside. Given an uncertainty in our compass bearing, and in the distance we have walked, what is our uncertainty in position in the N–S, E–W direction? This requires a transformation from polar to rectangular coordinates. Put y in the N–S direction, x in the E–W direction, and if the distance walked is r in a compass direction θ. Then

$$y = r \cos(\theta),$$

$$x = r \sin(\theta).$$

Also,

$$\frac{dy}{dr} = \cos(\theta),$$

$$\frac{dy}{d\theta} = -r\sin(\theta),$$

$$\frac{dx}{dr} = \sin(\theta),$$

$$\frac{dx}{d\theta} = r\cos(\theta).$$

Thus

$$\sigma_y^2 = \cos^2(\theta)\sigma_r^2 + r^2\sin^2(\theta)\sigma_\theta^2,$$

$$\sigma_x^2 = \sin^2(\theta)\sigma_r^2 + r^2\cos^2(\theta)\sigma_\theta^2$$

gives the relationship between the uncertainties in (x, y) and those in (r, θ).

Chapter 13

Instrumentation for Photon Detection in Space

J.L. Culhane

13.1 Introduction

Spacecraft carrying scientific instruments are now enabling a wide range of studies in the basic physical sciences. Since the launch of *Sputnik 1* in 1957, the pursuit of physical science in space has developed in three broad areas that may be simply characterised as follows.

First, the remote observation of the environment beyond the Earth, from the Solar System to the entire Universe, has provided an enormous expansion of the boundaries of traditional pre-space-age astronomy. The principal value of placing astronomical instruments in space lies in the ability to eliminate the absorbing and distorting effects of the Earth's atmosphere on the incoming radiation and, at visible wavelengths in particular, to eliminate the background radiation which the atmosphere emits. Over the past half century, use of space platforms has led to huge advances in astronomical science and it is fair to say that this discipline can not now be practised meaningfully without access to space.

Secondly, the ability to place instrumented spacecraft anywhere in the Solar System has allowed the *in situ* study of the Moon, the planets, and other Solar System bodies such as comets and asteroids. It has even been possible for humans to land on the Moon's surface, to place instruments there for longer duration observations and to return with samples of lunar material. While the engagement of human beings on a planetary surface has so far been confined to the Moon, instrumented spacecraft have travelled throughout the Solar System. The interactions of the solar wind plasma stream with the planets and other bodies has led to an enormous expansion of our understanding of plasma physics in a planetary context. In addition we have now established that the solar wind defines a region of influence around the solar system — the heliosphere, whose boundaries with the rest of the galaxy are now on the verge of being delineated by *in situ* observations with the pair of *Voyager* spacecraft.

The final broad area of physical science enabled by space platforms involves the ability to remotely observe the surface of the Earth — oceans, land and ice. Clearly work in all of these areas has previously relied on *in situ* observations and it is probably fair to say that the earth science community has required a significantly longer time to become convinced of the advantages of the space perspective for their discipline. However, it is now clear that the global view, coupled with the uniform observational sampling that results from the use of space platforms in polar and other near-Earth orbits, has led to dramatic advances in the earth sciences, particularly in cases where remote areas of the Earth's surface are involved.

Astronomy remains very largely in the domain of basic science. However, the use of remote sensing observations of the Sun, coupled with *in situ* observations of plasmas and charged particles in the near-Earth environment, has an increasing practical role in enabling an understanding of the risks posed to near-Earth communications and global positioning spacecraft by the dramatic outbursts from the Sun that are associated with mass ejections and flares. Damage can even result on the Earth's surface when terrestrial power distribution systems become overloaded following the high induced magnetic fields that can result from these large solar events. The embryonic applied field emerging from the recognition of the importance of the Sun–Earth coupling in these areas has come to be known as space weather.

Observations of the Earth's surface from space have even wider practical applications not only to meteorology, or traditional weather studies, but also to the much longer-term aspects of climate change. The exchange of energy between oceans, land and atmosphere and the impact of energy flow within the oceans are poorly understood phenomena that can have a considerable impact of the Earth's climate if existing balances should change. Changes in the atmospheric concentration of carbon dioxide and ozone provide dramatic examples of phenomena with the potential for causing climate change on uncomfortably short timescales. The long-term stability of polar ice masses and alterations to the cyclic changes in sea-ice thickness pose similarly dramatic threats to the stability of the Earth's climate, again on timescales as short as a hundred years. Key aspects of all of these phenomena require a global perspective for their complete understanding which only observations from space can deliver.

Successful observations from space pose significant challenges for innovation and reliability that lead to opposing pressures on instrument designers. Operation must be guaranteed for long periods in a uniquely hostile environment where the principal difficulties arise from the temperature and radiation extremes encountered and by the high vibration loads experienced during rocket vehicle launches. Often there are serious constraints on mass and power and the need to provide unusual operation conditions such as milli-Kelvin (mK) operating temperatures for sensitive photon detectors. It will not be possible in this brief chapter to discuss all of the relevant aspects of space instrumentation. In particular a discussion of space plasma physics and of Earth observation techniques is beyond the scope of this chapter. Thus the focus will first be on the requirements of space astronomy and on the space instruments used throughout the electromagnetic spectrum.

13.2 The Electromagnetic Spectrum

The quantum description of electromagnetic radiation emphasises its wave–particle duality. Thus energy is transferred in packets or quanta called photons and the energy of a photon of frequency ν (E_ν) is given by

$$E_\nu = h\nu = \frac{hc}{\lambda}, \tag{13.1}$$

where h is Planck's constant, c is the velocity of light and λ is the wavelength. When photons of high energy (short wavelength) are involved, the particle or photon nature of the radiation is most conveniently stressed. Alternatively, at long wavelength (low energy), the wave description becomes more appropriate. However, Equation (1) always applies. The names given to the different regions of the spectrum are indicated in Table 13.1 along with the appropriate wavelength, frequency and energy ranges.

More specialised units are typically used in the different ranges. Thus, for X- and gamma rays, photon energy units are often used where the electron volt (eV) is common. Here 1 eV $= 1.6 \times 10^{-19}$ J and the X-ray range is ~ 100 eV to 100 keV. While energy can still be important, particularly when considering interactions with detectors, UV and optical photons are more usually described by wavelength where the overall range stretches from 10 nm to 700 nm. Units of nm are used in the near-IR, giving way to μm for longer wavelengths. However, as we approach and enter the microwave range, mm and cm wavelength units give way to frequency units — THz, GHz and MHz. Finally, in the radio spectrum, both frequency and wavelength units are in common use.

Photons of electromagnetic radiation can interact with matter in a variety of ways, depending on their energy, as indicated schematically in Figure 13.1 (see also Chapter 8). At lower photon energies, the photoelectric effect (a) involves the removal of a bound electron by the incoming photon (ionisation) where the

Table 13.1   Regions of the electromagnetic spectrum.

	Wavelength (m)	Frequency (Hz)	Energy (J)
Radio	$> 1 \times 10^{-1}$	$< 3 \times 10^9$	$< 2 \times 10^{24}$
Microwave	$1 \times 10^{-3} - 1 \times 10^{-1}$	$3 \times 10^9 - 3 \times 10^{11}$	$2 \times 10^{-24} - 2 \times 10^{-22}$
Infrared	$7 \times 10^{-7} - 1 \times 10^{-3}$	$3 \times 10^{11} - 4 \times 10^{14}$	$2 \times 10^{-22} - 3 \times 10^{-19}$
Optical	$4 \times 10^{-7} - 7 \times 10^{-7}$	$4 \times 10^{14} - 7.5 \times 10^{14}$	$3 \times 10^{-19} - 5 \times 10^{-19}$
UV	$1 \times 10^{-8} - 4 \times 10^{-7}$	$7.5 \times 10^{14} - 3 \times 10^{16}$	$5 \times 10^{-19} - 2 \times 10^{-17}$
X-ray	$1 \times 10^{-11} - 1 \times 10^{-8}$	$3 \times 10^{16} - 3 \times 10^{19}$	$2 \times 10^{-17} - 2 \times 10^{-14}$
Gamma-ray	$< 1 \times 10^{-11}$	$> 3 \times 10^{19}$	$> 2 \times 10^{-14}$

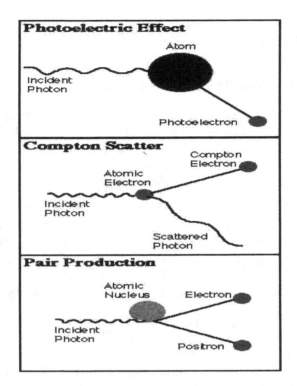

Figure 13.1   Schematic illustration of the processes for photon interaction with matter, (a) photoelectric absorption, (b) Compton scattering, (c) pair production.

photon energy, $E_\nu > E_B$, the electron binding energy. When E_ν has increased above E_B for electrons in the target atoms, the photons instead undergo Compton scattering with electrons (b). Here the photon energy is reduced — unlike in the case of Thomson scattering — and the electron emerges with the energy lost by the incoming photon, expressed as a wavelength change as

$$\frac{\lambda' - \lambda}{\lambda} = \left(\frac{h\nu}{m_e c^2} \right) (1 - \cos\alpha) , \qquad (13.2)$$

where α is the angle between the scattered electron and the incoming photon. This process operates at hard X-ray and lower gamma-ray energies. Finally, when the photon energy has increased to more than twice the electron rest energy, or $2 m_e c^2 = 1.022$ MeV, the process of pair production can occur (c). Here the incident photon can produce an electron–positron pair in the Coulomb field of a nucleus. In practice, use of pair production in gamma-ray detectors becomes practical for $E_\nu > 20$ MeV. All three of the above processes are important for photon detectors, depending on the photon energy involved. See Longair (1992) for a discussion of their behaviour with E_ν.

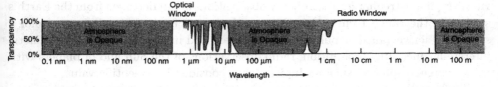

Figure 13.2   The transmission of the Earth's atmosphere for photons of the electromagnetic spectrum as a function of wavelength.

Photoelectric absorption has a crucially important role in the Earth's atmosphere, as is apparent in Figure 13.2. Here the transmission or transparency of the atmosphere is plotted against wavelength. Although the plot extends only to 0.1 nm at the short wavelength end — a photon energy of 12.4 keV — the atmosphere remains opaque to all higher-energy photons.

While X-rays with $E_\nu > 50$ keV can penetrate to ~ 30 km above the Earth's surface, where they can be studied by balloon-borne detectors, practically speaking it is necessary to place instruments in spacecraft that remain outside the atmosphere for long periods so that the radiation which cannot penetrate to the Earth's surface may be studied. Thus Figure 13.2 indicates the compelling need for observations of electromagnetic radiation from space. Furthermore, even the narrow though important visible spectral range (400–1000 nm) is affected by atmospheric turbulence or 'seeing' conditions. These limit the useable angular resolution of ground-based optical telescopes to ~ 1 arc sec. While this can be reduced by about a factor of 10 using current active or adaptive optics techniques, such performance is at present available only in the IR range at $\lambda > 1$ μm. We will see later that the Hubble Space Telescope (HST), with angular resolution of ~ 0.03 arc sec, operating at visible wavelengths in near-Earth orbit with sky background per resolution element reduced by ~ 1000 and free from atmospheric turbulence, has enormously advanced our understanding of the Universe.

13.3   Photons in Space

The previous section has made it clear why it is necessary to observe the Universe from above the Earth's atmosphere. We will now outline the scientific rationale for studies in different spectral ranges and describe the main features of several large astronomical space missions presently or soon operating throughout the spectrum from the radio to the gamma-ray range.

13.3.1   *Radio*

Although Figure 13.2 indicates that the atmosphere is opaque for radio emission at $\lambda > 20$ m, this region has not recently been studied with space-based antennae.

However, mainstream radio astronomy observations are undertaken from the Earth's surface in the range 1 cm $< \lambda <$ 20 m and play an important role in cosmology, studies of active galaxies, pulsars, variable stars and in observations of the galactic interstellar medium. In addition, radio observations of the Sun and of planets with active magnetospheres, such as Jupiter, have considerable scientific value.

Single dish radio telescopes have poor angular resolution due to their operation at long wavelengths. Thus for a telescope of aperture D operating at wavelength λ, the full width at half maximum of the angular response curve is $\theta_{\mathrm{FWHM}} \sim \lambda/D$. For a telescope with $D = 100$ m operating at $\lambda = 10$ cm, $\theta_{\mathrm{FWHM}} \sim 3.5$ arc min. However, if we assume that the properties of a radio source do not change on timescales comparable to that of Earth rotation, then provided that the amplitude and relative phase of signals arriving at two or more separate antennae can be measured, an angular resolution related to the separation of the antennae will be obtained. This is known as aperture synthesis. One example that employs this principle is the Very Large Array (VLA) in New Mexico, which consists of 27 radio telescopes set on three 20 km arms radiating from a common origin and separated by 120°. The signals received by each individual telescope are connected to a central site where they are correlated. This array can yield a performance approximating that of a single dish of aperture 40 km. The technique, first developed by Martin Ryle and colleagues at Cambridge, can provide good maps of extended sources with excellent angular resolution; ~ 0.5 arc sec at 10 cm in the case of the VLA. For a more detailed account of this technique, see Longair (1992) and references therein.

In a variant of this approach, known as very long baseline interferometry (VLBI), antennae situated at intercontinental distances from each other, e.g. US West Coast to Europe or ~ 9000 km, receive radio signals from a source and record them on high speed recorders, with the data being related to highly accurate time standards at each location. Correlation is then done off-line with data tapes from each receiver. Although the quality of extended source mapping is comparatively poor due to the smaller number of antennae, angular resolutions of ~ 2 mile arc sec are obtainable. Japan's ISAS has significantly enhanced the capability of the Earth-based VLBI network by placing a radio telescope in orbit. Named *Halca*, the spacecraft was launched on an M-V rocket and placed in an elliptical orbit with an apogee of 21,000 km. This facility has enabled a factor of ~ 3 enhancement in angular resolution together with better mapping capability for extended sources. An impression of the spacecraft in orbit with its antenna deployed is given in Figure 13.3(a), while features of the telescope and spacecraft are shown in Figure 13.3(b). The main reflector has diameter ~ 8 m and consists of a gold-coated molybdenum mesh suspended between six extensible masts. The secondary reflector of the Cassegrain configuration, diameter ~ 1 m, is also shown deployed. Radio waves are focussed by the two reflectors into a 2.5-m-long feed horn. Given the potential power of this technique, more ambitious missions are being planned in Russia, the USA and Japan, with spacecraft apogees of $\sim 100,000$ km.

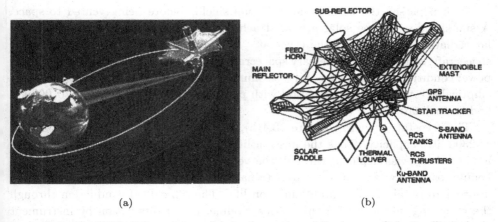

(a)   (b)

Figure 13.3   The ISAS *Halca* mission. (a) An impression of the spacecraft in its elliptical orbit. The increase in available baseline over those available between the Earth-based telescopes is apparent. (b) Diagram identifies the main elements of the deployed telescope and spacecraft. (Courtesy of ISAS.)

13.3.2   *Sub-mm/IR*

Extending from about 1 to 1000 μm in wavelength, this spectral range has gained enormously in importance in recent years. Observations have emphasised the study of cooler objects, star-forming regions and interstellar gas and dust along with the spectroscopy of planetary atmospheres in the solar system. As telescope sensitivities have increased, the desire to study proto-galaxies and evolving active galaxies at increasingly high redshifts has further stressed the value of IR observations since familiar visible range spectral features will show up at IR wavelengths for high redshift objects. Finally, studies of the anisotropy of the 3 K cosmic background radiation, whose spectrum peaks at $\lambda \sim 150$ μm, are providing the principal observational information about the nature and origin of the Universe. In the future, it is likely that searches for the existence of terrestrial planets associated with nearby stars will also emphasise observations in this spectral range.

Space IR telescope performance is limited by the temperature of the optical elements and the associated detector systems. Thus a major factor in the sensitivity increase of IR telescopes and instrumentation has stemmed from the ability to operate both instruments and telescope mirrors at temperatures in the range below 10 K. A good example is the ESA *Infra-red Space Observatory (ISO)* mission, launched by an *Ariane 4* from Kourou in November 1995. The schematic diagram in Figure 13.4 shows both telescope optics (60 cm aperture) and detectors surrounded by a superfluid helium cryostat. Some of the detectors (IR bolometers) were coupled directly to the helium tank so as to run at $T \sim 1.7$ K. Other detectors and the optical elements were cooled using the boil-off gas. The gaseous helium was passed through the optical support structure, where it cooled the telescope and the other instrument detectors.

It then passed the baffles and radiation shields before being vented to space. A sun-shield, with the solar arrays attached to the outside, prevented direct solar radiation from heating the cryostat. The service module, at the base of the spacecraft, contained the necessary spacecraft sub-systems, including data handling, power conditioning, telemetry and telecommand and attitude and orbit control. The elliptical orbit had a 24-hour period with perigee ~ 1000 km, apogee $\sim 70,000$ km and an inclination of $5°$.

The cryostat, shown in Figure 13.4(b), is surrounded by a liquid helium tank of 2300-litre capacity. Thus the mission lifetime was set by the rate at which the liquid volume was depleted through the venting of gaseous helium to space. The helium loss rate was in turn due to external heat sources that had to be minimised in order to obtain an acceptable mission life. These included conduction through the mounting interface with the service module, power dissipation by instrument electronics internal to the cryostat and external radiation sources, e.g. Sun, Earth. With a predicted mission lifetime of 18 months, *ISO* in fact operated cold for 28 months. However, use of cryostats with consumable coolants (e.g. liquid helium, solid hydrogen) leads to a considerable mass penalty, ground handling problems and, above all, limited mission life. Thus much effort is being devoted to developing other methods of cooling, which involve use of electrical power to drive closed cycle systems, rather than consumable cryogens. We will discuss some of these approaches later.

Two major missions to be launched together on an *Ariane 5* in 2007 are the ESA *Herschel* and *Planck* spacecraft. The former is a far-IR and sub-mm observatory that will undertake imaging and spectroscopy in the range of 60–670 μm. It will carry a passively cooled 3.5 m telescope mounted behind a sunshade while the instruments will be housed in a superfluid helium cryostat. Since the orbit is around the Earth–Sun Lagrange 2 (L2) point, which is ~ 1.6 M km from the Earth on the opposite side to the Sun, the telescope will run at a stable temperature of ~ 60 K. Hence the liquid helium is required only for instrument cooling. However, the mission lifetime is still limited to ~ 3–5 years. With the advent of more capable launch vehicles, space astronomy missions will make increasing use of the L2 orbit to benefit from passive cooling of optics and freedom from the Sun, Earth and Moon as background sources. The spacecraft is shown in Figure 13.5(a), where the telescope primary and secondary mirrors, cryostat and hexagonal service module may be seen.

The *Planck* mission (Figure 13.5(b)), which will share the *Ariane 5* launch vehicle with *Herschel* and will therefore also be placed in orbit around the L2 point, has a more specialised purpose, namely the study of temperature fluctuations in the cosmic microwave background radiation. Operating in the wavelength range of 350–10,000 μm, the radiation is collected by a 1.5 m Gregorian mirror and fed to arrays of radio receivers and bolometers. Given the location at L2, the mirror and first stage of the cooling system will run at 60 K. For the *Planck* focal plane

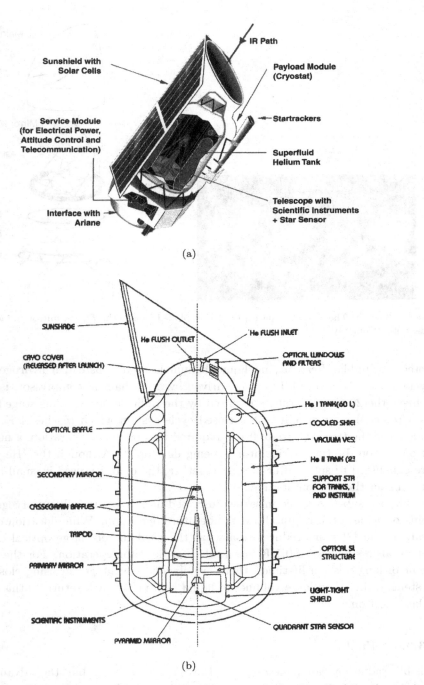

(a)

(b)

Figure 13.4   (a) Schematic diagram of the ESA *ISO* spacecraft showing all of the the mission elements. (b) Cryostat and telescope design showing the key elements requiring cooling. (Courtesy of ESA.)

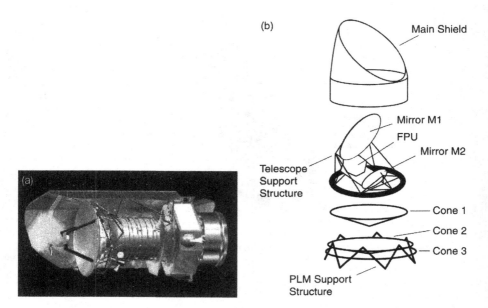

Figure 13.5   (a) The *Herschel* spacecraft. (b) Exploded view of the *Planck* mirror and focal plane assembly. (Courtesy of ESA.)

detectors, unlike *Herschel*, no liquid helium is used. Instead a hydrogen closed cycle Joule Thomson (J-T) cooler, driven by a mechanical compressor, is used to achieve the 20 K temperature required by the radio receivers. This stage also acts as a pre-cooler for a helium J-T closed cycle system which reaches 4 K. Finally, to achieve the temperature of 0.1 K required for bolometer operation, a novel type of open loop dilution refrigerator is being developed. Although the ^{3}He and ^{4}He are exhausted to space after flowing together, the quantity used is small and does not impact the mission lifetime.

The IR/sub-mm space missions described above rely heavily on cryogenic systems to achieve the required low temperature operation. While location of spacecraft at the L2 point reduces somewhat the need to cool large optical elements, temperatures as low as 0.1 K are needed for detector operation. For the achievement of long mission lifetimes, there are obvious advantages in using closed cycle systems rather than large volumes of liquid helium. We will return to this point in a later section.

13.3.3   *Visible*

Visible radiation can penetrate the Earth's atmosphere, but the advantages of space for large aperture telescopes operating in the visible and near/IR ranges were pointed out in Section 13.2. The NASA–ESA Hubble Space Telescope (HST) provides a prime example. Launched by the space shuttle in April 1990 into a circular orbit of 593 km and 28.5° inclination, the length of the telescope and spacecraft was $\sim$ 16 m, with a total mass of 11,100 kg. A schematic of the telescope is shown in Figure 13.6(a). The configuration is Cassegrain with Richey–Chretien optics and a primary aperture of 2.4 m. Four focal plane instruments are mounted behind the primary mirror. The mirror figures and surface roughness permit diffraction-limited operation at 250 nm with a resolution of $\sim$ 0.03 arc sec and an ability to register objects as faint as $m_v \sim 27$. Depending on the instrument in place at the focus, radiation in the range of 100 nm to 2.5 μm may be studied though observations at much longer wavelengths are possible. The diffraction-limited performance of the telescope requires extraordinary precision from the spacecraft attitude control system. With the fine guidance sensors locked on bright stars within 10–14 arc min of the object to be studied by the telescope, the pointing stability is $\leq$ 0.007 arc sec. Stars of $m_v > 15$ are suitable and a special catalogue of all available stars was prepared before launch by the Space Telescope Science Institute (STScI), whose staff are responsible for HST scientific operation and observation planning.

Although the low near-Earth orbit restricts access to the sky and greatly complicates observational planning, it confers one enormous advantage. The HST was designed so that it could be revisited periodically by the space shuttle for servicing, boosting of the orbit to ensure an extended mission life and changing of focal plane instruments. While four such servicing missions have now been made and a fifth and final visit is planned for July 2003, the first servicing mission, which took place in December 1993, was absolutely crucial for the future of the HST. Shortly after the spacecraft launch, it was discovered that there was a small but significant error in the figure of the primary mirror. The resulting spherical aberration meant that the telescope could never be perfectly in focus, and instead of the anticipated diffraction-limited performance, a situation existed where $\leq$ 15% of the encircled

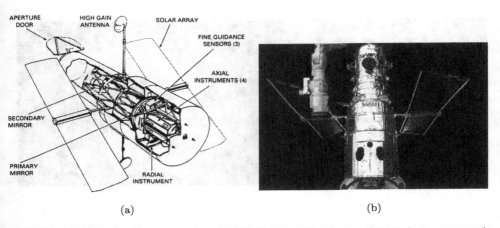

(a)   (b)

Figure 13.6 (a) Schematic diagram of the HST, showing the telescope layout, instrument positioning and fine guidance sensors. The large solar arrays are deployed. (b) The HST mounted in the space shuttle cargo bay during a servicing mission. (Courtesy of STSci and NASA.)

energy from a point source fell within a circle of diameter 0.2 arc sec. During the first servicing mission, one of the existing instruments, the High Speed Photometer, was removed to make way for a set of corrective optical elements which fed corrected images to the remaining three instruments. Following this revision, the HST has begun to observe at close to its designed diffraction-limited performance. It is planned to operate the mission to 2010. Thus Hubble, which has already revolutionised many branches of astronomy, will continue to have a dramatic impact on the subject for many years to come.

Proposed to operate in the IR spectral range, the Next Generation Space Telescope (NGST; recently renamed the James Webb Space Telescope), which NASA and ESA plan to launch in 2010, will be the successor in space of the HST. The NGST will require several important technological innovations to advance the subject substantially beyond the HST. With emphasis shifted from visible to IR — the NGST will observe in the 0.6–28 μm range — low temperature operation at $T \sim 50$ K is essential. Thus the telescope will be placed in orbit around the L2 point, like the *Herschel* and *Planck* missions. The proposed primary mirror diameter of 6.5 m allows detection of objects 10–100 fainter than those visible to the HST while ensuring diffraction-limited performance with resolution ≤ 0.1 arc sec at $\lambda \sim 2$ μm. However, available launchers, e.g. *Ariane 5*, limit the payload mass at L2 to ~ 5500 kg. This requires a primary mirror mass of ≤ 15 kg/m^2, which can only be achieved with a folding segmented mirror whose figure is actively controlled by cryogenic actuators. In addition, a huge deployable sunshade must be erected to reduce background and ensure low temperature operation for the optics. Advances are also required in methods of feeding light from individual sources in the focal plane to the spectrometers and in the development of IR detector arrays. Figure 13.7 shows two concepts at present under study.

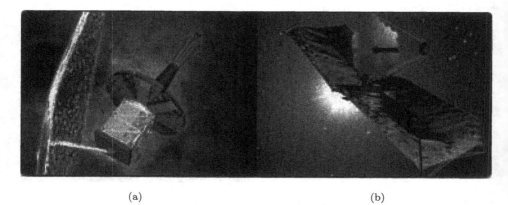

(a)   (b)

Figure 13.7   Proposals for the implementation of the NGST. (a) The Lockheed Martin concept; (b) the Ball Aerospace/TRW concept. (Courtesy of NASA.)

13.3.4 *X-ray*

In the X-ray and extreme UV spectral range, wavelength units (nm) give way to energy units (keV). This region equals the IR/sub-mm in scientific importance since it is concerned with the study of very energetic objects over the widest observable redshift range. In binary star systems, the process of gravitational accretion of matter from ordinary stars by the strong potential wells of compact galactic objects — neutron stars and black holes — is important in generating very hot X-ray-emitting material. The fast-spinning neutron stars created in supernova explosions have large magnetic fields and accelerate electrons to very high energies so that pulsed X-ray synchrotron emission results from their interactions with the magnetic field. All these objects are now observed in distant normal galaxies. Other extragalactic X-ray sources include clusters of galaxies that contain large volumes of hot ($T \sim 10^8$ K) gas and give rise to extended sources of X-ray emission. Active galactic nuclei (AGNs), e.g. quasars and Seyfert galaxies, are highly luminous X-ray sources where the emission is believed to result from the accretion of material from the host galaxy by a massive black hole. Studies of the latter two object categories at high redshifts have important implications for understanding the nature and origin of the Universe. Finally, it has been established that normal stars are X-ray emitters usually with higher luminosity than that of the Sun.

Since X-rays of energy 0.1–10 keV are absorbed by quite small amounts of material and have refractive indices $n \leq 1$, conventional optical systems are not appropriate. Thus the subject began with the use of large area gas-filled X-ray detectors equipped with mechanical collimators which gave a very poor angular resolution of $\sim 1°$. However, with $n \sim 0.995$ for a typical metal, optical elements that rely on total *external* reflection may be used. If $n = 1 - d$, then, by Snell's law of refraction, for angles of *grazing* incidence $\theta < \theta_c$ with $\cos \theta_c = 1 - d$ or $\sim (2d)^{1/2}$, the rays will undergo external reflection. Here the value of θ_c depends on λ and on the atomic number (Z) of the reflector. For high values of Z, reflectivity falls off less steeply with grazing incidence angle, so metallic coatings of gold or nickel are often used. Since X-rays have very short wavelengths — typically 1 nm — it is necessary to figure and polish the glass substrate with extreme care. For surface roughness in particular, a value of ~ 0.2 nm rms is presently achievable, which allows angular resolution of ~ 0.5 arc sec.

To ensure that the Abbe sine condition is obeyed so as to avoid severe coma for off-axis rays, it is necessary that a practical X-ray imaging system involves two reflections. A simple grazing incidence paraboloid of revolution, as proposed by Giacconi and Rossi in 1960, will act as a radiation collector whose field of view can be limited by placing an aperture at the focus (Figure 13.8(a)). This has the advantage of allowing a very small detector with low background counting rate to be used. A system of this kind was flown on the NASA *Copernicus* mission and marked the first in-orbit use of X-ray reflection optics for X-ray astronomy (Bowles *et al.*, 1974).

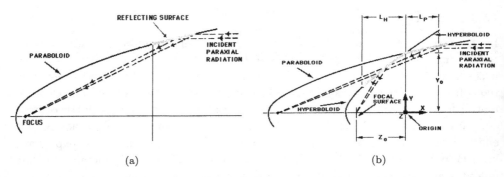

Figure 13.8   Optical diagrams for grazing incidence X-ray reflection. (a) Simple paraboloid of revolution with single reflection. (b) Paraboloid confocal and coaxial with a hyperboloid of revolution. Following two reflections, incident X-rays are focussed to form an image.

However, no image is formed. Wolter, in 1952, proposed a configuration that does produce images (Figure 13.8(b)). Known as the Wolter Type I, it involves the use of a paraboloid followed by a confocal and coaxial hyperboloid. Although there is some off-axis aberration, fields of view of 30–40 arc·min are achievable. Such telescopes have been used to good effect in making X-ray images of the Sun with ~ 3 arc sec resolution. The present state of the art in X-ray astronomy is represented by two observatory class missions — NASA's *Chandra* and ESA's *XMM-Newton*, which will now be described.

The *Chandra* spacecraft carries four pairs of mirrors arranged in a nested Wolter Type I configuration, as shown in Figure 13.9(a). The mirror elements are made from low expansion glass. They have been figured and polished to very high accuracy and then coated with iridium ($Z = 77$) to ensure good reflectivity over a range of grazing angles from 27 to 51 arc min. The maximum element aperture is 1.2 m and the focal length 10 m. The on-axis angular resolution of 0.5 arc sec is the best yet achieved in an X-ray optical system. Images are registered with two kinds of X-ray camera based on microchannel plate and CCD technology.

We will discuss CCDs in a later section. In addition there are two objective transmission gratings, made of gold, placed behind the mirrors. The system operates in the range of 0.1–10 keV. The spacecraft is illustrated in Figure 13.9(b), with its solar arrays deployed. The aperture cover door also acts as a sunshade when open and enables the telescope to be pointed to within 45° of the Sun. *Chandra* was launched by the space shuttle and later used its own propulsion system to achieve an elliptical orbit of 10,000 km by 140,000 km with 28.4° inclination. The spacecraft spends 85% of its 64-hour orbit above the Earth's trapped particle belts and so can observe individual X-ray sources for up to 55 hours.

The ESA *XMM-Newton* mission involves a different approach, namely the fabrication of the optical surfaces by replication. While the optical configuration remains Wolter Type I, the figure and smooth surface are produced on the outside of a solid mandrel. This is then coated with gold, after which a solid nickel shell is built

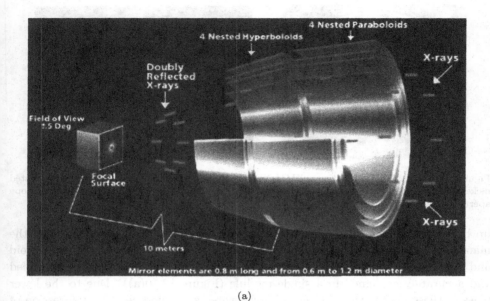

(a)

(b)

Figure 13.9 The *Chandra* mission. (a) The X-ray mirrors nested in Wolter I configuration. The 10 m focal length is measured from the junction of the paraboloid and hyperboloid arrays. (b) *Chandra* spacecraft with arrays deployed. Key elements of the payload are indicated. (Courtesy of SAO and NASA.)

Figure 13.10   (a) One of the three *XMM-Newton* X-ray mirror modules containing 58 nested nickel shells. (b) *XMM-Newton* spacecraft showing the three mirror module and optical telescope aperture. (Courtesy of ESA.)

up by electroforming. The shell with its gold coating is then separated from the mandrel under carefully controlled conditions. Each shell includes a paraboloid and hyperboloid section and the thickness is such that 58 shells can be assembled and accurately coaligned in a single module (Figure 13.10(a)). Due to the lower mass of these optics, the spacecraft carries three such modules. An impression of the spacecraft is shown in Figure 13.10(b). The three X-ray mirror modules can be seen with protective doors open. *XMM-Newton* also includes an optical/UV telescope of 0.3 m aperture which can detect objects of $m_v \sim 24$, an equivalent performance to that of a 4 m telescope on the Earth. *XMM-Newton* was launched on 10 December 1999 on an *Ariane 5* vehicle and, like *Chandra*, placed in a highly elliptical orbit with perigee 7,000 km and apogee 114,000 km. The orbital period is ~ 48 hours, of which ~ 40 hours is available for uninterrupted observation of a target source.

The layout of the X-ray instrumentation for two of the three telescope modules is shown in Figure 13.11. Arrays of gold-coated reflection gratings are placed behind the telescope so as to intercept and disperse 40% of the incoming X-rays on a metal oxide semiconductor (MOS) CCD array. The X-ray energy resolution of these devices is sufficient to allow photons diffracted in different grating orders to be identified even though the dispersed spectra are superimposed. For the third telescope module, there are no gratings and all of the X-ray flux goes to a novel pn CCD detector which has high quantum efficiency for X-rays up to 10 keV.

The angular resolution at < 5 arc sec is significantly coarser than that of *Chandra*. However, the effective aperture (area multiplied by reflectivity) is more than ten times greater at $E_\nu \sim 6$–7 keV — a key energy range for study of emission lines of neutral and highly ionised Iron. Thus *XMM-Newton* has the primary goal of measuring the X-ray spectra of distant objects and complements *Chandra*, which emphasises angular resolution but lacks the effective area.

X-ray images of the Sun have been obtained for more than 10 years with the *Yohkoh* grazing incidence telescope launched in 1991 (Figure 13.12). Given the

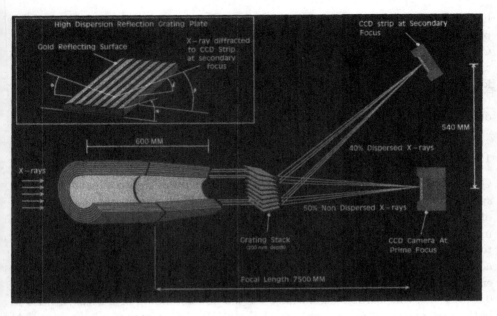

Figure 13.11　A stack of ruled gratings (detail in inset) forms spectra with resolution $E/\Delta E \sim$ 100–2000 on *XMM-Newton*. (Courtesy of ESA.)

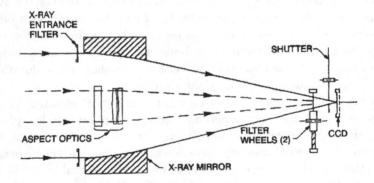

Figure 13.12　Solar X-ray telescope flown on the ISAS *Yohkoh* mission. (Tsuneta *et al.*, 1991; with kind permission of Kluwer Academic Press.)

comparatively large solar X-ray flux, a single pair of Wolter I type reflectors provides sufficient collecting area. The mirrors are figured and polished on low expansion Zerodur with a surface roughness of ~ 0.38 nm rms. Angular resolution is limited by the 2.46 arc sec pixel size of the CCD — a 1024×1024 pixel front illuminated device responding in the range of 0.3–4.5 nm. The temperature of the emitting plasma can be estimated by taking the ratio of signals obtained with different thin filters (Tsuneta *et al.*, 1991). There is also a thin entrance filter to eliminate visible light to which the CCD would respond. In order to maximise use of the available pixels, exposure is controlled with a high-speed mechanical shutter rather than by CCD frame transfer. An on-axis white light camera obtains solar images, which are registered on the same CCD, thus allowing features in the X-ray images to be identified with structures seen at visible wavelengths. Largely due to the X-ray telescope, *Yohkoh* has revolutionised our understanding of the solar corona with observations throughout a solar cycle.

13.3.5   *Gamma-ray*

Gamma-ray photons arise from nuclear rather than atomic processes and so strictly speaking the relevant energy range begins at 0.5 MeV. However, many of the targets of interest are also hard X-ray emitters and hence gamma-ray missions also include instruments sensitive down to $E_\nu \leq 10$ keV. The compact galactic objects referred to in Section 13.3.4 are in some cases also gamma-ray emitters. Similarly, the acceleration of electrons to relativistic energies in pulsars can lead to gamma-ray emission. The ability to detect emission lines from excited nuclei at energies above 0.5 MeV has considerable significance for the understanding of nucleosynthesis in both stars and supernovae. Images of the Galaxy in gamma-ray lines, e.g. ^{26}Al at 1.8 MeV, allow us to identify nucleosynthesis sites. For $E_\nu > 100$ MeV, the collisions of cosmic-ray protons with nuclei in galactic gas clouds produce emission from the decay of the resulting neutral pions. Going beyond the Galaxy, AGNs are also gamma-ray emitters, and in particular the Compton mission (see below) has identified blazars — a class of AGN — as being highly luminous above 100 MeV. Finally, the phenomenon of gamma-ray burst sources has assumed considerable significance since it became clear that, at least for the short burst duration, these might be the most luminous objects in the Universe.

Hard X-ray and gamma-ray photons are not reflected or refracted. In addition, given the steepness of high-energy photon spectra, comparatively few photons are available at the highest energies — particularly for extragalactic objects. Gamma-ray astronomy therefore requires large area detector systems with massive absorbing elements so that the photons may be converted to a useful signal from which energy and directional information may be obtained. The NASA *Compton Gamma-Ray Observatory* (*CGRO*) provides a good example of the approaches needed to operate at the highest energies (Figure 13.13(a)). Launched on the space shuttle in April 1991, with a mass of ~ 17 tonnes, it was placed in a near-Earth orbit. Sensitive in the energy range 30 keV to 30 GeV, it was, after the HST, the second of NASA's Great Observatory series and at the time the most massive astronomy spacecraft yet flown. For photons above 100 keV the Compton effect becomes important, while for $E_\nu > 30$ MeV it is necessary to rely on pair production for photon detection.

The EGRET instrument produces images in the 30 MeV to 30 GeV range using high-voltage gas-filled spark chambers. High-energy gamma rays enter the chambers

(a) (b)

Figure 13.13   (a) The NASA *CGRO* spacecraft; (b) the ESA *INTEGRAL* spacecraft. (Courtesy of ESA and NASA.)

and produce electron–positron pairs, which cause spark trails. The path of the particles is recorded, allowing the determination of the direction of the original gamma ray. A NaI crystal beneath the spark chambers measures the original gamma-ray energy. The instrument is surrounded by a scintillator shield for background rejection and includes a complex array of time-of-flight scintillation detectors to distinguish downward- from upward-moving particles. The space between the spark chambers contains thin plates made of a material with high atomic number to allow pair production to occur. Within a 20° field of view, source positions are measured with a precision in the range 5–30 arc min. This massive ($\sim$ 2 tonnes) and complex instrument still represents the state of the art for gamma-ray detection at $E_\nu > 30$ MeV.

For the energy range of 0.8–30 MeV, a large Compton Telescope (COMPTEL) is used to measure photon energy and arrival direction. It comprises two large scintillator arrays (4200 cm^2 upper, 8600 cm^2 lower) separated by a distance of 1.5 m, with each array surrounded by an anticoincidence shield. The incoming photons are Compton-scattered in the upper array and fully absorbed in the lower, with the locations and deposited energies of both interactions being measured. The instrument has a 1 sr field of view on the sky and can determine source positions with 5–30 arc min accuracy. This is also a large instrument with a weight of 1.5 tonnes. There are two remaining instruments. The Oriented Scintillation Spectrometer Experiment (OSSE) consists of four NaI scintillation detectors, sensitive to energies from 50 keV to 10 MeV. Each of these detectors can be individually pointed. This allows observations of a gamma-ray source to be alternated with observations of nearby background regions. An accurate subtraction of background contamination can then be made. Finally, the Burst and Transient Spectrometer Experiment (BATSE) is an all-sky monitor sensitive from about 20 to 600 keV. It comprises eight identically configured detector modules, each module containing two NaI(Tl) scintillation detectors. The eight planes of the large area detectors are parallel to the eight faces of a regular octahedron, with the orthogonal primary axes of the octahedron aligned with the coordinate axes of the *Compton* spacecraft. Thus this instrument has an all-sky view and is optimised for the detection of gamma-ray bursts.

The ESA *INTEGRAL* mission (Figure 13.13(b)), launched on 17 October 2002 on a Russian *Proton* rocket, emphasises energy and spatial resolution in the energy range of 15 keV to 10 MeV. Weighing 3600 kg, the spacecraft was placed in a highly elliptical orbit with perigee 10,000 km, apogee 153,000 km and inclination 51.6°. The orbit allows about 90% of its 72-hour duration to be spent above the charged particle belts. Thus uninterrupted source observations of up to 65-hour duration are possible.

The Spectrometer observes gamma-ray point sources and extended regions in the 20 keV–8 MeV energy range with an exceptionally good energy resolution of 2.2 keV (FWHM) at 1.33 MeV. This is accomplished using an array of 19 hexagonal high purity germanium detectors cooled by a Stirling cycle system to a temperature of 85 K. A hexagonal coded-aperture mask is located 1.7 m above the detection plane. It provides a fully coded field of view of 16° with an angular resolution of 2°. In order to reduce background radiation, the detector assembly is shielded by an anticoincidence system that extends around the bottom and side of the detector almost completely up to the coded mask. The aperture, and hence contribution by diffuse cosmic gamma rays, is limited to $\sim 30°$. A plastic scintillator veto system below the mask further reduces the 511 keV background from electron–positron annihilation.

The Imager provides 12-arc-min source identification and a response to continuum and broad lines over the 15 keV–10 MeV energy range. A tungsten coded-aperture mask, 3.2 m above the detection plane, is optimised for high angular resolution. As diffraction is negligible at gamma-ray wavelengths, the angular resolution is limited by the spatial resolution of the detector array. The Imager design takes advantage of this by utilising a detector with a large number of spatially resolved pixels, implemented as physically distinct elements. The detector uses two planes: one 2600 cm^2 front layer of CdTe pixels and a 3100 cm^2 layer of CsI pixels. The CdTe and CsI arrays are separated by 90 mm, which allows the paths of the photons to be tracked in 3D as they scatter and interact with more than one element. Events can be categorised and the signal-to-noise ratio improved by rejecting those which are unlikely to correspond to gamma-ray photons. The aperture is restricted by a lead shielding tube and shielded in all other directions by an active bismuth germanium oxide (BGO) scintillator veto system.

There are two further supporting instruments that extend the mission's photon energy range to X-ray and visible wavelengths. The X-Ray Imager will make observations simultaneously with the main gamma-ray instruments and provide images with arc min angular resolution in the 3–35 keV energy band. The photon detection system consists of two high-pressure imaging microstrip gas chambers (1.5 bar, 90%

xenon + 10% methane), at a nominal gas gain of 1500. Each detector unit views the sky through its coded-aperture mask located at ~ 3.2 m above the detection plane. The Optical Monitoring Camera (OMC) consists of a passively cooled CCD (2055 × 1056 pixels; imaging area 1024 × 1024 pixels) working in frame transfer mode. The CCD is located in the focal plane of a 50 mm aperture lens including a Johnson V-filter to cover the 500–850 nm wavelength range.

13.4 Photon Detection

For detection, photons must be absorbed in a medium — gas, solid or liquid — and the creation of detectable electric charge or field, heat or light must result. At low photon energies (radio, sub-mm, IR) where the wave nature of the radiation is important, *coherent* detectors, which respond to the electric field strength of the incoming signal, are used. They can preserve phase information. *Thermal* detectors simply absorb the radiation and the resulting absorber temperature increase changes its electrical properties in a measurable way. Primarily useful for IR/sub-mm radiation, they have recently shown potential as single photon X-ray detectors. *Photon* detectors respond directly to the incoming quanta, which have sufficient energy to release charge in a solid or gaseous absorber. The charge may be directly collected or produce secondary effects, e.g. visible light in a scintillator. Crucial to the useful detection of photons are quantum efficiency — the fraction of the photons converted to signal, and noise — the uncertainty in the output signal.

13.4.1 *Low energy photons — radio, sub-mm, IR, visible*

For radio, sub-mm and far-IR photons, coherent and thermal detectors are commonly used, while in the visible range, the charge coupled device (CCD), which relies on production of charge in an array of silicon elements or pixels, is the current system of choice. We will also comment briefly on superconducting tunnel junctions (STJs; see Section 13.4.2), which have potentially important applications for non-dispersive spectroscopy in the UV and visible ranges.

13.4.1.1 *Hetrodyne receivers*

Among coherent detectors, hetrodyne receivers are the most relevant for astronomy. An incoming signal is added to a local oscillator signal. The amplitude of the summed signal is modulated at the difference or 'beat' frequency. If this 'mixing' is done in a nonlinear circuit element, e.g. a diode, then ideally $I \propto V^2$ or $\propto E^2$, where E is the electric field strength, and the mixer produces an output proportional to the power of the incoming signal. A diagram showing the key elements of a hetrodyne receiver system is shown in Figure 13.14. The incoming radiation field is combined with the field generated by the local oscillator and fed to a secondary

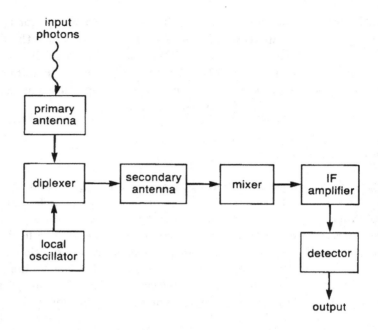

Figure 13.14   Elements of a hetrodyne receiver system.

antenna — often a wavelength-matched wire surrounded by reflectors to improve its transfer efficiency. Arrangements of this type are called quasi-optical. Feed-horns coupled to waveguides are also commonly used for $f \leq 800$ GHz. The mixed signal will be amplitude-modulated at the intermediate frequency $f_{IF} = |f_S - f_{LO}|$, where the signal at f_{IF} contains spectral and phase information for the incoming signal at f_S provided that the local oscillator signal (f_{LO}) remains steady in frequency and phase and that f_{IF} remains within the system bandwidth. The output of the mixer is passed to an intermediate frequency (IF) amplifier and then to a rectifying and smoothing circuit called a 'detector' which is not a photon detector but rather an array of filters tuned to different frequencies with spectrometers — digital autocorrelators or acousto-optic spectrometers — on their outputs.

The hetrodyne technique involves the down-conversion of a very high incoming frequency, at which there may be no available amplifiers, to a lower frequency (f_{IF}) that can be handled by microwave signal-processing techniques with the possibility of obtaining high spectral resolution. Current mixer devices include Schottky diodes, superconductor–insulator–superconductor (SIS) devices and hot electron bolometers (HEBs). A detailed discussion of these can be found in Rieke (1994) and references therein. The Shottky device is a classical semiconductor diode but with structure optimised for high frequency operation by means of a small contact between a metal and a semiconductor. Its performance can be significantly enhanced by operation at 70 K. The SIS device employs superconductors to produce a strongly non-linear circuit element. With the superconducting band gap ~ 1000 times smaller than that in a semiconductor, the SIS devices operate with lower local oscillator power and greater sensitivity than the Schottky diode. However, they must be operated below the critical temperature for superconductivity ($T \cong 5$ K). SIS devices are not effective for $f > 1200$ GHz but recently the use of fast HEB mixers has been allowing operation up to ~ 2000 GHz. For IF amplification, special transistors — high electron mobility transistors, HEMTs — are fabricated in gallium arsenide. With small electrode structures, e.g. gate ≤ 1 μm, operation up to $f \sim 10^{11}$ Hz can be achieved.

The High Frequency Instrument (HIFI) on ESA's *Herschel* mission (de Graaw *et al.*, 2000) provides a good example of a space hetrodyne system. It will cover the band 480–1250 GHz with SIS mixers while HEB systems will be used in the 1410–1910 GHz band. For spectral analysis, the lower frequency range will use a wide band acousto-optical spectrometer with an instantaneous IF range of 4 GHz and a resolution of 1 MHz. In the high frequency band, a high resolution digital auto-correlator spectrometer will cover an IF frequency range of either 1 GHz (resolution 200 kHz) or 500 MHz (100 kHz resolution). Most observations will involve beam switching to measure background with a separation of ~ 3 arc min between target and reference field and a chopping frequency of ~ 1 Hz.

13.4.1.2 *Bolometers*

These thermal detectors, so far principally used in the IR and sub-mm bands, absorb photons and measure the resulting rise in temperature through the changes in electrical properties of very sensitive thermometers. In the thermal model in Figure 13.15(b), the detector is connected by a link of thermal conductance G to a heat sink at temperature T_S. If the detector absorbs a constant power P_{in} which raises the temperature by T_{in}, then by definition, $G = P_{in}/T_{in}$. If we assume that a fluctuating photon power $P_\nu(t)$ is deposited in the detector and absorbed with efficiency ε, then

$$\varepsilon P_\nu(t) = \frac{dQ}{dt} = C\left(\frac{dT_{in}}{dt}\right), \tag{13.3}$$

where C is the heat capacity given by $dQ = CdT_{in}$. The total power being absorbed is then

$$P_\nu(t) = P_{in} + \varepsilon P_\nu(t) = C\left(\frac{dT_{in}}{dt}\right) + GT_{in}. \tag{13.4}$$

If $P_\nu(t)$ is applied to the detector as a constant step function at $t \geq 0$ when $P_\nu = P_1/\varepsilon$, then the solution of (4) is

$$T_{in}(t) = \frac{P_{in}}{G} + \frac{P_1}{G}(1 - e^{-t/(C/G)}) \tag{13.5}$$

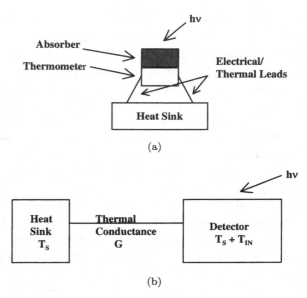

Figure 13.15   Bolometer schematics. (a) Simple view of operational elements; (b) thermal model.

and the thermal time constant of the detector is $\tau_T = C/G$. So for $t > \tau_T$, $T_{\text{in}} \propto P_{\text{in}} + P_1$, if T_{in} can be measured, then the system can measure the input power. The elements of a bolometer system are shown in a simple sketch in Figure 13.15(a). The radiation absorber is attached to a thermometer which is usually a small Si or Ge chip that has been doped to yield a large temperature coefficient of resistance and to have an impedance suitable for matching to a low noise amplifier — usually a JFET. Doping can be carried out either by ion implantation (Si) or by exposing high purity material (Ge) to a flux of neutrons in a reactor. For the latter technique — nuclear transmutation doping, NTD — the dominant doping reaction occurs when Ge atoms capture neutrons and decay to Ga. The device is called a composite bolometer since absorber and thermometer are separate items, so each can be optimised separately. The connecting leads both establish the thermal conductance to the heat sink and carry current to the resistor. The high impedance low noise amplifier measures the voltage, which responds to changes in resistance and hence to changes in temperature. So, from (13.5), changes in input power are measured.

The detailed theory of bolometer operation is given by Rieke (1994). Key performance parameters are noise equivalent power (NEP) and electrical time constant $\tau_e = C/(G - \alpha P_D)$, where $\alpha(T)$ is the temperature coefficient of resistance and P_D is the power dissipated in the detector. Thus the frequency response could be improved by increasing G. However, for the two principal components of electrical noise, Johnson and thermal noise, $\text{NEP}_J \propto G T^{3/2}$ and $\text{NEP}_T \propto G^{1/2} T$. Hence it is necessary to trade sensitivity and speed of response by varying G, with $(\text{NEP})\, \tau_e^{1/2}$

as a useful figure of merit, while keeping C as low as possible. Since G is $\propto T^2$, operating at the lowest possible temperature is clearly an advantage.

We have discussed the ESA *Herschel* and *Planck* missions in Section 13.3.2. Both employ semiconductor bolometer-based instruments to cover parts of their wavelength range — 250–670 μm for the *Herschel* Spectral and Photometric Imaging Receiver (SPIRE) and 115–1000 μm for the *Planck* High Frequency Instrument (HFI). Griffin (2000) discusses the application of bolometers to these missions. The devices used, which represent the present state of the art in the field, are NTD Ge 'spider web' composite bolometers developed by the Caltech/JPL group (Mauskopf *et al.*, 1997). The absorber, which is a metalised silicon nitride structure, is shown in Figure 13.16. The thermometer element, a small NTD germanium crystal (Haller, 1985), can be seen in the middle of the web structure. The spacing in the web pattern is much smaller than the wavelength of the incoming radiation, so the web acts like a plane absorber. Besides drastically reducing the overall heat capacity, the web minimises the cross section for interaction by cosmic ray particles. This bolometer design has achieved NEP = $1.5 \cdot 10^{-17}$ W/$\sqrt{\text{Hz}}$ with a time constant $\tau_e = 100$ ms at 300 mK, while at 100 mK — the operating temperature of the *Planck* bolometers — NEP = $1.5 \cdot 10^{-18}$ W/$\sqrt{\text{Hz}}$ with $\tau_e = 65$ ms. Comparison of these figures illustrates the importance of low temperature operation.

Given that an individual bolometer (a few mm) is small compared even to the telescope diffraction spot (tens of mm) and that the first pass absorption efficiency is only 50% for an absorber in free space, the device is usually mounted in an integrating cavity and coupled to the incident beam from the telescope through a single mode feedhorn. Griffin (2000) discusses the optimum packing of an array of bolometers and their feedhorns to sample as much as possible of the telescope focal plane. However, there are significant gaps in coverage and typically 16 separate pointings are required to create a fully sampled map of the telescope field of view.

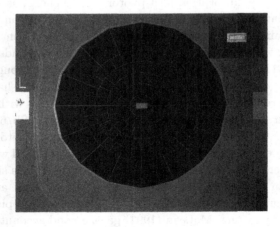

Figure 13.16   Enlarged image of the JPL-Caltech bolometer. (Courtesy of J. Bock.)

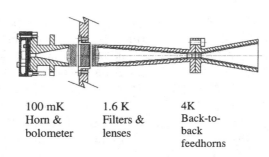

100 mK	1.6 K	4K
Horn &	Filters &	Back-to-
bolometer	lenses	back
		feedhorns

Figure 13.17   The *Planck*/HFI feedhorn-coupled bolometer. (Courtesy of ESA.)

A major advantage of this approach lies in excellent rejection of stray light and EM interference together with the elimination of sidelobes by careful feedhorn design. This is especially important for the *Planck* mission, which will undertake sensitive measurements of temperature anisotropies in the cosmic background radiation. The triple feedhorn structure to be used in the *Planck* high frequency instrument is shown in Figure 13.17. While it is possible in principle to eliminate feedhorns and use an array of square bolometers to give full image sampling, issues of vulnerability to stray light and interference remain. However, for the future, transition edge sensors (TESs) with SQUID-based multiplexers (Chervenak *et al.*, 2000) may allow the successful fabrication of large filled arrays. Single detectors have already been built that show superior values of (NEP) $\tau_e^{1/2}$ at 300 mK. We will discuss the TES devices in the context of X-ray observations, where they are also showing considerable promise.

13.4.1.3   *Charge-coupled devices (CCDs)*

CCDs are photon-detecting pixel arrays that make use of intrinsic photoconduction. In a semiconductor, the absorption of a photon whose energy is greater than the band gap of the material will lift an electron into the conduction band and create an electron–hole pair that can conduct measurable electric current. The commonly used element is Si, whose band gap energy of 1.12 eV corresponds to a long wavelength cut-off of 1.11 μm. Use of CCDs has been extended throughout the visible into the X-ray range up to a photon energy of $\sim$ 10 keV and has revolutionised astronomy. For IR response there are intrinsic semiconductors with lower band gap energies and longer cut-off wavelengths, e.g. Ge/1.85 μm, InSb/6.9 μm. The addition of impurities to Si will produce an extrinsic photoconductor in which lower energy photons can excite electrons from energy levels associated with the impurity atoms that lie closer to the conduction band. Impurity elements such as B and Sb added to silicon can extend the IR response to $\lambda \le 30$ μm, while addition of impurity atoms to germanium can push this to $\lambda \le 130$ μm. The topic of IR arrays is a specialist and complex one. McLean (1997) gives a good account of these devices and their application in astronomy.

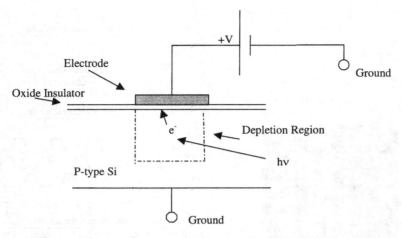

Figure 13.18   Diagram of the MOS capacitor structure. Charge created in the depletion region by photon absorption is stored under the electrode for later transfer.

For the operation of CCDs, the key element is the metal-oxide–semiconductor structure, which behaves like a capacitor (Figure 13.18). When a positive voltage is applied to the p-type silicon at the electrode, the majority carrier holes are repelled and a depletion region is swept free of charge. Incoming photons produce electron–hole pairs of which the electrons are attracted to the insulator under the electrode where they are held pending transfer. The charge storage capability depends on the applied voltage and on the size and depth of the depletion region, which are set by the electrode dimensions, and the resistivity of the Si.

A practical device consists of an array of MOS capacitors arranged in groups of three for the readout that is commonly used. A sheet of p-type silicon is covered with an insulating layer of silicon oxide, on top of which is placed three sets of electrode strips for parallel transfer of collected charge (Figure 13.19(a)). Following exposure, each element or pixel of the array will have acquired a charge proportional to the input photon flux. An appropriate application of phased voltage waveforms (clocking) will then cause the charge to be transferred in the parallel or y direction. Each set of three transfer electrodes defines a row, with the individual pixels in a row being defined by column isolators or channel stop diffusions. These are heavily doped, and therefore conducting, regions that inhibit field penetration. Hence the charge from each pixel is transferred only in the y direction towards the serial output register. Here the electrode orientation and charge flow are appropriate for charge transfer in the x direction. Charge is transferred in parallel between rows but, after each row transfer, the serial register charges are shifted out in the x direction to the output amplifier. This process is continued until all the pixel charges resulting from a single exposure to a scene have been cleared.

The diagram in Figure 13.19(b) indicates how charge is transferred from one pixel to the next. Each pixel has three electrodes and the first schematic shows

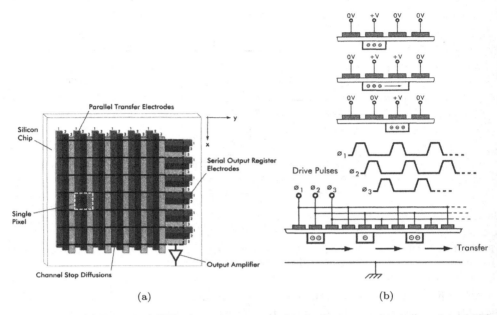

(a)   (b)

Figure 13.19   (a) Format of CCD electrode structure showing storage and output register MOS elements. (b) Three phase clocking scheme for transfer of charge between electrodes and from pixel to pixel.

the transfer of charge from one *electrode* to the next. The voltage on an adjacent electrode is first raised to the same value as that on the initial electrode. This results in a sharing of charge. The initial electrode voltage is then set to zero and the disappearance of the related potential well results in the transfer of all the charge to the adjacent electrode. During the exposure, the middle electrode on each pixel is held high to ensure collection of the liberated charge. Steps i and ii must be repeated three times in order to complete a pixel-to-pixel transfer. The necessary clock waveforms and the *pixel* transfer are shown in the lower part of the schematic. Pixel dimensions are typically in the range of 12–25 μm. McLean (1997) describes the use of a buried channel, an additional n-type region that enables charge transfer to take place some distance below the plane of the electrodes. Use of this structure avoids the charge trapping that can occur due to crystal defects near the surface. McLean also discusses the processes of reading out and digitising the charge packets from each pixel as they emerge from the output register.

When first developed, CCDs were illuminated from the front and so incoming photons had to pass through the electrodes and the insulating layer. Although these were kept thin and the electrodes in particular were made from heavily doped silicon, front illumination leads to reduced quantum efficiency and in particular to an inability to respond to UV photons. While the front-illuminated QE can be improved by 'virtual phase' operation (Janesick *et al.*, 1981; see also McLean, 1997) with the use of only one physical electrode per pixel, a more direct approach which extends the response to EUV and soft X-ray wavelengths involves illuminating from the back side of the detector. Figure 13.18 shows that below the depletion region, there is a layer of p-type Si of undetermined thickness. Back side illumination requires thinning of the detector, usually by etching away the intervening Si so that the physical depth of the device is as close as possible to the depletion depth d. Here $d = (2\kappa\varepsilon_0\mu\rho_s V)^{1/2}$, with κ the dielectric constant, ε_0 the permittivity of free space, μ the charge carrier mobility and ρ_s the resistivity of silicon. V, the applied voltage, must be chosen so that d is always just less than the physical depth of the device. For typical CCDs, $\rho_s \sim 10$–25 Ω cm, which leads to $d \sim 3$–10 μm. We will see later that use of higher resistivity silicon can allow higher values of d and the efficient detection of X-ray photons with $E_\nu \leq 15$ keV.

It is crucial during transfer to preserve the value of charge collected in each pixel. Since CCD formats are now of order 4 k×4 k pixels, the charge in a pixel remote from the readout amplifier must undergo many transfers before being read out. The key figure is the charge transfer efficiency (CTE) between pixels. CTE values > 0.99999 are required to keep the output charge loss at an acceptable level of $< 1\%$. Using the buried channel approach helps maintain CTE. More generally, dark current also affects CCD operation, with thermal generation in the neutral bulk silicon in the depletion region and from surface states at the Si–SiO$_2$ interface. The latter represents the largest current source by a factor of $\sim 10^2$–10^3, where the wide range is due to the effect of processing variability. Operation in inversion mode allows holes from the channel stops to neutralise the surface states. Use of multi-phase pinned (MPP) architecture, which involves the addition of a boron implant below the phase 3 electrode, greatly reduces the impact of surface states. These techniques are well described by McLean (1997). Finally, operation at low temperature has a dramatic impact on noise sources, as may be seen from Equation (13.6), where the pixel generated dark current as $f(T)$ is given by

$$N_\mathrm{D} = 2.55 \cdot 10^{15} N_\mathrm{o} d_\mathrm{p}^2 T^{1.5} \exp\left(\frac{-E_\mathrm{G}}{2kT}\right) \quad \text{e/s/pixel}, \tag{13.6}$$

with N_o the dark current in nA/cm^2 at room temperature, d_p the pixel size in cm, T the operating temperature in K and E_G the band gap energy. CCDs are generally operated below 223 K while even lower temperatures, $T \leq 190$ K, will ameliorate the effect of radiation damage by ensuring that traps related to damage-induced defects in the crystal remain filled.

13.4.2   *Higher energy photons — EUV and X-ray*

A brief discussion of high energy X- and gamma-ray detection is included in Section 13.3.5 above. Here we will focus on the energy range of 0.03–10.0 keV. X-ray astronomy in particular has in the past relied on proportional counters and microchannel plates where, for $E_\nu \sim 1$ keV, the former combined a measure of non-dispersive energy and position resolution ($E/\Delta E \sim 5$ and $\Delta x \sim 200$ μm) while the latter provided no energy resolution but $\Delta x \sim 20$ μm. Proportional counters are disappearing from use while microchannel plates, though employed very effectively on the *Chandra* mission, will for astronomy be confined to niche applications for $\lambda \geq 50$ nm. While their application as particle detectors (ion and electron) in space plasma physics missions is of growing importance, they will not be discussed here. Fraser (1989) gives a good account of these systems.

13.4.2.1   *X-ray and EUV application of CCDs*

The development of back thinned and illuminated CCDs has greatly extended the device response to higher energy photons. Visible light photons produce one hole-electron pair, and so with readout noise ≤ 5 e$^-$, *many* photons must be integrated in a pixel to yield a detectable signal. At X-ray and EUV energies significantly above the silicon band gap, 3.65 eV is required to create a hole–electron pair, with the remaining energy above E_G going to heat the crystal lattice. This compares favourably with the ~ 30 eV needed to create an ion–electron pair in a gas-filled proportional counter, and so the absorption of a *single* 100 ev photon liberates ~ 30 electrons and allows a degree of non-dispersive energy resolution to be achieved. However, all of the charge created by the incoming photon must be collected for measurement at the output amplifier. This aim can be compromised since bare silicon develops a 3 nm native oxide layer in air. Together with the uncertainty in etching away the correct depth of silicon to match the depletion depth, the device may have a field-free entry region of ~ 30 nm in which some of the charge created by an absorbed photon is either lost to recombination or diffuses laterally so that it is not captured in the depletion region of a single pixel. Such charge loss or spreading will degrade both energy and position resolution, particularly for photons with $E_\nu \leq 500$ eV. Several techniques has been tried with the aim of allowing an electric field to be sustained in this entry region so that any charge created may be guided into the appropriate depletion region. Implantation of accelerated ions or pulsed laser annealing of the surface in a gas of appropriate ions for implantation has been successfully used to allow the establishment of an electric field in this region. The data in Figure 13.20 show measured and calculated quantum efficiencies for the wavelength range for photons with $\lambda \leq 60$ nm or $E_\nu > 20$ eV (Lemen, 2002). This device, manufactured by EEV, has $d \sim 8$ μm and a back side ion implantation. The solid line shows the calculated QE for $1 < \lambda < 1000$ nm. Measured points for the visible range also show good agreement with the model. The two points indicated in black are from weak and poorly resolved EUV lines and are probably subject to large systematic errors. Similar devices will be employed in the EUV Imaging Spectrometer (EIS) on the ISAS *Solar-B* spacecraft which will register solar spectra from a multilayer-coated diffraction grating in the 15–30 nm range (Culhane *et al.*, 2002).

At higher X-ray energies, the depth of the depletion region where photons are absorbed sets the quantum efficiency (QE). The QE values for the two CCD cameras

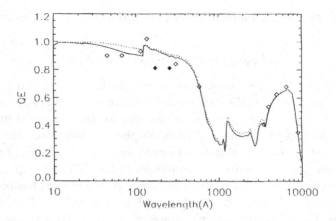

Figure 13.20 EEV Type 42/10 measured QE and model calculations (10 A = 1 nm). (Courtesy of J. Lemen.)

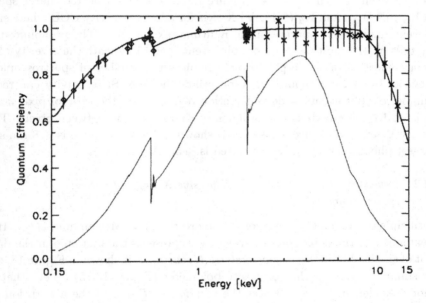

Figure 13.21 Quantum efficiency for the *XMM-Newton* PN (*upper*) and MOS (*lower*) CCDs. (Courtesy of ESA.)

on the *XMM-Newton* mission are compared in Figure 13.21. One is based on MOS technology and is front-illuminated with $d = 40$ μm (Turner *et al.*, 2001) while the other employs an array of PN junctions and is back-illuminated with $d = 300$ μm (Struder *et al.*, 2001). Because of its smaller d value, the MOS QE begins to fall at $E_\nu > 4$ keV while the PN QE remains high up to 10 keV. At lower energies, the MOS QE values are less than for the PN due to the absorbing effect of the insulating oxide layer and charge transfer electrodes (thickness $\sim$ 1–2 μm of Si).

For CCDs used in a single photon per pixel mode, the energy resolution is given by

$$\Delta E_{\text{FWHM}} = 2.36 \cdot \omega \sqrt{N^2 + F \cdot E/\omega}, \qquad (13.7)$$

where $\omega = 3.65$ eV, N is the system noise in rms electrons, which is principally read-out amplifier noise assuming the dark current contribution has been suppressed, and $F \sim 0.12$ is the Fano factor for Si which accounts for the statistical uncertainty in the process of heat loss to the lattice following photon absorption. Hence the read-out noise N must be minimised if useful resolution is to be achieved for low energy photons. Values ≤ 5 e$^-$ rms are now available for the CCD on-chip output amplifiers, and for sampling times of ~ 20 μs per pixel, $N \sim 2$ e$^-$ rms has been achieved. Energy resolution for a noise-free system and for two values of readout noise is plotted against E_ν in Figure 13.22(a). However, for $E_\nu \leq 700$ eV, loss of charge in the initial absorption process will significantly degrade the energy resolution. This occurs because of the splitting of charge between neighbouring pixels. The impact can be seen even at higher energies. Figure 13.22(b) shows collected charge spectra of Mn K$_\alpha$ (5.9 keV) and K$_\beta$ (6.4 keV) X-rays for a low resistivity CCD where single pixel events have been separated from multiple pixel events. The pronounced low energy tails on the multiple pixel or split event spectra indicate the severity of the problem. Selection of single pixel events reduces the number of spectroscopically valid events by ~ 2. Although an appropriate choice of Si resistivity can reduce the number of split events, a deep depletion region where the ratio of pixel size to depth is $\ll 1$ can exacerbate the problem of charge sharing between pixels. Thus, for the EUV sensitive device whose QE is shown in Figure 13.20, d is ~ 8 μm since the largest photon energy to be registered is ~ 50 eV.

13.4.2.2 *Cryogenic detectors for non-dispersive X-ray and optical spectroscopy*

The principle of bolometer operation has more recently (McCammon *et al.*, 1984) been extended to the detection of X-rays and represents the first step in the development of single photon *microcalorimeters*. A schematic is shown in Figure 13.23(a) and is similar to that of the composite bolometer (Figure 13.15) except that the thermometer element now measures the increase in T due to the absorption of a *single* X-ray photon. The ideal absorber should ensure that as much as possible of the incoming X-ray energy is converted to heat described as phonons or quantised vibrations of the crystal lattice. For $T \leq 1$ K, mean phonon energy (kT) is $\leq 10^{-4}$ eV, so the statistical fluctuation in phonon number will be ~ 1000 times less than for the charge created following photon absorption in a silicon CCD where $\omega = 3.6$ eV. Thus the possibility emerges for achieving non-dispersive energy resolution similar to that of a Bragg spectrometer, which has substantially lower throughput for incoming photons. In an ideal situation, where all other noise sources can be made small, the energy resolution will be set by the phonon number fluctuations

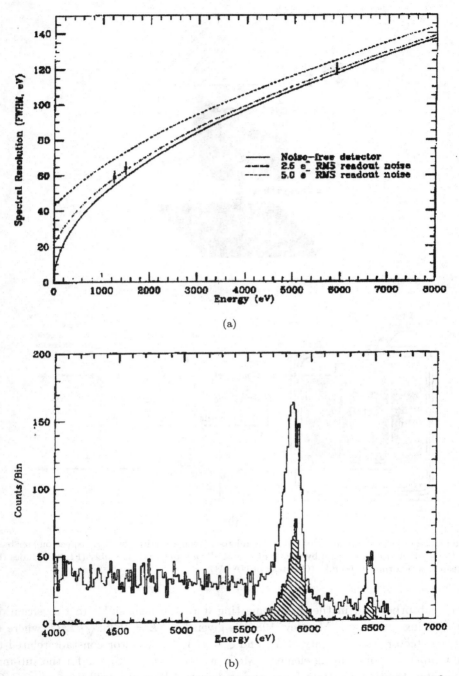

(a)

(b)

Figure 13.22   (a) Energy resolution vs E_ν for zero, 2.5 and 5 electron rms readout noise figures. (b) Mn K_α (5.9 keV) and K_β (6.4 keV) spectra for single (hatched) and multiple pixel events. Data are from an MIT Lincoln Laboratories CCID 7 device. (Lumb *et al.*, 1991)

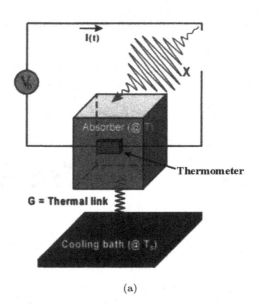

(a)

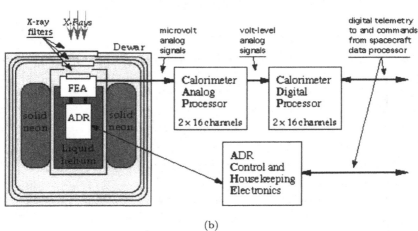

(b)

Figure 13.23   (a) The elements of the X-ray micro-calorimeter. (b) The X-ray spectrometer built for the ISAS *Astro E2* mission by the GSFC group. The Front End Assembly (FEA) includes the detector, which runs at 60 mk. (Courtesy of NASA.)

in the absorber and in the link connecting it to the heat sink. In this situation the limiting resolution (Moseley, 1984) is given by $\Delta E = 2.36\xi\sqrt{kT_S^2 C}$, where C is the detector heat capacity at T_S and $\xi \sim 1$–3 is a detector constant related to the temperature-measuring element (McCammon *et al.*, 1986). As for the sub-mm bolometer, the time constant for restoring detector thermal equilibrium $\tau_T = C/G$, so in the single photon application this will be the fall time of the temperature pulse that follows X-ray absorption.

The microcalorimeter to be flown on the ISAS *Astro-E2* mission in 2005 is shown in Figure 13.23 (Kelley *et al.*, 2000). Placed at the focus of a nested array of thin shell grazing incidence mirrors, it has 32 separate detectors covering a field of view of ~ 10 arc min^2. Pumped solid neon and liquid helium cryogens establish temperatures of 17 K and 1.3 K while an adiabatic demagnetisation refrigerator holds $T_S \sim 60$ mK. Thermometer elements are ion-implanted silicon and absorbers are ~ 8 μm HgTe, giving 95% QE at 6 keV. Absorber choice is a compromise between good QE and low heat capacity. For detector heat capacity $C \sim 2 \cdot 10^{-13}$ J, the limiting resolution at 60 mK is $\Delta E \sim 6$ eV, while the value achieved is ~ 10 eV. The thermal link conductance $G \sim 2 \cdot 10^{-11}$ W/K establishes the single photon pulse fall time at 6 ms. Although this limits the photon counting rate to ~ 10 events/s to minimise pileup, a range of selectable neutral density X-ray filters is included to increase dynamic range.

Another approach to microcalorimetry uses a transition edge thermometer combined with an absorber to form a transition edge sensor (TES). The device, first developed by the NIST group (Irwin, 1995), uses a superconducting layer biased in temperature to lie within the narrow transition range between normal and superconducting states. This situation is shown in Figure 13.24(a), where the parameter α characterises the sharpness of the transition. The sensor (Figure 13.24(b)) is biased by a current flowing through the supeconducting layer, which is made from two materials (e.g. Al/Ag), one of which is superconducting. Use of a second element in close proximity allows the tuning of the superconducting transition temperature. Typically used bilayers include Al/Ag, Al/Cu, Ir/Au Ti/Au, Mo/Au and Mo/CU, while the transition temperature ranges from 15 mK to 1 K. The silicon nitride membrane provides a thermal coupling to a heat sink (not shown) whose temperature must be significantly below the transition value. As the bilayer cools, its

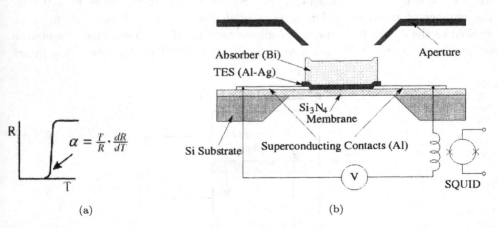

$$\alpha = \frac{T}{R} \cdot \frac{dR}{dT}$$

(a)   (b)

Figure 13.24   (a) Plot of R against T, showing the transition from the superconducting to normal resistivity state. (b) TES schematic; the Si substrate is connected to a heat sink which is at a temperature significantly below the transition temperature.

resistance falls and heating in the layer (V^2/R) increases. A stable operating point, and corresponding resistance value, is reached when the Joule heating in the layer equals the heat flowing to the sink. Thus the TES stabilises within its resistivity transition range due to an effect called negative electrothermal feedback.

Following energy deposition by an absorbed X-ray photon, temperature and resistance increase, leading to a decrease in current and Joule dissipation while the heat flow to the sink remains the same. The X-ray energy is thus proportional to the transient current decrease but, because of the electrothermal feedback (loop gain ~ 15), the pulse decay time is much faster than the natural relaxation time τ_T and so the output counting rate limit increases to ~ 500 events/s. The same trade-off as for the bolometer-based system must be made between X-ray absorption and thermal capacity. Bismuth of 1.5 μm thickness is used in the NIST device shown in Figure 13.24(b). The low impedance of the TES ($\sim$ mΩ) allows coupling to a low noise, low power superconducting quantum interference device (SQUID) amplifier. The energy resolution so far achieved by the NIST group is ~ 2 eV at 1.5 keV (Woolman *et al.*, 2000), but the need to trade resolution with pulse fall time must be kept in mind. For future X-ray astronomy applications, e.g. Constellation-X, XEUS, the need to develop arrays of TES sensors will require significant effort. A number of possible schemes for both array construction and readout are being studied by the SRON group (Bruijn *et al.*, 2000).

Although STJs have achieved energy resolution of ≤ 10 eV for 1 keV photons, interest in their application has shifted to single photon detection in the visible/UV range, so we will focus on that aspect here. Peacock *et al.* (1998) and Verhoeve (2000) provide detailed discussion. The STJ consists of two superconducting layers separated by a thin insulating barrier (Figure 13.25). Absorption of a photon is followed by a series of fast processes in which the photon energy is converted to free electrons by breaking Cooper pairs. For operating temperatures of $\sim 0.1\ T_c$, the number of thermal carriers is very small but a magnetic field, B, must be applied to the device to suppress Josephson current. If the absorption of a photon of wavelength λ creates N_o photons, then $N_o(\lambda) = 7 \cdot 10^5/\lambda E_{Ta}$, where λ is in nm

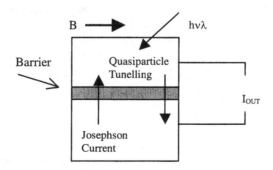

Figure 13.25   Operation of the STJ as a photon detector.

and $E_{\text{Ta}} \sim 0.664$ meV is the energy gap for tantalum ($T_c \sim 4.5$ K). A 500 nm photon will create ~ 2000 electrons.

Applying a DC bias across the two superconducting films and the thin barrier forms the tunnel junction and enables the transfer of quasiparticles from the upper to the lower film by the process of quantum-mechanical tunnelling. Following an initial tunnelling, a particle can tunnel back to the upper film and then tunnel forwards again. This process can be repeated several times before the loss of the particle. Hence on average each quasiparticle will contribute $\langle n \rangle$ times to the signal, with $\langle n \rangle$ in the range of 1–100, depending on the construction of the STJ. Multiple tunnelling can be enhanced by adding thin layers of lower gap material, e.g. Al, on each side of the barrier so that the quasiparticles are forced to remain close to the barrier and away from possible surface trapping sites. If we include the fluctuations in $\langle n \rangle$ (Goldie, 1994), then for a perfectly symmetrical junction and in the absence of noise components, the resolution in nm is

$$\delta\lambda_{\text{Ta}} = 2.8 \cdot 10^{-3}\lambda^{3/2}E_{\text{Ta}}^{1/2} \left(F + 1 + \frac{1}{\langle n \rangle} \right)^{1/2}, \tag{13.8}$$

where $F \sim 0.22$ is the Fano factor appropriate to tantalum. Thus for large $\langle n \rangle$, the limiting resolution $\delta\lambda_{\text{Ta}} \sim 30$ nm for $\lambda = 500$ nm or $\lambda/\delta\lambda_{\text{Ta}} \sim 15$. In practice there will be at least two further contributions to fluctuation, namely analogue electronic noise — independent of E_ν — and detector spatial non-uniformity — $\propto E_\nu$. It is the latter effect in particular that limits the energy resolution for X-ray photons. Results obtained by the Estec group (Verhoeve, 2000) are given in Figure 13.26. The plot is in energy units but extends from 246 nm to 1983 nm, with the measured resolving powers for tantalum ranging from 25 down to 4 at the long wavelength end. Measured resolution, electronic noise and device-limited resolution are shown. The predicted tunnel-limited resolutions are also plotted for Molybdenum ($E_G = 0.139$ meV) and Hafnium ($E_G = 0.02$ meV). The response of a tantalum STJ as measured at Estec is shown in Figure 13.27. The first order wavelength of the monochromator is set to 2000 nm and the signal is shown to be resolved above the noise level. The detected spectrum extends to ninth order at a wavelength of 22 nm. The STJ is mounted on a sapphire substrate and is illuminated from the back through the sapphire. It is operated at $T \sim 0.3$ mK, which is $< 0.1T_c$ for tantalum. The device quantum efficiency is shown in Figure 13.28 for the UV range of 140–240 nm. Measured points, taken at the BESSY/PTB synchrotron facility, agree well with the calculated values. The drop in sensitivity at 140 nm is due to the short wavelength cut-off of sapphire. Use of a MgF$_2$ substrate would extend the response to 115 nm.

A 6×6 element array has been built and used for several astronomical observations, including the observation of the Crab Nebula's optical pulse profile (Perryman *et al.*, 1999) and the measurement of QSO redshifts to $z > 4$ (de Bruijne *et al.*, 2002). An 18×50 element array is being developed. A diagram showing the construction

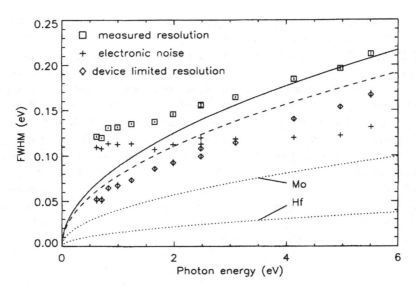

Figure 13.26   Measured and calculated energy resolution in eV plotted against E_ν. (Courtesy of P. Verhoeve.)

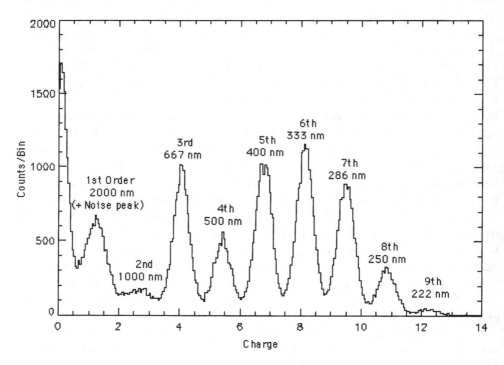

Figure 13.27   Tantulum STJ response to input from a grating monochromator, with intensity plotted against collected charge. (Courtesy of ESA.)

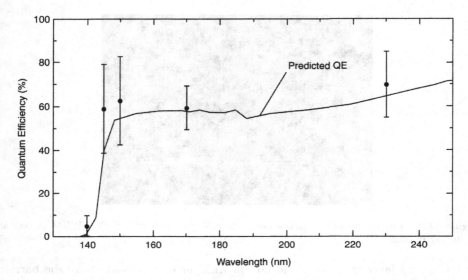

Figure 13.28 Calculated and measured quantum efficiency for the Estec tantalum STJ. (Courtesy of ESA.)

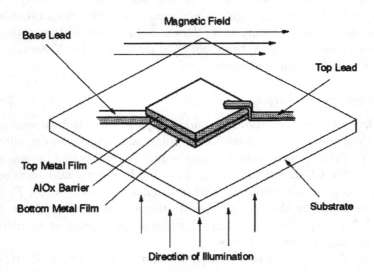

Figure 13.29 Construction of the tantalum STJ. The magnetic field suppresses Josephson current. (Courtesy of ESA.)

of a single STJ is given in Figure 13.29. The device has a 20 μm $\times$ 20 μm cross section, with the lower film, mounted on the sapphire substrate, being 100-nm-thick epitaxial tantalum on top of which is a 30 nm quasiparticle trapping layer. The two tantalum films sandwich an aluminium oxide barrier through which the electrons tunnel. The 1-nm-thick barrier is formed by controlled oxidation of the

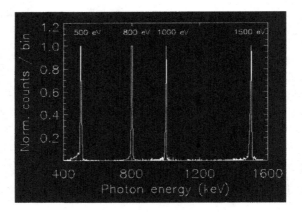

Figure 13.30   Energy resolution at energies < 2 keV. Below 1 keV, 6 eV FWHM is typical. (Courtesy of ESA.)

first trapping layer while a second 30 nm Al layer is deposited above the barrier. The construction is completed by a deposit of 100 nm of poly-crystalline tantalum. Bias is applied through the leads and the collected charge signals are read out using a charge-sensitive preamplifier operated at room temperature. Although the energy resolution at $E_\nu > 1$ keV is significantly worse than that demonstrated for the TES device, the curves in Figure 13.30 show a useful measure of resolution at lower energies, with values of ~ 6 eV being obtained for 0.8 and 1 keV X-rays. However, the resolution rapidly worsens above 2 keV.

13.4.2.3   *Cryogenic techniques*

It is clear from the material above that many applications require operation at very low temperatures. While use of the L2 orbit and large sunshades can allow passive cooling to $T < 60$ K, future missions may require active cooling, as was employed for the ESA *ISO*. CCDs require cooling, to ~ 180 K while ~ 60 K is necessary for the large germanium spectrometers used in gamma-ray spectroscopy. Even greater problems are posed by the need to operate the new generation of non-dispersive visible and X-ray spectrometers and IR/sub-mm bolometers at temperatures $\leq$ 300 mK and ranging down to 10 mK.

To date, missions that have required detector operation at $T < 4$K, e.g. *ISO*, have employed consumable cryogens such as liquid helium as the primary cooling system. This approach will still be used in a number of future missions, e.g *Herschel*. While helium boils at 4.2 K, reduction of the vapour pressure above the liquid can significantly reduce this figure. Furthermore He does not solidify but enters a 'superfluid' state below its triple point of 2.2 K. In space, even lower temperatures may be obtained by managing the pressure above the liquid through control of the pumping speed to the outside vacuum. Thus a heat sink at 1.8 K can be provided in *Herschel* as the base for a further cooling system to hold detectors at 0.3 K.

Use of orbits that are distant from the Earth can ensure operation with a relatively cold spacecraft, which in turn allows maintenance of very low operating temperatures with smaller liquid helium cryostats. The NASA *Space IR Telescope Facility (SIRTF)*, to be launched in 2003, will be placed in an Earth-trailing heliocentric orbit in which it will drift behind the Earth as the latter orbits the Sun. With a drift rate away from the Earth of 15 Mkm/year, the spacecraft will run at a temperature of $\sim$ 40 K. This allows the use of a relatively small liquid helium cryostat (950 kg) that enables the telescope (0.85 m aperture) and focal plane detector systems to be operated at 5.5 K and 1.4 K respectively for a mission lifetime of five years. In contrast, an *ISO*-like system would require 5700 kg of helium to achieve the same temperatures and mission life. In addition, remoteness from the Earth allows greater operational flexibility, with $\geq$ 30% of the sky being instantaneously visible to the *SIRTF* telescope.

Another approach, employed in the ISAS *Astro-E2* mission (Figure 13.23(b)), involves the use of two stages of cooling with consumable cryogens. To reach the final microcalorimeter operating temperature of 60 mK, it is necessary to have a starting temperature for the ADR of $\sim$ 1.5 K. If this were to be achieved with only a liquid helium cryostat, $\sim$ 3000 litres of helium would have been required. Instead an annular vessel containing 170 kg of solid neon surrounds the helium and ADR systems. With a melting point of 24.6 K, pumping of this enclosure to space allows a temperature of 17 K to be maintained for a lifetime of two years. In this situation, some 30 litres of liquid helium, pumped to achieve a temperature of 1.3 K, is sufficient for a lifetime of five years. While solid hydrogen has a melting point of 14 K and a latent heat five times greater than that of neon, safety considerations precluded its use on *Astro-E2*. A solid hydrogen cryostat was flown on NASA's *Wide-field IR Explorer (WIRE)* mission, where 4.5 kg of the material was used to cool the IR telescope optics to 13 K and the detectors to 6.5 K. The planned lifetime was four months but unfortunately the hydrogen was lost early in the mission due to the premature opening of the cryostat entrance cover.

The difficulties and limitations of operating with large volumes of consumable cryogen have led to consideration of the possibility of cryogen-free operation. The Stirling cycle cryocooler has been used in a number of space missions, e.g. ESA *ERS-1, -2, Envisat*, to cool Earth-looking IR detectors from 300 K to $\leq$ 80 K. It is a closed cycle system which comprises two elements: (i) a pressure oscillator working at ambient temperature in which a reciprocating piston transmits mechanical energy to the enclosed gas and generates a pressure oscillation in the refrigerator cold finger, and (ii) a cold finger cylinder in which a displacer containing a regenerator matrix separates two volumes at ambient and cold temperatures. When reciprocating in the cylinder, the displacer forces gas from one volume to the other. Appropriate relative phasing of the pressure oscillations and the displacer motion generates the cooling effect. The cooling capacity is given by $Q_c = f \int \alpha V_c dP$, where f is the operating frequency, $\alpha = T/V (dV/dT)_P$ is the expansivity, P is the pressure in the cooler and

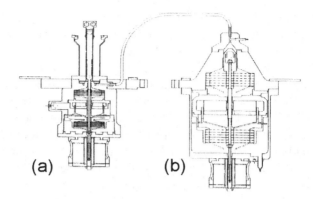

Figure 13.31   The Oxford/RAL split Stirling cycle cooler. (a) Compressor; (b) displacer with cold finger. (Courtesy of T. Bradshaw.)

V_c is the variable expansion volume. The 80 K system developed at Oxford and RAL for use on the ERS spacecraft (Werrett *et al.*, 1985) is shown in Figure 13.31. It is a split Stirling cycle machine operating with a helium gas fill. Moving coil motors drive both compressor and displacer. Linear position sensors are used to measure position and their outputs are used to control relative phase. Use of a drive system on each element rather than relying on pressure pulse transmission to a free piston allows greater cooling efficiency and the maintenance of balance in the mechanisms to minimise residual momentum.

While the system illustrated in Figure 13.31 will achieve $T \geq 80$ K, two-stage Stirling systems have been designed to reach $T \sim 20$–30 K. For temperatures below 20 K, mechanical cyclic coolers become less efficient and systems based on Joule–Thomson expansion of appropriate gases have been developed. A closed cycle system proposed by Orlowska *et al.* (1990) is shown in Figure 13.32. Here the Stirling compressors and a two-stage displacer achieve temperatures in the 20–30 K range. The prototype described by Orlowska *et al.* can provide 200 mW of heat lift at 30 K with a 20 K base temperature. Two Joule–Thomson compressors, with axes in line to minimise uncompensated momentum, then drive [4]He gas which has been pre-cooled to $T \sim 25$ K, through a J–T orifice. The gas will first pass through the inner tube of a series of concentric heat exchangers, where it is cooled by low-pressure return gas from the J–T expansion. A prototype of this system has achieved a base temperature of 3.6 K.

Another closed cycle system is the sorption refrigerator (Figure 13.33) which is used to cool the bolometers in the *Herschel* SPIRE instrument to 0.3 K. In this case, the cold reservoir is maintained by pumped [4]He from the *Herschel* cryostat. However, the reservoir could be maintained by another closed cycle system. The sorption pump contains gaseous [3]He in contact with microporous activated charcoal arranged to have a large ratio of surface area to volume. With the heat switches S1 open and S2 closed, the pump is heated to $T \sim 30$ K. This drives the gas to

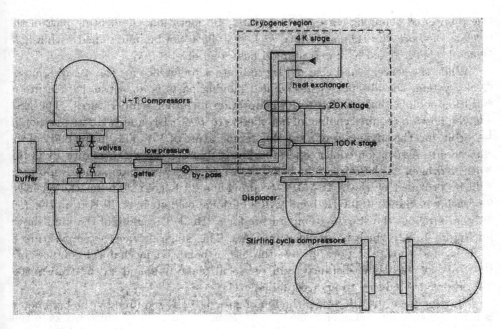

Figure 13.32 Schematic diagram of a two-stage Stirling system coupled to a Joule–Thomson expansion stage to achieve 4 K. (Courtesy of T. Bradshaw.)

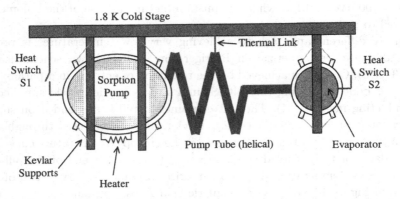

Figure 13.33 Operation of the ^{3}He sorption refrigerator, a closed cycle system that works between $T < 2$ K and $T \sim 300$ mK. (Courtesy of ESA.)

the evaporator, where it is liquefied at 1.8 K, and for zero g operation, held by surface tension forces in a 'sponge' of sintered copper or aerogel. With the heater off, S1 closed and S2 open, vapour is pumped back to the cold sorption stage. This reduces the vapour temperature in the evaporator to $T \sim 0.3$ K, as required by the bolometers. When the transfer of ^{3}He to the sorption stage is complete, the evaporator temperature will begin to rise and the cycle must begin again by heating the pump and changing states of S1 and S2. In operation, the heating part of the cycle is of short duration, so a duty cycle of $\sim$ 95% can be maintained with a hold time of $\sim$ 40 h at 0.3 K in extracting a heat load of ≤ 10 μW.

While the scarcity and cost of ^{3}He indicate a preference for the use of a closed cycle system, the absence of mechanical moving parts is an attractive feature of sorption cooling that has led to its adaptation for other temperatures. The ESA *Planck* mission requires cooling of bolometers to $T \sim 0.1$ K and, with its position in L2 orbit, has the advantage of a starting temperature of $T \leq 50$ K in the body of the spacecraft. A sorption cooler, operating in the manner shown in Figure 13.33 but with hydrogen as the refrigerant being transferred between pump and evaporator, will be provided by the NASA JPL to provide a 20 K cold stage for the receivers operating in the 30–100 GHz band. An additional cold stage at 18 K from the JPL system provides the starting temperature for a ^{4}He J–T system, of the kind shown in Figure 13.32, which is provided by RAL. This stage delivers a temperature of ≤ 4 K, and though it does involve the use of balanced mechanical compressors, they are located some distance from the focal plane unit and so the transmitted vibrations are reduced to an acceptable level.

Starting from the 4 K stage, the final figure of 0.1 K required by the bolometers is achieved with a novel open cycle system (Benoit *et al.*, 1997) which works in the manner of a J–T system but exhausts to space a mixture of ^{3}He and ^{4}He at the very low rate of 12 μ mole/s. This process is analogous to the one operating in a dilution refrigerator but works without the presence of gravity. It provides a cooling power of ~ 100 nW at 0.1 K while the mission lifetime for use of the bolometers is constrained to ≤ 3 years.

A totally different approach to achieving very low temperatures is provided by the Adiabatic Demagnetization Refrigerator (ADR) shown schematically in Figure 13.34(a). Cooling is achieved by the reduction of entropy in a paramagnetic salt. This is done by placing the salt (S) in a strong magnetic field supplied by a superconducting magnet (M). The salt container or pill is suspended on supports with very low thermal conductance, e.g. Kevlar, and is connected through a heat switch (W) to a sink at a temperature set by the entropy–temperature curve of the salt. The detector to be cooled (D) is attached to the other end of the pill. Note that the entropy–temperature diagram comprises a family of curves, two of which are shown in Figure 13.34(b), with magnetic field H as a parameter.

The ADR operation cycle is shown in Figure 13.34(b). Starting at A, the initial magnetic field is zero and the salt is at the heat sink temperature, T_i. A magnetic field of ~ 1 T is then applied while the switch W remains closed. Thus, at B, the strong field aligns the paramagnetic salt molecules, which reduces their entropy or thermal motion. The switch H is then opened and the salt cools quickly to a substantially lower temperature while the magnetic field is adjusted to H_f (C). Heat input from the detector will then raise the entropy as the molecules come out of alignment with the field, but this change is managed by progressively further

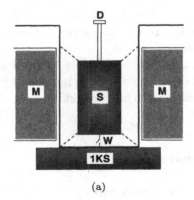

(a)

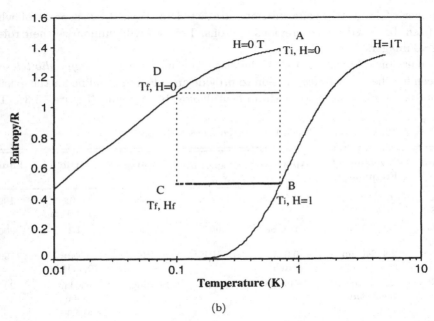

(b)

Figure 13.34   (a) ADR schematic showing the salt pill (S), magnet (M), heat switch (W), detector mounting position (D) and 1 K cold sink. (b) Entropy–temperature curves for a paramagnetic salt. The ADR operating cycle (A–B–C–D) is also shown.

reducing the magnetic field so that the temperature, T_f, remains constant. When the magnetic field finally reaches zero, the salt pill cannot extract any more thermal energy, but the heat from the detector is now in the form of magnetic entropy due to the random orientations of the salt molecules (D). Brief application of a high field, $H = \delta$, will realign the molecules and result in a temperature increase of the salt, making the detectors temporarily inoperable (A′). Closure of W will remove the excess thermal energy to the sink and leave the pill and detector at T_i after which the cycle — now A′–B–C–D — can begin again.

The hold time for such a system can be ~ 24 h for a typical detector heat load while about 1 h is required to recycle the system. Given the magnetic properties of suitable salts, it is necessary to use two ADRs operating in series to cover an appropriate temperature range. Use of a double ADR (Hepburn *et al.*, 1996) allows cooling from a 4 K heat sink, which can be provided by mechanical or sorption coolers, to temperatures as low as 10 mK. In addition lower magnetic field values can be employed in the d-ADR system.

13.5   Imaging and Spectroscopy

13.5.1   *Imaging at longer wavelengths*

A number of large telescope missions are listed in Table 13.2, aspects of which have been discussed in the previous sections. Here we will summarise their role as sensitive imagers.

As mentioned in Section 13.3.1, the major role in space for a large *radio telescope* has been for the ISAS *Halca* mission to provide an enlarged baseline for the existing ground-based VLBI network with an 8 m deployable antenna (Figure 13.3). This

Table 13.2   Large space telescopes.

Mission	Wavelength/ Frequency	Aperture	$T_{aperture}$	Cooling	Orbit	Launch
Halca (ISAS)	1.6, 5, 22 GHz	8 m, mesh	–	Passive	560× 21,400 km	1997
Herschel (ESA)	60–670 μm	3.5 m, SiC	60 K	Passive	L2	2007
ISO (ESA)	2.4–240 μm	0.6 m, Zerodur	10 K	^{4}He	1000× 70,500 km	1995
IRIS (ISAS)	1.8–26 μm 50–200 μm	0.7 m, SiC	6 K	Stirling/ ^{4}He	650 km polar, sun synch.	2004
SIRTF (NASA)	3.0–180 μm	0.85 m, Be	5.5 K	Passive/ ^{4}He	Heliocentric, Earth-trailing	2003
NGST (NASA)	0.6–28 μm	6.5 m, segmented, Be or glass composite, active figure control	50 K	Passive	L2	2010
HST (NASA)	0.15–2.5 μm	2.4 m, quartz	–	Passive	593 km, LEO	1990
Solar-B (ISAS)	0.48–0.65 μm	0.5 m, glass	–	Passive	650 km, polar, sun synch.	2005

system allows baselines of up to 30,000 km to be achieved to ground-based radio telescopes and is yielding angular resolutions of ~ 0.5 milli-arcsec compared to the 2 milli-arcsec previously available from ground-based VLBI. In addition the larger number of baselines available with *Halca* allows much enhanced u–v plane coverage and hence improved mapping of extended sources. Significant technical difficulties are posed by the need to transmit radio source signal and phase information to the ground station on a 128 Mbps link and to receive on board from the ground a highly stable hydrogen maser reference frequency for phase transfer. In addition the orbit must be known with high accuracy — 3–10 m is being achieved with *Halca*.

For the *IR range*, it is interesting to compare *ISO* and *SIRTF*, which undertake imaging in similar wavelength ranges. *ISO* ceased operation in mid-1997 following the expected depletion of its ^{4}He cryogen, whereas *SIRTF* is due for launch in 2003 and has a planned lifetime of ~ 5 years. Thus the differences mainly follow from the rapid pace of technical development in the IR astronomy field. Both telescope arrangements, of Ritchey–Chretien design, are shown in Figure 13.35. Because the *ISO* orbit is closer to the Sun while that of *SIRTF* is always behind the Earth, much greater care is required in the design of baffles for *ISO*. The deeper orbit allows the *SIRTF* spacecraft structure to run at ~ 40 K and so the telescope and instruments require a much smaller volume of ^{4}He. The *SIRTF* telescope mirrors are made from Be, which has a low heat capacity and expansion coefficient, and are lighter than the *ISO* mirrors, which are made from low expansion quartz. Both telescopes have similar wavelengths of 5–6 μm for diffraction-limited operation, as has Japan's *IR Imaging Survey* (*IRIS*) mission, which also employs a Richey–Chretien optical configuration with lightweight SiC mirrors. In a near-Earth orbit, it relies on Stirling cycle cryocoolers and 170 litres of liquid helium to provide a mission lifetime of ~ 550 days. The Stirling coolers will allow continued near-IR operation even after the liquid He cryogen has been exhausted.

While the *SIRTF* primary mirror area is only a factor of 2 larger than that of *ISO*, it is running significantly cooler and in an orbit where there is much less restriction on the pointing direction. In addition, advances in IR array technology have meant that for *ISO*, in the range 2.5–5.0 μm, a 6×6 arc min^2 field of view was imaged on an array of 32×32 pixels of 12 arc sec/pixel, whereas with *SIRTF*, for four bands in the range of 3.6–8 μm, a 5.1×5.1 arc min^2 field will be imaged on a 256×256 pixel array with 1.2 arc sec/pixel.

The *IRIS* mission is primarily designed to conduct an all-sky survey in four bands in the far-IR range (50–200 μm). The first IR all-sky survey, undertaken by the *IRAS* mission in 1985, had a sensitivity of ~ 1 Jy and a positional resolution of ~ 3 arc min. The *IRIS* survey, using arrays of unstressed and stressed Ge:Ga detectors, should be completed in the first year of the mission. It will have ~ 10–30 mJ sensitivity, depending on wavelength along with positional resolution in the range of 30–50 arc sec. The spacecraft can also be operated in pointed mode and has three IR arrays for the 1.8–26 μm range. These are InSb (512×412 pixels) and

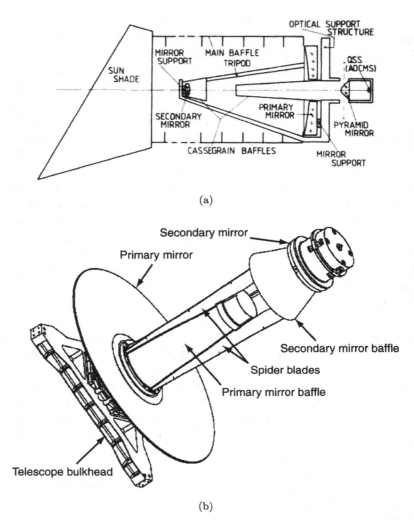

(a)

(b)

Figure 13.35   Optical layouts of (a) *ISO* 0.6 m telescope (courtesy of ESA) and (b) SIRTF 0.85 m telescope (courtesy of NASA).

two Si:As impurity band conduction (256 × 256) devices. For a 500 s observation, the 5σ sensitivities vary through the band from $\sim$ 1–150 μJ.

The *Herschel* and *NGST* missions will be placed in orbit about the L2 point (Figure 13.36). L2 is located $\sim 1.5 \cdot 10^6$ km from the Earth, remote from the Sun and beyond the Moon's orbit. There is a similarly located point (L1) in the direction of the Sun and the *SOHO* spacecraft has been in orbit around L1 since 1995. Both are gravitational potential saddle points in the Earth–Sun system but are unstable on a time scale of $\sim$ 25 days. This necessitates small station keeping manoeuvres to maintain the spacecraft in orbit and at an essentially constant distance from

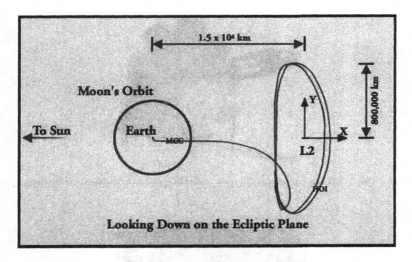

Figure 13.36   A quasi-stable orbit about the second Lagrange point (L2).

the Earth. Both spacecraft require large sunshades for the total blocking of both directed and reflected solar radiation. Given the additional ability of the telescopes and other payload elements to radiate to deep space, mirror temperatures in the range of 50–60 K can be maintained.

Herschel will carry two photometry systems for the 60–210 μm and 250–500 μm ranges. Both employ ^{3}He sorption coolers (Figure 13.33) connected to the ^{4}He cryostat to allow bolometers (13.4.1.2) to run at 0.3 K. The Photodetector Array Camera and Spectrometer (PACS, Poglitsch *et al.*, 2000) employs two filled silicon bolometer arrays of 32 × 64 pixels and 16 × 32 pixels. The former can register either in the range of 60–90 μm or 90–130 μm, selectable by filter interchange, while the latter always responds from 130 to 210 μm. The common field of view is 1.75 × 3.5 arc min^2, with corresponding pixel sizes of 3.3 or 6.3 arc sec. The 5σ point source sensitivity for a 1 h observation is $\sim$ 2–3 mJy, with the lower figure needing on-array chopping of the input signal. For the 250–500 μm range, a field of view of 4 × 8 arc min^2 is observed simultaneously at 250, 350 and 500 μm by arrays of 139, 88 and 43 feedhorn-coupled spider web bolometers (Figures 13.16 and 13.17) with beam sizes of 17, 24 and 35 arc sec (Griffin *et al.*, 2000). For these detectors, the 5σ point source sensitivity for a 1 h observation is $\sim$ 2.5–3 mJy, but since the beam is not completely filled, a 16-point 'jiggle' pattern scan is needed to survey the entire field of view with a sensitivity reduction of 3–3.5.

The *Hubble Space Telescope* (*HST*) imaging performance (0.03 arc sec/m_v $\sim$ 27) and unique serviceability in orbit have been discussed in Section 13.3.3. Although it should continue in operation to 2010, its successor, *NGST*, will concentrate on performance in the IR spectral range. Indeed it is a measure of *HST*'s extraordinary success that five of the seven major astronomical telescope missions

Figure 13.37   The 6.5 m *NGST* primary mirror with its segments deployed. (Courtesy of NASA.)

Figure 13.38   Low expansion glass mirror, reaction structure and actuators. (Courtesy of Kodak Corp.)

listed in Table 13.2 will emphasise the range $\lambda \geq 1$ μm in order to study objects at even higher redshifts.

The *Next Generation Space Telscope* (*NGST*) mission, to be launched in 2010, was described briefly in Section 13.3.3. The proposed 6.5 m diameter primary mirror is too large to fit in the available rockets and hence must be of segmented construction (Figure 13.37), with deployment on arrival in L2 orbit. In addition the mass restriction to < 15 kg/m^2 for the primary mirror means that traditional high stability glass structures cannot be used and hence active wavefront sensor control of the mirror figure will be essential. Several approaches are being studied for fabricating the mirror segments. These involve thin reflecting layers — Beryillium, glass composite or thin low expansion glass — mounted on lightweight graphite composite reaction structures. A large number of linear actuators are mounted to bear on the underside of the mirror surface and are driven so as to optimise segment figure and segment-to-segment alignment (Figure 13.38). On arrival at L2 and following cool-down to < 50 K, the segments will be in a misaligned state with their figures deformed. A series of steps is then executed, beginning with coarse alignment and focussing of the segments using the image of a bright isolated star. The wavefront is then sensed with successively greater precision and a phase map produced which is used iteratively to compute the actuator settings, leading to a wavefront error of 0.1 μm at 1.0 μm. It is hoped that coarse alignment and phase correction will be done only once while fine phasing will be repeated periodically throughout the mission. Since it will be necessary to use the focal plane instruments, e.g. the Near IR Camera (NIRCam, $\lambda \leq 5$ μm) for wavefront sensing, the modules involved must be fully redundant. Development of precision actuators to run at $T < 50$ K is seen as a significant technical challenge. For the long wavelength detectors ($\lambda > 5$ μm) that need to run at $T \sim 6$ K, sorption-pump-driven coolers using hydrogen and helium to get to 6 K would seem preferable since the need for precise telescope pointing makes it difficult to use cyclic mechanical coolers or compressors. However, should the spacecraft temperature run at ~ 30 K, a consumable cryogen solution may be possible.

The principal imaging instrument proposed is the NIRCam, which will be sensitive in the range of 0.6–5.0 μm. It will cover the wavelength bands 0.6–2.3 μm and 2.4–5.0 μm, with 4096^2 and 2048^2 arrays with the likely detector choice being HgCdTe since these devices can be adequately cooled by passive radiation to deep space. A field of view of 2.3×4.6 arc min^2 may be observed in the two bands simultaneously, leading to pixel sizes of 0.034 and 0.068 arc sec/pixel. Camera sensitivity is at a level of 1 nJy for a 50,000 s observation, as can be seen from Figure 13.39, where the spectra of two strongly redshifted galaxies at $z = 5$ and

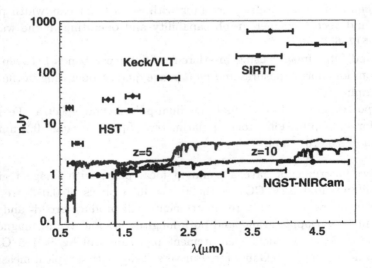

Figure 13.39   Sensitivity (nJy) of a NGST Near IR Camera design (Rieke *et al.*, 2001) for $z \leq 10$ galaxies (5σ, 50,000 s). (Courtesy of M. Rieke.)

$z = 10$ are also shown. Sensitivities for the band are indicated for the *HST* and *SIRTF* imagers while the performance of the ground-based 10 m Keck telescope is also plotted. Band pass and tuneable filters are used to provide a measure of spectral resolution and are discussed by Rieke *et al.* (2001). It is anticipated that the proposed NIRCam will view the formation of the earliest galaxies and connect them to the known Hubble sequence. Use of coronagraph discs in the focal plane should allow the study of the planetary formation process around nearby stars.

Unlike the other large telescopes described in Table 13.2, the ISAS *Solar-B* mission is targeted at high spatial resolution studies of the Sun. The 0.5 m optical telescope provides diffraction-limited images of angular resolution 0.2 arc sec at $\lambda = 500$ nm, corresponding to a distance of 150 km on the Sun's surface. For ground-based telescopes, atmospheric seeing limits the resolution to ≥ 1 arc sec, so the arguments for operation in space are similarly compelling to those used in support of *HST*. The sun-synchronous orbit will allow uninterrupted viewing of the Sun for eight months of the year. While the astronomical telescopes require cooling to reduce thermal noise, the *Solar-B* telescope needs to retain its figure and alignment while observing the Sun continuously. Two views of the design are shown in Figure 13.40. The main support truss structure is made from a carbon fibre composite to ensure strength and low thermal expansion while the optical elements are supported on invar mounts. The primary mirror is radiatively coupled to surrounding cold plates while a substantial part of the heat from the primary is reflected out through the side of the telescope by the 'heat dump' mirror. The cold plates and the secondary mirror radiate to space.

The incoming radiation is fed through the collimator lens and in parallel to:

- Narrow-band tuneable birefringent filter with ~ 0.01 nm bandwidth providing imaging and vector magnetograph capability and operating in the wavelength range of 517–657 nm;
- Broad-band filter imager using interference filters, passbands between 0.3 and $1 \cdot 10^{-3}$ nm for short exposures and high image quality, operating in the range of 388–670 nm;
- Spectro-polarimeter; obtains dual-line dual-polarisation spectra (Fe I 630.1/.2 nm) for high precision Stokes polarimetry. Spectral range 0.2 nm with a resolution $\sim 2.5 \cdot 10^{-3}$ nm.

The first two instruments have a common focal plane with image registration by a back-illuminated E2V CCD with 2048×4096 pixels of 0.053 arc sec and frame transfer operation. The three instruments will produce broad- and narrow-band filter images, Dopplergrams and both longitudinal and vector magnetic field measurements. Magnetic field measurement precision will be ~ 1–5 Gauss for longitudinal field and 30–50 Gauss for transverse field, with a typical measurement time of ~ 5 min.

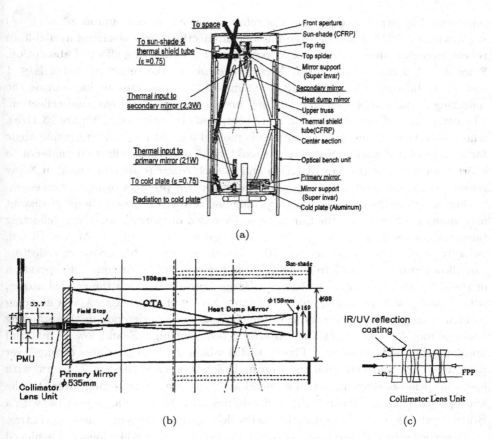

(a)

(b)   (c)

Figure 13.40   *Solar-B* 0.5 m telescope mounting (a) truss structure, primary and secondary mirrors, heat dump mirror and thermal radiation paths. (b) Solar radiation paths, operation of heat dump mirror and connection to focal plane instruments through collimator lens and polarisation modulator unit (PMU). (c) Collimator lens unit. (Courtesy of ISAS.)

13.5.2   *Grazing incidence systems and multilayers*

The concept of X-ray reflection at grazing incidence has already been discussed in Section 13.3.4, where we found that the value of critical angle for total external reflection depended on the X-ray wavelength λ and on the atomic number (Z) of the reflector. Considering the photon–matter interaction in more detail, we find that when incoming X-rays, with energy greater than the binding energy of the outer electrons, impinge on a typical metal, they encounter a 'quasi-free' electron gas. The value of the refractive index is characteristic of the number density of free electrons $\sim n_a f$, where n_a is the number density of atoms and f is the number of free electrons per atom — the atomic scattering factor. However, the electrons are not truly free and so some of the incoming photon energy will be absorbed. This situation can be represented by expressing the index of refraction as a complex number $n^* = (1 - \delta) - i\beta$, where $\delta \propto \lambda^2$ represents the number of effectively free electrons available to reflect incoming photons and $\beta \propto \lambda^2$ represents the competing effect of absorption. Since δ and β are both small at X-ray wavelengths, the refractive index is ≤ 1 and, as we have seen in Section 13.3.4 (Figure 13.8), rays must be incident on the reflecting surface at angles $< \theta_c$ if they are to undergo total external reflection. The competing effects of reflection and absorption are indicated in Figure 13.41(a), where reflectivity from an ideal surface is plotted against normalised grazing angle for a range of β/δ values. However, for real high Z materials, reflectivity extends to shorter wavelengths (Figure 13.41(b)) and so such coatings are often used in X-ray reflectors. Discontinuities in these plots result from the presence of absorption edges.

For wavelengths $\lambda \leq 50$ nm, high angular resolution observations of distant astronomical objects and the Sun have so far relied on *grazing incidence* reflecting telescopes, several of which have been described in Section 13.3.4. At $\lambda \sim 10$ nm, reflectivity at normal incidence is $\sim 10^{-4}$, but the use of *multilayer-coated* reflectors can allow figures of $\geq 30\%$ to be achieved in the range of 10–50 nm. The operation of these systems is similar to that of a Bragg crystal, in which the crystal atoms, with lattice plane spacing d, reflect incoming X-rays of wavelength λ at a glancing incidence angle θ and the condition for constructive interference in the outgoing beam is met for $n\lambda = 2d\sin\theta$. Alternate layers of high Z and low Z material are deposited on a substrate (Figure 13.42) where d is the thickness of one layer pair. Rays are reflected from the thin high Z layers while the low Z layers, with low absorption coefficient, act to suspend and separate the reflecting layers by the appropriate distance. Much higher reflectivities are achieved than is possible with a Bragg crystal, which relies on individual widely spaced atoms as reflecting centres.

In accordance with the Bragg relation, operation at normal incidence is achieved for $d = \lambda/2$. However, the uniformity of substrate and layer thickness that is available limits operation to $\lambda \geq 5$ nm. Roughness and uniformity are often characterised by the Debye–Waller parameter, σ, where achievable reflectivity $R_a = R_{\text{Theor}} \exp(-4\pi\sigma/\lambda)^2$. Thus, for an acceptable response at $\lambda \sim 5$ nm, $\sigma \sim 0.2$ nm is required. This is possible for the substrate, but for layer boundaries, $\sigma \sim 0.6$–0.7 nm is currently achieved due to the formation of amorphous interlayers. It is also essential that the atoms in the alternate layers do not diffuse into each other or combine to form new compounds at the interface. At near normal incidence the bandwidth of the reflectance curve $\Delta\lambda \sim \lambda/N_{\text{LP}}$, where N_{LP} is the number of layer pairs. Increasing N_{LP}, besides reducing the bandwidth, increases the peak reflectivity as more layers contribute to the diffracted signal, but eventually a limit is reached due to the absorption or radiation in the structure.

For the wavelength range of 13–30 nm, good results have been obtained with Mo/Si layer pairs since the absorption coefficient of Si drops significantly for wavelengths ≥ 12.4 nm — the location of the Si L absorption edge. Results from Windt (1998) are shown in Figure 13.43 for a structure with 40 Mo/Si layer pairs. The plot

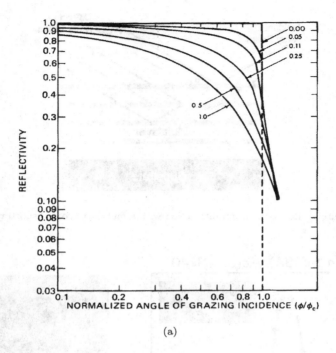

(a)

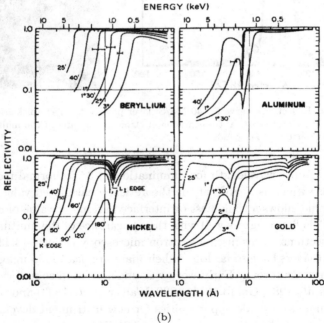

(b)

Figure 13.41 (a) Reflectivity plotted against normalised grazing incidence angle for a range of β/δ values (Hendrick, 1957). (b) Reflectivity against X-ray wavelength with grazing incidence angle as a parameter (Giacconi *et al.*, 1969).

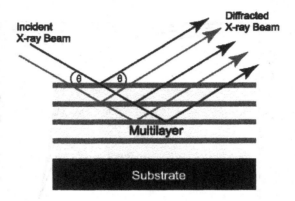

Figure 13.42   X-ray beam interaction with a multilayer-coated substrate.

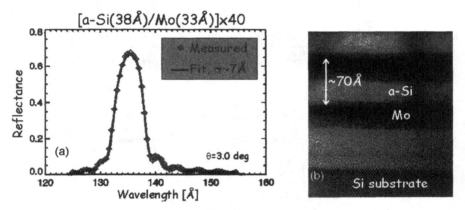

Figure 13.43   (a) Reflectance curve for 40 Mo/Si layer pairs with layer thickness and derived σ value indicated. (b) Cross-sectional transmission electron micrograph of the multilayer structure (Windt, 1998). (Courtesy of D. Windt.)

of reflectance against wavelength for illumination at 3° to normal incidence shows a peak response which is ~ 85% of the theoretically calculated value. A fit to the reflectance profile allows a roughness or interface parameter value of $\sigma \sim 0.7$ nm to be deduced. This figure agrees well with the appearance of the multilayer structure seen in cross-sectional transmission electron microscopy (Figure 13.43(b)).

Mo/Si multilayers have so far found their main application in imaging and spectroscopy of the solar corona. Early work was undertaken in sounding rocket flights (Underwood *et al.*, 1987; Golub *et al.*, 1990; Walker *et al.*, 1993) and the rocket programmes continue to provide opportunities for new instrument development. More recently the *Solar Heliospheric Observatory* (*SOHO*) and *Transition Region and Coronal Explorer* (*TRACE*) missions have carried multilayer telescopes operating in the 15–30 nm range.

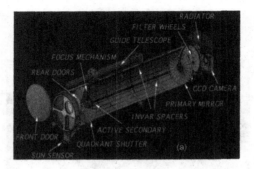

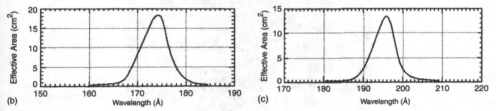

Figure 13.44   (a) *TRACE* EUV telescope. Three of the four mirror quadrants have Mo_2C/Si coatings. (b) Multilayer response at 17.3 nm and (c) at 19.5 nm. The third quadrant response is centred at 28.4 nm. (Courtesy of Lockheed and NASA.)

The *TRACE* EUV telescope (Tarbell *et al.*, 1994), a 0.3 m Cassegrain system, is indicated in a schematic diagram in Figure 13.44(b), along with other features of the instrument. The primary and secondary mirrors are sectored in four quadrants, three of which have Mo_2C/Si multilayer coatings, which show a somewhat better performance than Mo/Si. A quadrant shutter allows one sector at a time to view the corona and register images of the appropriate pass band on the focal plane CCD detector, which is cooled by connection to a radiator. The fourth quadrant is designed to view wavelengths ≥ 120 nm where multilayers are not required. The filter wheel system is used to select pass bands of interest in this range. The telescope has a field of view of 8.5×8.5 arc min^2 and an angular resolution of 1 arc sec, while the 1024×1024 pixel CCD samples at 0.5 arc sec/pixel. The TRACE spacecraft was launched in April, 1998 into a sun-synchronous orbit of 800 km and at the time of writing, is continuing to obtain high resolution EUV images of the solar corona.

For $\lambda \leq 4$ nm or $E_\nu \geq 0.3$ keV, interface imperfections characterised by $\sigma \sim$ 0.7 nm make it difficult to achieve useful reflectance values near normal incidence. However, it is still possible to operate suitable multilayers at grazing incidence and to extend their response to higher energies. Windt (1998) has compared the calculated reflectivity of a single iridium layer with measured values at 8 keV for a range of W/Si multilayers with $N_{LP} = 100$ and 2 nm $< d <$ 9.9 nm. He has demonstrated very significant reflectivity enhancements in narrow energy bands for grazing incidence angles in the range of 1–2.5°.

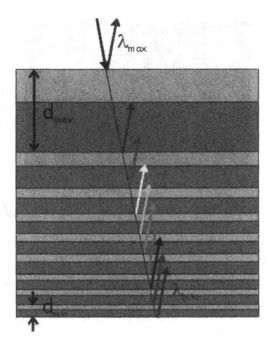

Figure 13.45   Schematic of the operation of a depth-graded multilayer structure. (Courtesy of D. Windt.)

For energies > 10 keV, development of depth-graded X-ray multilayer coatings can offer enhanced reflectivity at grazing incidence over a broad energy range. In these structures, the bilayer thickness varies continuously with depth such that each bilayer effectively reflects a different wavelength, so by adjusting the distribution of bilayer thicknesses, broadband reflectance can be achieved (Figure 13.45). For X-ray energies at $E_\nu > 100$ keV, pairs of materials may be chosen to permit values of $N_{LP} \geq 1000$ to ensure a broad energy response. Use at higher energies with large N_{LP} is possible since the layer absorption coefficients will be small. However, even at grazing incidence, interlayer roughness and diffusion can significantly degrade reflectance. As for the normal incidence cases discussed above, layer interface quality is characterised by the parameter σ while theoretical reflectance at each interface is determined by the optical constants of the materials.

Windt *et al.* (2002) have recently reported the development of W/SiC graded-thickness multilayers optimised for response in the range of 140–170 keV. They first determined values of σ for periodic multilayers using 8.04 keV Cu Kα radiation at grazing incidence. The W/SiC multilayers had $N_{LP} = 300$ and d values in the range of 0.65–2.15 nm. Values of σ between 0.34 and 0.23 nm were obtained, with the latter being the smallest interface parameter yet reported. Depth-graded structures were then fabricated from W/Sic and W/Si, and measured reflectances in the 140–170 keV range are shown in Figure 13.46. Optical constants were measured for

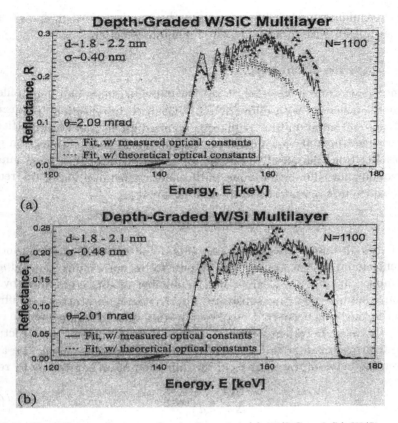

Figure 13.46 Hard X-ray reflectances for depth-graded (a) W/SiC and (b) W/Si multilayers. (Windt *et al.*, 2002)

single layer films and compared with the predictions from theory. Although the grazing incidence angle of only 6.8 arc min is very small, a useful reflectance of $\sim$ 20–30% was obtained. The ability to collect and focus radiation at energies of 20–200 keV is likely to increase sensitivities by factors of $\sim$ 1000 with respect to those of present large area detector systems and to lead to important advances in high energy astronomy.

13.5.3 *Spectroscopy*

We have mentioned briefly in Section 13.4 the hetrodyne receiver systems that are used for high-resolution observations of line emission from molecules at far IR/sub-mm and radio wavelengths. While we have discussed IR and visible imaging techniques in Section 13.5.1, to cover spectroscopic techniques at these wavelengths is beyond the scope of this chapter. We will, however, deal with the methods used for high resolution observations at UV, EUV and X-ray wavelengths by introducing dispersive techniques and briefly summarising the non-dispersive methods that have already been dealt with in Section 13.4 on photon detection.

13.5.3.1   *Dispersive spectroscopy*

Dispersing a spectrum in wavelength so that individual lines and line profiles may be resolved is achieved with diffraction gratings used in reflection or transmission or by diffracting radiation from regularly organised atoms in a crystal lattice. The multilayers described above can also be used for spectroscopy. Diffraction grating action results from the effective passage of radiation through a large number of parallel equidistant slits of the same width, and a generalisation of the treatment of double slit interference leads to the grating equation

$$d(\sin \theta + \sin \alpha) = n\lambda, \tag{13.9}$$

where d is the distance between adjacent slits, i and θ are the angles for incidence on and reflection from the grating with respect to the normal direction and n is the order number. In practice a large effective number of slits is produced by ruling grooves on a suitable reflecting substrate or by a holographic process, and diffraction occurs in the outgoing beams reflected from the thin smooth regions between rulings. For X-ray wavelengths where reflection is less efficient, transmitting structures of narrow slit arrays made with high Z metal are also sometimes used. The angular separation $\Delta\theta$ between wavelengths in the diffracted beam is given by the relation

$$\frac{\Delta\theta}{\Delta\lambda} = \frac{n}{d} \cos \theta. \tag{13.10}$$

Overlapping of spectra in different orders can occur given that

$$d(\sin \theta + \sin \alpha) = \lambda_1 = 2\lambda_2 = 3\lambda_3. \tag{13.11}$$

Use of (10) can establish that that the resolution $\lambda/\Delta\lambda = nN$, where N is the number of slits usually specified by the number of rulings per unit length for a reflection grating.

The simple grating operation concepts described above refer to plane gratings. Curved gratings (e.g. concave) which focus as well as diffract are usually employed in practice to reduce the number of optical elements in the spectrometer. A system of this kind is shown in Figure 13.47 — the Paschen mount, which shows how the angle i and thus the range of diffracted wavelengths can be controlled by moving the slit around a circle called the Rowland circle. However, in practice configurations like this are no longer used, and for space instruments, greater complexity is required to fit within the inevitable mass and volume constraints.

A high-resolution normal incidence spectrometer used for solar coronal studies on the ESA/NASA *SOHO* mission is shown in Figure 13.48 (Wilhelm *et al.*, 1995). The instrument operates in the range of 33–160 nm though in practice the sensitivity

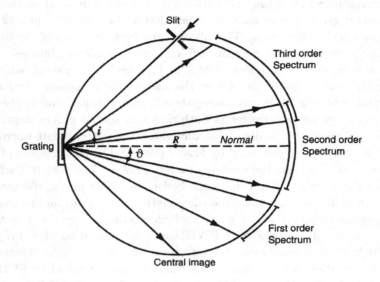

Figure 13.47 Simple normal incidence concave grating arrangement.

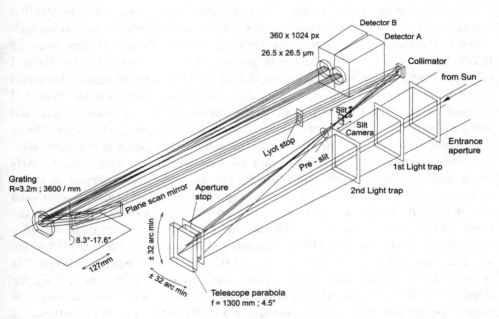

Figure 13.48 Layout of the SUMER normal incidence EUV spectrometer on *SOHO*. (Wilhelm *et al.*, 1995; with kind permission of Kluwer Academic Publishers.)

for $\lambda < 50$ nm is low due to the radiation undergoing three normal incidence reflections before reaching the grating. Radiation from the Sun is incident on the primary mirror — an off-axis parabola made from SiC that has a scan range of ± 32 arc min in two perpendicular directions. This allows light from any part of the Sun to be focussed on the entrance slit. A second off-axis parabola then collimates the beam and feeds it to a plane scan mirror, which in turn feeds the grating, which has a ruling density of 3600 lines/mm. Given the sizes of the detectors, comparatively short spectral ranges (2.5–5.0 nm) are registered simultaneously and so observation of the complete wavelength ranges in both first and second orders requires rotation of the scan mirror in the dispersion direction. The detectors are microchannel plates whose operation is discussed by Fraser (1989). The combination of the slit and detector pixel sizes sets the resolving power $\lambda/\Delta\lambda \sim 17{,}700$–$38{,}300$, which corresponds to 0.002 nm at 50 nm in second order and 0.004 nm at 160 nm in first order. The spatial resolution along the slit (length ~ 6 arc min) and perpendicular to the dispersion direction is ~ 1 arc sec, which is again set by the detector pixel size since the optical elements have FWHM angular resolution of ≤ 0.7 arc sec. With its high spectral resolution, the spectrometer can determine Doppler shift line-of-sight velocities to $\sim \pm 1$ km/s. The emission lines studied by SUMER are mainly produced in the plasma temperature range of $2 \cdot 10^4$–$2 \cdot 10^6$ K, so studies of the chromosphere and transition region are emphasised.

For studies at shorter wavelengths, though having some overlap with the SUMER wavelength coverage, *SOHO* also carries a spectrometer operating at grazing incidence in the range of 15–80 nm. Known as the Coronal Diagnostic Spectrometer (CDS, Harrison *et al.*, 1995), its layout is shown in Figure 13.49(a). The Wolter I telescope configuration shown in Figure 13.8(b) is preferred for imaging applications but it imposes a higher convergence on the reflected rays than does the Wolter II configuration (Figure 13.49(b)), in which the second reflection from the hyperboloid surface is external so a beam of lower divergence is fed to the spectrometer. Hence this configuration is used in CDS with the components being gold-coated zero-dur. The stop placed immediately after the telescope selects two separate parts of the annular output beam, which are fed simultaneously to a normal incidence spectrometer (NIS) and a grazing incidence spectrometer (GIS) through a common slit. In the NIS, following diffraction by two toroidal gratings which respond in the ranges of 30.8–38.1 nm and 52.3–63.3 nm, the resulting spectra are registered on a microchannel plate intensified CCD. A number of slits may be selected, all of which are 4 arc min in length. Spatial coverage of the Sun in the dispersion direction is provided by a scanning mirror which can cover a range of 4 arc min in 2 arc sec steps. The GIS uses a spherical grazing incidence grating with four microchannel plate detectors placed around the Rowland circle. Since the spectra are astigmatic, pinhole entrance slits of 2, 4 and 8×50 arc sec are used to feed a 1000 lines/mm grating. These slits can be moved perpendicular to the dispersion direction while the scan mirror covers the Sun in the dispersion direction. The telescope field of

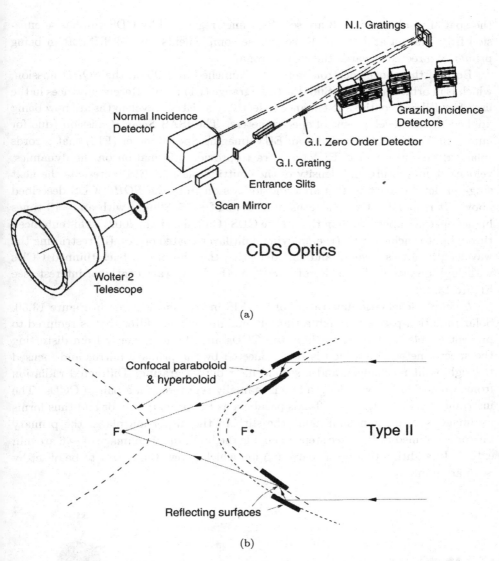

Figure 13.49 (a) Optical elements of the CDS system. A Wolter type II imaging telescope feeds the normal and grazing incidence spectrometers. (b) Schematic diagram of the Wolter II configuration. (Harrison *et al.*, 1995; with kind permission of Kluwer Academic Publishers.)

view is 4×4 arc min, with the entire instrument being mounted on actuators so that all parts of the Sun may be covered. Although Doppler line-of-sight velocities of ~ 35 km/s can be measured by CDS, the main purpose of the instrument is the determination of plasma temperature and density by line intensity ratio methods. Given the greater difficulty associated with grazing incidence operation, spectral resolving powers of $\lambda/\Delta\lambda \sim 700$–4500 are available over the wavelength range while the spatial resolution is ~ 3 arc sec. The lines registered by CDS emphasise emission from plasma at $T > 10^6$ K, so its use complements that of SUMER in being primarily directed towards the solar corona.

Both of the above spectrometers were launched in 1995 on the *SOHO* mission, which is in orbit around the Sun–Earth Lagrange (L1) point. Recent advances in the use of multilayer coatings to enhance reflectivity at EUV wavelengths are now being applied to the development of spectrometers. The ISAS *Solar-B* mission (due for launch in August 2005) includes an EUV Imaging Spectrometer (EIS) that records solar extreme ultraviolet (EUV) spectra to provide information on the dynamics, velocity, temperature and density of the emitting plasma. EIS represents the next stage of development in this area and follows on from the *SOHO* CDS described above. It responds at wavelengths in a range of ~ 15–30 nm, with an ~ 10 times higher effective aperture than that of the CDS (Culhane *et al.*, 2002). This enhanced throughput is achieved by (i) employing multilayer-coated optics, (ii) restricting the wavelength ranges covered and (iii) detecting the photons in back-thinned CCDs with high quantum efficiencies of $\sim 80\%$ at the EUV wavelengths of interest (see Figure 13.20).

A simple schematic illustration of the EIS instrument is given in Figure 13.50. Solar radiation passes through a thin aluminium entrance filter that is required to prevent visible light from reaching the CCDs and thermal energy from distorting the spectrometer. Radiation is then collected by the primary mirror and focused through a slit mechanism and a second filter onto a grating. Diffracted radiation from two designated wavelength bands is finally registered on a pair of CCDs. The instrument uses a stigmatic off-axis parabola as the primary mirror and thus forms an image ~ 4 arc min high along the slit. In the dispersion plane, the primary mirror is scanned in 1 arc sec steps to enable the build-up of an image of ~ 6 arc min extent. In addition there is a coarse motion which allows the mirror to be offset by ~ 15 arc min.

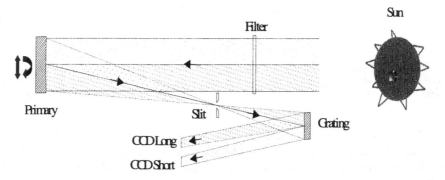

Figure 13.50   Schematic diagram of EIS. Solar radiation enters from the left and passes through the filter to the primary mirror. Following passage through the slit to the grating, it is diffracted to the two E2V CCDs, which respond in the band 15–21 and 25–29 nm.

The structure is fabricated from composite panels rather than from Al alloy as used in CDS, to ensure low mass and a very low overall thermal expansion coefficient. It has an overall length of 3.2 m, with the other two dimensions being ~ 0.5 m. The thin (150 nm) entrance filter is contained in a vacuum housing which will be maintained at ~ 0.1 atm pressure until the spacecraft has achieved orbit. This protects the very thin filter from acoustic loading during the launch phase. The primary mirror, of 150 mm aperture, has each half of its area coated with a different thickness Mo/Si multilayer which is optimised for the two different wavelength ranges. Two slits and two slot apertures are mounted on a four-position mechanism. There is a high-speed shutter for exposure control and a redundant thin aluminium light-rejection filter behind the slit. The grating, with a uniform ruling density of 4200 lines/mm and multilayer coatings, which match those of the mirror, is equipped with a focus mechanism for final in-orbit adjustment. Finally, the two selected EUV wavelength bands are registered by two back-illuminated CCDs of high quantum efficiency for the chosen wavelengths. These devices, produced by E2V, have a format of 2048 (dispersion) × 1024 (along slit) 13.5 μm pixels.

The mirror and grating combination has a spatial resolution capability of 2 arc sec while the sampling is at 1 arc sec for each 13.5 μm CCD pixel. The multilayer coatings are optimised for two wavelength ranges, namely 17–21 nm and 25–29 nm. Emission lines from transition region (0.1–1.0 MK), coronal (1–3 MK) and solar flare (6–25 MK) plasmas are included. For data of good statistical quality, $\lambda/\Delta\lambda \sim 9000$–15,000 allows the measurement of velocity from line centroid displacement with a precision of ± 3 km/s, while non-thermal line width measurement will be to an accuracy of ± 20 km/s following subtraction of thermal Doppler and instrumental profiles. Under control of the shutter, exposure times can range from < 50 ms (suitable for solar flares) through 3–10 s for strong active region lines up to 20–30 s for quiet Sun emission.

We discussed the Bragg law of diffraction in Section 13.5.2 on multilayers where the outer electrons in a continuous strip of high Z atoms reflect the radiation. Prior to the development of multilayer coatings, diffraction from the individual atoms in a crystal lattice provided a means of reflecting X-rays that exhibited narrow profiles with angle because of the regularity of the crystal structures. This property was extremely valuable for achieving very high wavelength resolution, but given the relatively small number of diffracting centers in the lattice, Bragg spectrometers have low photon throughput. Thus, apart from some early examples of their use on sounding rocket flights for supernova remnant studies (Zarnecki and Culhane, 1977), they have not been used for the study of distant X-ray sources. However, major advances have been made in our understanding of the solar corona through Bragg spectrometer studies of the line emission from highly ionised ions.

Given the Bragg law, $n\lambda = 2d\sin\theta$, where d is the distance between lattice planes in an appropriate crystal, there are two methods of displaying an incoming spectrum in wavelength, which are indicated in Figure 13.51. Conventional

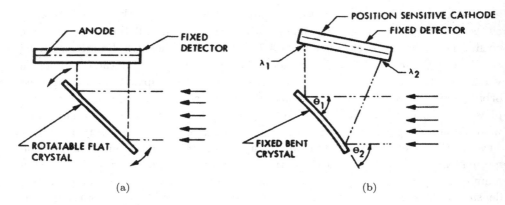

Figure 13.51   Two Bragg spectrometer geometries used in solar X-ray studies. (a) Plane crystal rotated through a range of Bragg angles with a detector large enough to respond to the required range of wavelengths. (b) Curved crystal presenting a range of Bragg angles to the incoming beam with the spectrum registered by a position-sensitive detector.

Bragg spectrometers scan in wavelength by rotating a flat crystal so that a range of angular positions (θ) converts to a range of wavelengths according to Bragg's law (Figure 13.51(a)). Acton *et al.* (1980) flew a spectrometer of this type on the NASA *Solar Maximum Mission* (*SMM*) as part of the X-ray polychromator. These systems give broad wavelength coverage but, in sampling only a single wavelength at a time, do not provide a time-coherent set of line intensities for solar coronal structures that are highly time-variable. To overcome this limitation, the Bent Crystal Spectrometer (BCS; Figure 13.51(b)) was developed for space use (Catura *et al.*, 1974; Rapley *et al.*, 1977). Since the sensitivity of these spectrometers $S \propto 1/\Delta\theta$, where $\Delta\theta = \theta_2 - \theta_1$, is the range of incident Bragg angles, the appropriate choice of wavelength range is important. The *SMM* X-ray Polychromator also included a BCS component, which responded in eight separate wavelength intervals using curved germanium crystals to register the emission lines from Fe and Ca ions with $\lambda/\Delta\lambda$ in the range of 4000–10,000.

More recently, a simpler system, but with 10–60 times greater sensitivity than that of the SMM spectrometers, has been flown on the ISAS *Yohkoh mission* (Culhane *et al.*, 1991). Results from SMM indicated that observations crucial to understanding the energy release process in solar flares needed to be obtained early in flare with high time resolution. Emission from the resonance lines of four ion species, S XV, Ca XIX and Fe XXV, which have helium-like configurations, and Fe XXVI, which is hydrogen-like, is studied by four germanium crystals mounted in the spacecraft, as shown in Figure 13.52. Radiation passes through the entrance apertures and is incident on four bent crystals, from which it is diffracted into gas-filled proportional counter detectors that register the arrival position and hence the Bragg angle for each photon. X-ray line spectra are thus obtained in a common integration period, which can be as short as 1 s, for each of the four

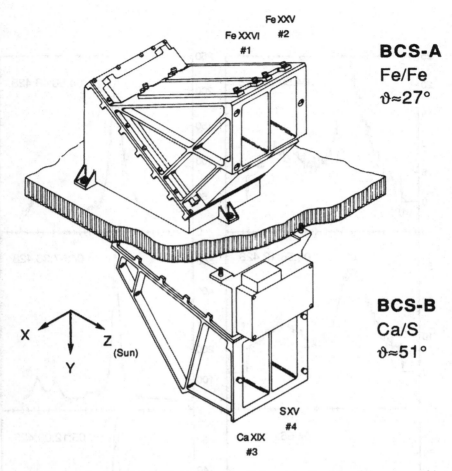

Figure 13.52   The four *Yohkoh* BCS channels mounted on opposite sides of the spacecraft's centre panel. Wavelength ranges are paired for roughly equal Bragg angles.

wavelength ranges. Examples of the Ca XIX resonance line spectrum are shown in Figure 13.53 for a solar flare observed in December 1991. While the spectrum obtained in the declining or plasma cooling phase of the flare shows a symmetric line profile (Figure 13.53(e,f)), during the early impulsive phase, a blueshifted profile (Figure 13.53(a,d,c,d)) indicates plasma upflow into the corona at $v \sim 400$ km/s following the deposition of non-thermal energy by electron beams in the lower solar atmosphere. Integration times as short as ~ 6 s for panels (a)–(c) allow the short duration impulsive phase of the flare, which lasts for ~ 60 s, to be studied in detail.

Properties of the four spectrometer channels are listed in Table 13.3. The way in which the crystal is cut determines the value of the space between the atomic planes — the 2d parameter from Bragg's law. Taken together with the radius of curvature imposed on the crystal, this determines the wavelength range covered by

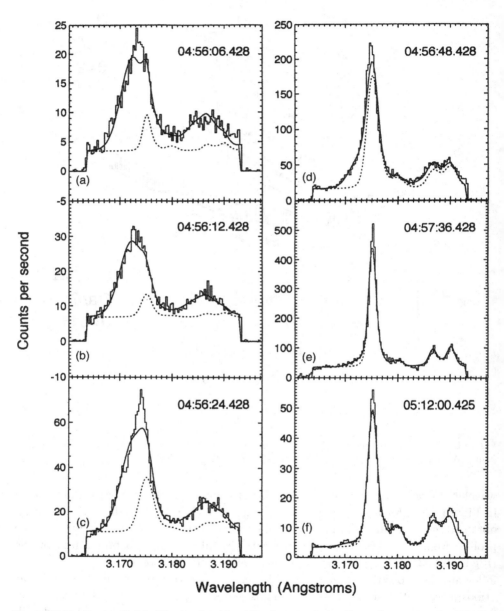

Figure 13.53   In panels (a)–(d), the dotted profile shows emission from plasma at rest, with the bulk of the emission (solid line) due to upflowing gas. For panels (e) and (f), the profile is entirely due to stationary plasma.

Table 13.3 Properties of the *Yohkoh* Bragg Crystal Spectrometer.

Channel ID	Crystal/ Cut	Crystal radius (m)	Ion	Wavelength range (nm)	$\lambda/\Delta\lambda$	Resolution (nm$\times 10^{-5}$)
BCS-A						
1	Ge/220	13.64	Fe XXVI	0.17636–0.18044	4700	3.8
2	Ge/220	10.20	Fe XXV	0.18298–0.18942	3500	5.3
BCS-B						
3	Ge/220	9.60	Ca XIX	0.31631–0.31912	6000	5.3
4	Ge/111	4.56	S XV	0.50160–0.51143	2700	18.6

each of the channels. X-rays from the Sun enter along the z direction through thin Kapton aluminised filters (not shown in the figure) used for heat rejection.

While medium-to-high-resolution dispersive spectroscopy has been a tool for solar coronal observations since the first use of Bragg spectrometers in the late 1960s, use of the technique for distant cosmic X-ray sources has finally come of age with its deployment on the *XMM-Newton* and *Chandra* missions. These have been discussed in outline in Section 13.3.4 and we will now examine two contrasting approaches, namely reflection and transmission grating spectroscopy.

We discussed briefly the *XMM-Newton* Reflection Grating Spectrometer array in Section 13.3.4, while the layout of the gratings is shown in Figure 13.11. Two of the three X-ray telescopes feed radiation to grating arrays, with half of the total reflected X-rays being passed to direct imaging MOS CCDs. The spectrometer is described in detail by den Herder *et al.* (2001) — see Figure 1 of that paper for a more detailed layout diagram. Nine back-thinned CCDs, each with 1024×384 format with 27 μm pixels and 30 μm depletion depth, are located on the Rowland circle and detect the dispersed spectra in a single photon counting mode. The arriving X-ray position on the CCD array is given by the dispersion equation

$$n\lambda = d(\cos\beta - \cos\alpha), \qquad (13.12)$$

where n is the spectral order with values $-1, -2, \ldots$, d is the groove spacing, and α and β are the angles made by incident and dispersed rays with the grating plane. First and second orders overlap on the CCDs, but the latter have sufficient energy resolution to allow the separate identification of photons in different orders. Each array contains 182 gratings, each measuring 10×20 cm. The grating substrates are of 1-mm-thick SiC and backed by five stiffening ribs. Gratings are replicated from a ruled master and are covered with a 200 nm gold reflecting layer. Groove density is ~ 646 grooves/mm at the centre and varies by $\pm 10\%$ from centre to each end to correct for aberrations associated with the converging beam. The support structure for the gratings is made from a monolithic billet of beryllium chosen for its low mass and good stability with temperature.

Two key performance measures, the RGS effective area and wavelength resolution, are shown plotted against wavelength in Figure 13.54. Effective area peaks at ~ 1.5 nm in first order and at ~ 1.0 nm in second order. It should also be remembered that there are two reflection grating modules, so for photon-limited observations, peak effective area in first order is ~ 140 cm^2 and ~ 60 cm^2 in second order. Over the wavelength range of 0.5–3.5 nm, $\lambda/\Delta\lambda$ varies from 60 to 500 in first order and from 80 to 800 in second order. These are approximate figures due to the difference between RGS1 and RGS2 values.

In contrast to *XMM-Newton*, the *Chandra* mission includes two transmission-grating spectrometers which, as the name suggests, consist of high Z metallic structures attached to a thin substrate and with spaces which transmit the X-rays. Two systems, the Low Energy Transmission Grating (LETG; Brinkmann *et al.*, 1997) and the High Energy Transmission Grating (HETG; Canizares *et al.*, 1985), can be put in place alternately behind the *Chandra* telescope array. To explain the principle of operation, we will concentrate on the HETG. The overall HETG system includes two sets of gratings, each with a different period. Of these, the medium energy gratings intercept the beam from the two outer shells while the high energy gratings intercept the beam from the two inner shells. Both grating sets are mounted on a single structure and so are used together. Diffracted radiation is registered by the CCDs of the *Chandra* Advanced Imaging Spectrometer optimised for spectroscopy (ACIS-S).

A schematic layout of the HETG in position behind the *Chandra* telescope is shown in Figure 13.55. X-rays from the mirrors are incident on the transmission gratings and are diffracted in one dimension through an angle β according to the grating relation

$$\mathrm{Sin}\beta = \frac{n\lambda}{d}\,, \tag{13.13}$$

where n is the order number, d is the spatial period of the grating lines and β is the dispersion angle. Undispersed radiation is registered at the zero order position. The enlarged views in Figure 13.55(a) show the medium and high energy grating facets and their relationship to the inner (higher energy) and outer (medium energy) mirror shells. The final enlargement on the extreme right of the diagram shows the grating structure at the level of the individual bars where the bar sizes are 0.2 μm and 0.1 μm for medium and high energy facets respectively. Facet construction is indicated in Figure 13.55(b). Here the gold bars, approximately rectangular in shape and with depth 360 nm (MEG) and 510 nm (HEG), are shown attached to thin polymide membranes coated with thin layers of gold and chromium.

The HETG support structure is a circular aluminium plate that can be swung into position behind the mirror array. A total of 336 individual grating facets, each $\sim 25 \times 25$ mm, are mounted on the structure. Each facet is positioned to bring the dispersed radiation to focus on the ACIS detectors. There are 192 facets on the outer two telescope annuli that constitute the MEG, with grating period

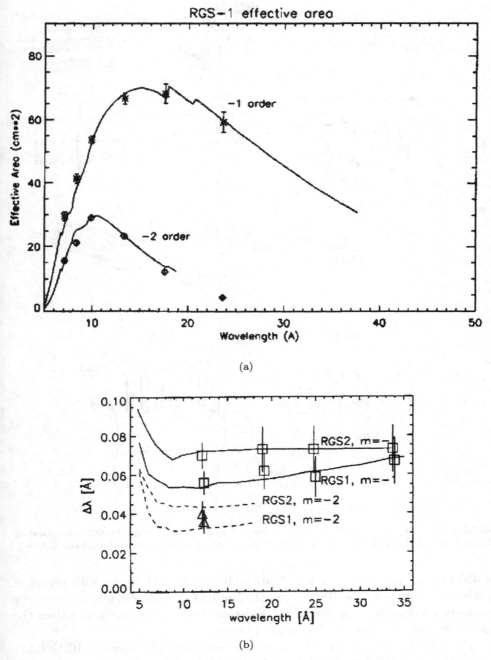

Figure 13.54   (a) RGS1 effective area plotted against wavelength for $n = -1$ and $n = -2$. (b) FWHM wavelength resolution plotted against wavelength for both orders. Values differ slightly for RGS1 and RGS2 (den Herder *et al.*, 2001). (Courtesy of ESA.)

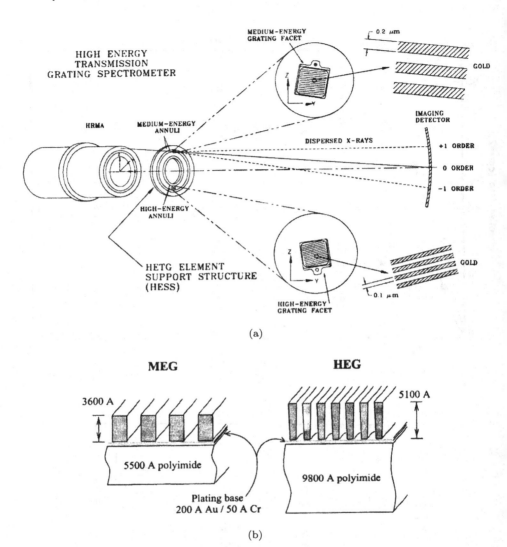

Figure 13.55   (a) Layout and operation of the *Chandra* High Energy Transmission Grating (HETG). (b) Detailed diagram of the HETG facet construction. (Courtesy of SAO and NASA.)

of 400.14 nm. The two inner annuli are addressed by 144 facets, with period of 200.08 nm, making up the HEG. The gold grating bars are electroformed onto the conducting substrate. The height and width of the bars are chosen to reduce the zero order and maximise the first order intensity.

The HETG system operates in two wavelength (energy) ranges: for HEG, 0.12–1.5 nm, and for MEG, 0.25–3.1 nm. For the HEG, the resolving power $\lambda/\Delta\lambda \sim$ 65–1070, while for the MEG, the range of values is $\sim$ 80–970. If the responses of all the orders are added together, the complete HETG system has an effective area ranging from 7 cm^2 at 2.5 nm to 25 cm^2 at 0.2 nm. While the peak value of 200 cm^2 at ∼ 0.8 nm is somewhat larger than that of the *XMM-Newton* RGS, the figures are significantly smaller at each end of the energy range.

13.5.3.2 *Non-dispersive spectroscopy*

For solar observations, the need for spectral resolution is such that the difference in application between imaging and spectroscopy is very marked. Thus with $\lambda/\Delta\lambda$ values ≥ 1000 required for meaningful study of the complex coronal emission line spectra and given the comparatively high intensities available for the Sun, emphasis has been concentrated on the design of dispersive systems with increasingly high throughput. However, in the EUV and X-ray ranges, it is difficult for dispersive spectrometers to match the angular resolution available with visible light instruments. So while it is not out of the question that the need for high spectral resolution observations of very small (≤ 0.5 arc sec) high temperature structures might suggest niche application for the new generation of non-dispersive instruments, this seems unlikely in the immediate future. In this short summary we will therefore concentrate on the general astronomical application of the various photon detectors discussed in Section 13.5. These have responses ranging from the red end of the visible spectrum (∼ 1.0 μm) to higher energy X-ray energies (∼ 0.1 nm) and provide useful wavelength or energy resolution in particular parts of the range.

Back-illuminated and -thinned CCDs can provide resolving power, $\lambda/\Delta\lambda$ in the range of 15–50 for $2.5 > \lambda > 0.1$ nm. Achieving these values in practice is not necessarily easy since subtleties of the CCD construction are involved in, for example, the charge sharing between neighbouring pixels when the photons are first converted to electron–hole pairs in the silicon. However, with quantum efficiencies of ∼ 0.8 achievable over the same wavelength range, these devices can yield ∼ 10–20 times greater effective collecting aperture than the best of the dispersive systems discussed above in the context of the *Chandra* and *XMM-Newton* missions. The use of CCDs therefore continues to be taken seriously for studies of faint quasars and their black hole nuclei at $z \geq 1$–2.

A major breakthrough in recent years has emerged from the application of the microcalorimeter concept to non-dispersive X-ray spectroscopy. These devices, which respond to the temperature increase that follows the absorption of a single photon, exhibit a much-reduced statistical uncertainty in signal size than is the case for the previous generation of non-dispersive photon detectors, which rely on direct conversion of photons to charge. The new detectors have the advantage that the photon-absorbing element can be physically distinct from the 'thermometer'. However, the need to minimise detector heat capacity and to run the devices at temperatures of $0.03 \leq T \leq 0.3$ K makes their use in space systems extremely challenging. This applies particularly to the need to construct device arrays with formats comparable to those of current CCDs or ∼ 2 k × 2 k pixels. Development of distributed readout systems, rather than relying as at present on an event processing chain for each individual pixel, is crucial to the ultimate success of these devices in astronomy. However, the available energy resolution is impressive. With a figure of 2 eV (FWHM) at 1.5 keV already achieved by the transition edge sensor, $\lambda/\Delta\lambda$ values 250–2500 for $\lambda \leq 2.5$ nm appear to be within reach.

The third category of detector that also works at low temperatures is the superconducting tunnel junction (STJ). Operated at $T \leq 0.1 \, T_C$, in materials (e.g. Ta) where the energy gap is 0.664 meV, absorption of 500 nm photon will lead to the creation of ~ 2000 electrons. While the first development activity for these devices was directed towards operation at X-ray wavelengths and an energy resolution (FWHM) of 6 eV has been achieved for 1 keV photons, difficulties in achieving uniform device surfaces appear to limit the performance for $E > 2$ keV. Thus, for the moment, the transition edge sensor offers greater promise at X-ray energies. However, the unique feature of STJs is their ability to provide useful non-dispersive wavelength resolution for $\lambda \leq 2 \, \mu m$. Measurements by the Estec group with Ta devices have yielded $\lambda/\Delta\lambda \sim 4$–25 for the range of $1980 > \lambda > 240$ nm and it is clear that performance should continue to improve through the UV and EUV to a figure of ~ 300 for 1 nm photons mentioned above. Availability of non-dispersive resolution with good quantum efficiency could transform optical astronomy in applications such as order sorting for very high resolution echelle spectrometers in a manner similar to that in which CCDs are being used for the *Chandra* and *XMM-Newton* dispersive spectrometers discussed above. In addition there is the potential for dramatic improvements in resolving power if future devices can be fabricated in Mo or Hf for which the energy gaps are 0.139 and 0.020 meV respectively.

13.6   Summary and Conclusions

In this chapter we have discussed a broad range of instrumentation for the observation of photons in space.

Section 13.2 contained a brief account of the electromagnetic spectrum, including a listing of the usual names for the spectral regions and a description of the processes for photon interaction with matter. The photoelectric absorption process is particularly important both for the detection of photons and for their interaction with the atmosphere of the Earth. The latter in fact supplies the reason why observations in most of the wavelength range must be undertaken from space since photon absorption by or interaction with the turbulence of the atmosphere effectively eliminates useful astronomical data taking for all but the radio spectrum.

In Section 13.3 we discussed the rationale for studies in the various spectral ranges from space-borne platforms. While the Earth's atmosphere transmits at radio wavelengths, placing a large radio telescope in deep Earth orbits with apogee 20,000–100,000 km allows the achievement of much longer interferometer baselines than is possible on the Earth's surface. The VLBI technique is crucial for achieving high angular resolution at the < 1 m arc sec level for observations of a wide range of objects from stars to distant active galaxies. The nature and performance of the Japanese *Halca*, the one major mission so far undertaken in this area, was outlined.

Sub/mm and IR observations require space-based observatories to avoid atmospheric absorption. Although there are a few narrow transmission windows, they are rendered variable by weather conditions even at high mountain observing sites. Space-based observations are therefore essential for consistent intensity measurements over a broad spectral range. This is particularly important for studies of the microwave background radiation, where advances in cosmology depend on the need to make subtle comparisons of all-sky intensity over a range of angular size scales. General-purpose space observatories are also important in this area given the growing emphasis placed on IR observations of high redshift (z) objects. Properties of the ESA *Planck* and *Herschel* missions which operate in these ranges were briefly discussed.

Even though visible and near-IR wavelengths are transmitted through the atmosphere, the effects of 'seeing' due to atmospheric turbulence and of the sky background make it necessary to go to space for the ultimate angular resolution and sensitivity. The *Hubble Space Telescope* (*HST*) has used these advantages to revolutionise optical astronomy while the future *Next Generation Space Telescope* (*NGST*), which, for reasons of achieving high sensitivity for very distant objects, will observe further into the IR range, will provide the next qualitative advance in the subject. Both of these missions were briefly discussed.

The *Herschel*, *Planck* and *NGST* missions will all be placed in orbit around the Lagrange L2 point, which is $1.5 \cdot 10^6$ km from the Earth in the anti-solar direction. Such an orbit simplifies operational planning by allowing optimum sky coverage with fewer restrictive Sun-avoidance constraints. It also makes simpler the operation of telescopes and detector systems at low temperatures. The growing importance of achieving very low operating temperatures is emphasised throughout the chapter.

Observations at X-ray wavelengths are mandated by atmospheric opacity. Along with the IR/sub-mm range, studies of X-ray emission represent a second key area of activity for future astronomers since they enable work on objects with extreme physical conditions due to the very high energies and temperatures involved in e.g. active galactic nuclei and supernova remnants. The need to operate imaging optical systems at grazing incidence is crucial for progress in this field, and so the difficulties in designing and producing these systems were briefly discussed. The two major observatories currently operating — NASA's *Chandra* and ESA's *XMM-Newton* — were described in outline. In addition dramatic advances in our understanding of the solar corona and the nature of solar flares have come about from the ten-year Japanese *Yohkoh* mission, which includes an imaging X-ray telescope. This instrument was also described.

The atmosphere also absorbs gamma rays. Detection and imaging at these energies are especially difficult due to the long photon stopping distances required and because reflection, even at grazing incidence, is not possible. Two missions were discussed in outline. The NASA Compton observatory has concluded its work and re-entered the Earth's atmosphere from its near-Earth orbit. The masses and sizes of the required instruments pose particular problems, with the 17-tonne Compton spacecraft being launched by a dedicated space shuttle flight. The instrumentation employed, particularly the two-tonne spark chamber, which is needed to respond at photon energies up to 30 GeV, has some similarity to systems that used to be employed with large ground-based particle accelerators. Although the scope of the discipline tends to be limited by the small number of available high energy photons, Compton has led to significant advances in the study of very high energy emission from certain quasars and in the understanding of the sources of gamma-ray bursts. The ESA *INTEGRAL* mission, launched in 2002, emphasises energy and spectral resolution in the somewhat lower energy range below 30 MeV but is expected to complement the achievements of Compton due to a substantially enhanced performance at these lower gamma-ray energies. The rather specialised detector systems required in this field were discussed in the context of these two missions and so were not treated further in the chapter.

In Section 13.4, the techniques for photon detection were discussed. At radio wavelengths, hetrodyne receivers allow both detection and high-resolution spectroscopy by mixing the incoming signal with the output of an appropriate local oscillator. For the sub-mm and far-IR, bolometers operate by measuring the temperature rise in an absorber that has been illuminated by incoming radiation. Detector noise depends on a high power of the temperature, so it is essential to operate these systems at temperatures of ~ 100 mK. The need to couple each pixel individually to the telescope by means of feedhorns makes it difficult to construct large arrays or to completely fill the telescope beam in the focal plane. Thus efforts are under way to develop large filled arrays using transition edge sensors.

For the visible and near-IR ranges, coupled arrays of semiconductors are in widespread use. Charge-coupled devices (CCDs), employed in large arrays with optical telescopes, consist of MOS capacitors, which allow the transfer of detected charge through the pixel array to one or more output amplifiers. Silicon CCDs are limited in their response to $\lambda \leq 1$ μm and so the IR arrays employ either intrinsic semiconductors with lower band gap energies than silicon or extrinsic photodetectors with impurity doping that creates energy levels lying closer to the conduction band. Use of these devices can extend the wavelength response to $\lambda \leq 130$ μm.

For EUV and X-ray photons, CCDs are also increasingly used though it is necessary to thin the devices so that the physical thickness of the photon-absorbing region matches the depletion depth in silicon as closely as possible. At X-ray wavelengths in particular, the collected charge is proportional to the energy of the absorbed photon and so for $\lambda \leq 2.0$ nm a useful measure of energy resolution becomes available without the use of a dispersive element such as a diffraction grating or Bragg crystal. For image registration, back-thinned CCDs can operate to $\lambda \leq 5.0$ nm with quantum efficiency (QE) > 50%. While the QE does not drop below 20–30% in the range up to 30 nm, microchannel plates, often used to provide intensified visible signals for CCDs, can yield somewhat higher QE values.

In the last decade, the principle of the IR bolometer has been applied successfully to X-ray detection but with the advantage that a very useful measure on non-dispersive energy resolution can be achieved for X-ray photons. Transition edge sensors, where superconductors, operated at just below the critical temperature for transition to normal resistivity, make the transition in response to the absorption of a photon, offer a promising new approach in this area. The resulting impulsive decrease in current is proportional to X-ray energy but, following the increase in resistivity and resulting current reduction, the device rapidly cools again to below T_c. The most promising wavelength range for application is below $\lambda \sim 1$ nm. As with bolometers, fabrication of arrays of these devices is crucial for their effective application in X-ray astronomy. Superconducting tunnel junctions, while also offering competitive energy resolution for wavelengths ≥ 1 nm, appear to have their most promising future applications in the UV and visible spectral ranges. Small arrays have already been used successfully on ground-based optical telescopes.

For IR and sub-mm and increasingly for X-ray wavelengths, detector operation at very low temperatures ($T \sim 0.1$–0.3 K) is of growing importance. This topic was discussed extensively in Section 13.4 and a range of approaches presented. These included the use of large volumes of liquid or solid cryogen, cyclic mechanical coolers, Joule–Thomson expansion coolers, the sorption pump and the adiabatic demagnetisation refrigerator. Cooling of instruments on spacecraft in near-Earth orbits is challenging due to solar radiation input and the proximity of the Earth. Thus, for several future astronomy missions, the Lagrange 2 quasi-stable orbit, which is around a point of balanced gravitational potential located 1.5 million km behind the Earth, will allow spacecraft structures to run at $T \sim 40$ K and provide an energetically favourable environment for further cooling of instrumentation.

In Section 13.5, imaging telescopes were discussed and a range of large space telescope missions ($d > 0.5$ m) that operate in the visible, IR, sub-mm and radio bands was described. The preponderance of IR missions is apparent, indicating that with increasing sensitivity, IR observations have become more relevant for the study of very distant objects. Among the exceptions are the 8 m deployable radio telescope (*Halca*) mission, which greatly extends the available baselines for VLBI studies of radio sources and the diffraction-limited 0.5 m solar telescope (*Solar-B*) mission, which can resolve to distances of 150 km on the Sun's surface. The physics of grazing incidence X-ray reflection was briefly reviewed and the development and application of multilayer coatings that allow near normal incidence operation in solar missions operating at EUV wavelengths were discussed. For photon energies greater than 10 keV, the use of graded depth multilayers is enabling the use of grazing incidence optical systems, which can focus X-rays. Operation at energies up to 180 keV has been demonstrated.

The techniques for high-resolution dispersive spectroscopy were presented. Reflection grating spectrometers are much used for high-resolution observations of the Sun's EUV spectra. Two spectrometers from the ESA/NASA *SOHO* mission, one operated at normal incidence and the other at grazing incidence, were described. For future work at EUV wavelengths, Mo/Si multilayers that greatly enhance reflectivity over narrow wavelength ranges will increasingly be used. As an example of this approach, a high-resolution imaging EUV spectrometer with multilayer coatings that will be flown on the Japanese *Solar-B* mission, was described. Bragg crystal spectrometers have been used in the past where observations with the highest wavelength resolution were required. The use of fixed curved crystals offers improved time resolution compared to plane scanning systems. As an example, the system used on the *Yohkoh* solar mission for study of the emission line spectra of highly ionised coronal S, Ca and Fe was discussed.

While dispersive spectrometers have mainly been used for high-resolution solar observations, the two major X-ray observatories, *XMM-Newton* and *Chandra*, are equipped with high-resolution reflection and transmission grating spectrometers respectively. The operation of these systems was discussed and their effective areas and wavelength resolutions were compared. For distant astronomical sources, the improved spectral resolution now being offered at X-ray wavelengths by the new generation of cryogenic detectors will probably lead to their adoption for future X-ray observatories. However, the weight, size and complexity of the large grating-based instruments have to be set against the complexities of the cooling systems needed to provide operating temperatures of ~ 0.1–0.3 K for non-dispersive detectors. In this connection, use of the L2 orbit behind the Earth will make the use of cryogenic detectors easier to achieve.

Chapter 14

Space Engineering

Alan Smith

14.1 Introduction

The development of spacecraft is a highly demanding activity. While reliability must be of the highest level, schedules must be adhered to and resources are limited. Scientific payloads add a further dimension — the need for internationally competitive performance.

Opportunities to provide scientific instrumentation for space flight present themselves from several national and international programmes, but most frequently from the European Space Agency, NASA and Japan.

This article explores the issues associated with the development of scientific instrumentation for satellites and space probes from the perspective of a university research group that has provided numerous instruments for space.

14.2 The Engineering Challenges of Space

Development of instrumentation for use in space poses a unique combination of challenges to the engineer and the project manager.

Expensive launch vehicles (rockets) are required to place a spacecraft in Earth orbit or beyond. Since a primary driver for the cost of the launch vehicle is the mass of the satellite, all aspects of the instrumentation that directly or indirectly (e.g. required electrical power) impact the mass are heavily constrained. In practice enormous lengths are taken to minimise mass and power consumption.

To minimise mass, very lightweight structures are used made from e.g. aluminium alloys or composite materials, and these are optimally designed. Where possible, instrumentation will be miniaturised, including the use of highly integrated, high-density electronics.

The need for low power tightly constrains the choice of electronic component. Typically CMOS devices are used. As we shall see below, other serious constraints on such components are also present.

Launch vehicles are not designed for specific, individual applications, at least not since the Saturn V/*Apollo* programme. The vehicle has to be chosen to best match the mission (and vice versa). Apart from simple mass considerations the launch vehicle will also constrain the physical volume of the spacecraft (although some adaptation of the faring of launch rockets is possible), especially in the case of the space shuttle. Possible orbits are also constrained, more so if the launch vehicle is to launch multiple satellites. A further, natural constraint comes about from the fact that the Earth is spinning and that the velocity of its surface is greatest at the equator. For this reason launch sites tend to be at low latitudes and most orbits are equatorial. Note that the orbit chosen will also influence the telemetry rates available to the instrument (how much data one can transmit to ground) since signal strength falls with the square of the distance and increased power implies increased mass.

To reach orbit the instrumentation must endure the experience of launch. During launch the instrument will be subjected to severe vibration and acoustic noise from the rocket motors. Mechanical shocks will also be present, caused by e.g. first stage separation. Therefore mechanical structures, and all instrumentation in general, must be designed to be not only low mass but also robust against mechanical stress. The need to provide a design margin to allow for uncertainty in the precise launch conditions increases these demands still further. Linear acceleration and depressurisation during the launch phase are also issues to be addressed when designing instrumentation for space.

When safely placed in orbit, another, quite different set of challenges are manifest by virtue of the space environment. In space one must deal with:

- The vacuum of space — cooling by convection is not present, contaminants can readily move from one part of an instrument to another, and high voltage discharge is an issue. Note that while the vacuum of space intrinsically has a very low pressure, out-gassing and local trapping of volatiles can lead to a significant ambient pressure (e.g. $> 10^{-3}$ mBar).
- The Sun's thermal radiation — typically a satellite will be illuminated by the Sun on one side ($T \sim 6000$ K) and by either the Earth ($T \sim 300$ K) or dark space ($T \sim 4$ K) on the other. This presents a serious thermal control problem that may be worsened by the need to orient the spacecraft in arbitrary directions while maintaining full irradiance of the solar panels.
- Ionising radiation — commercial electronic components are unsuited to a radiation environment. Components must be selected according to their resistance to radiation damage (radiation hardness) and further shielded where appropriate. Ionising radiation can also affect the properties of materials; for instance, some glasses will gradually become opaque.

- Atomic oxygen — at relatively low Earth altitudes solar radiation dissociates molecular oxygen and there is a relatively high concentration of atomic oxygen. This highly reactive species is swept up by satellites and can cause damage to delicate surfaces such as mirrors, solar panels or filters.
- Ultraviolet radiation — solar UV is also a hazard to many 'painted' surfaces such as thermal radiators which can discolour (and therefore lose their effectiveness) following illumination by the Sun. Solar UV will cause the polymerisation of organic contaminants which may build up on the sensitive surfaces of the spacecraft (e.g. radiators) and cause changes in thermal, electrical or optical properties.
- Dust — the phenomenon we call the zodiacal light is merely sunlight reflected from dust particles concentrated roughly in the plane of the ecliptic. Satellites passing through this cloud of dust may suffer damage to delicate surfaces and membranes. For missions to comets this is a major hazard since the density of dust is much higher around such objects, which are the origin of a significant part of the zodiacal dust.

Instruments deployed in space will have an anticipated life expectancy — typically 2–10 years. While in a very limited number of situations some refurbishment or repair of a satellite may be possible (e.g. the *Hubble Space Telescope*), in the vast majority of cases the satellite must be designed and constructed to operate without maintenance for its planned life. This naturally leads to the inclusion of fault avoidance measures such as the selection of only very high quality components and the de-rating of designs. Satellites also need to be fault-tolerant, i.e. they must be able to cope with faults as they develop on board. Redundant critical systems are commonplace; and increasingly spacecraft are being designed so that software can be uploaded to work around problems, etc.

The hostile environment mentioned above leads to progressive degradation of satellite systems (notably electronics, solar panels, radiators, and some scientific instruments). Mechanisms wear and the debris will ultimately affect their performance. Electronic components typically have a life expectancy that is related to their temperature of operation. Materials out-gas and the resultant contaminants build up in critical areas (such as the surface of mirrors). One must take all of these effects, and other more subtle ones, into consideration when designing the instrumentation. In particular, consideration must be given to the 'end of life' performance, which must continue to meet the needs of the mission.

Some satellite functions require the use of on-board consumables. Attitude control is often accomplished using hydrazine motors, and cooling of scientific instruments may involve the boil-off of liquid helium into space. When the consumables are fully depleted the associated function (and perhaps mission) is lost. Therefore the amount of such consumables sent into space must be carefully judged. Too little and the mission is cut short; too much and vital mass has been taken from other subsystems. Therefore the operational scenarios of the mission must be well understood during the early design phase and prudent, managed conservation of consumables must be a feature of the mission planning and operations.

Ultimately low-Earth orbits or even highly eccentric orbits with a low perigee will decay and the satellite will re-enter the Earth's atmosphere. Sometimes a trade-off between launch mass and lifetime has to be performed.

During the life of the spacecraft its behaviour is controlled through on-board and ground-based operations. Operations typically involve the reception of status and mission data and the transmission and execution of commands. Any operations plan must take into consideration:

- Remote operations — the only possible communication with (and therefore influence over) the spacecraft is via its telemetry uplink. If this fails, then the mission is lost. Moreover spacecraft should be considered as real-time systems that must be able to look after themselves between periods of contact and in the case of unscheduled loss of contact. This circumstance is unusual yet may dominate their design philosophy. An on-board computer will usually manage spacecraft operations locally with either direct or pre-loaded command sequences being executed.
- Electrical isolation — the vacuum of space contains energetic charged particles. When these impact on the spacecraft they tend to charge up the surface, which is unable to discharge to ground. Special measures may be necessary to deliberately discharge the spacecraft by means of an ion gun.

Often the instrumentation (e.g. solar panels, covers, booms, aerials, launch locks) is launched in a configuration, which, while providing the necessary resistance to vibrational damage, is inappropriate to its proper function. Deployment mechanisms are necessary and may be required to operate while out of ground contact. Complex operation sequences may be stored on board and executed automatically, yet with some margin of fault recovery.

14.3   Spacecraft Design

Complex systems may fail in complex and unexpected ways due to unforeseen emergent system properties or hidden internal interactions. While component level failures can be reduced to acceptable levels by the selection of high quality parts, minimising complex failure modes is more difficult. One approach is to reduce complexity through segregating, where possible, spacecraft functionality into separate sub-systems and then minimising the interaction between them. In systems engineering language this is referred to as 'diagonalising the matrix', i.e. in an interaction matrix of function vs. sub-system partitioning, most (but not all) of the non-diagonal elements are empty. This approach is very apparent in spacecraft design (with the exception of very small, highly integrated spacecraft where the boundaries between sub-systems are necessarily blurred).

Function Subsystem	Provide electrical power	Transmit data to ground	Provide stable pointing	Provide appropriate thermal environment	Acquire Science Data
Power	█				
Telecoms		█			
Data Handling and Control		█			█
Attitude Control			█		
Structure					
Thermal				█	
Science Payload					█

Figure 14.1   A partitioning matrix for a scientific satellite.

Shown in Figure 14.1 is a typical, high-level matrix of this form for a generic, scientific satellite.

The essential sub-systems in any spacecraft include:

Power. This sub-system comprises: solar panels which convert solar energy into electrical power; batteries which act as intermediate storage of electrical power and are essential for spacecraft operation prior to solar-panel deployment and/or Sun acquisition and during eclipses; power distribution, conversion and regulation electronics; current monitors and switches. Batteries have a finite lifetime, which needs to be carefully managed through appropriate charge and discharge cycles. Electrical noise is often an issue on spacecraft and can be either conducted or transmitted. Power converter noise must be regulated and filtered, and noise between sub-systems via the power sub-system or ground lines must be controlled. Synchronisation of the power converter clocks is commonplace.

Telecommunications. Data must flow both to (up-link) and from (down-link) the spacecraft. The telemetry capacity in either direction is governed by the power of the transmitter, the sensitivity of the receiver and the distance between them. Most satellites utilise pre-existing hardware. Naturally, more distant spacecraft are therefore likely to have lower telemetry rates.

Data handling and control. The spacecraft will be managed by an on-board computer system that will monitor its 'health', receive (from the telecommunications sub-system) and execute commands, acquire and store data, and pass mission data to the telecommunications sub-system. This sub-system will comprise both processor units and data storage units. For mass storage, tape-recorders have been replaced by solid-state memories in recent years. The need for radiation tolerance and high reliability, together with a general reduction in availability of MIL standard parts, now places severe constraints on parts selection and hence design and performance. The growth in scientific data volume has been very large and so it is increasing necessary to perform onboard data compression prior to transmission. Loss-less compression algorithms can provide only modest compression ratios (depending upon the structure within the data) and very often a much more radical approach is needed using lossy compression. For scientific data, square-root compression is favoured since it adds no artefacts to the data. For a greater compression ratio more exotic routines (e.g. wavelet or Jpeg) are used. Ultimately, very high compression is possible if one accepts that data analysis that would normally be performed on the ground must be performed on board with the resultant transmission of data products rather than raw data.

Attitude control. Most scientific satellites must control their attitude (pointing direction) in space. Usually various constraints are associated with attitude control. These include the need for illumination of solar panels (which are seldom movable relative to the main spacecraft since this is expensive), and exclusion angles to prevent solar illumination of sensitive instrumentation (Sun avoidance). Attitude control involves attitude determination (through magnetic, Sun, star and Earth limb sensors typically) and attitude correction (using e.g. hydrazine thrusters, magneto-torquers and reaction wheels). Attitude control is handled through a feedback process to provide the necessary stability.

Structure. The various sub-systems must be held within an appropriate, lightweight yet stiff structure. Often this is separated between a service module (comprising most of the major sub-systems) and a payload module (including the various scientific instruments). The structural element of these modules is a subsystem in its own right although its function is relatively passive. Significant cost savings can be afforded by re-use of service module design elements. However, re-use of payload module elements is more problematic owing to the unique nature of every scientific payload. Very lightweight structures are possible through the use of composite materials (Formula 1 racing cars have composite structures). While enormous savings are possible in mass, problems still remain regarding the stability, out-gassing and thermal conductivity of such materials. Nevertheless even very contamination-sensitive EUV instruments are now being flown within composite structures.

Thermal. It is usually necessary to control the temperature of payload and service module elements. Some level of control is immediately afforded through the use of thermal blankets which reflect the sunlight yet retain infrared, thermal radiation. If further control is needed a passive radiator directed towards deep space is an effective solution. Where fine-tuning is necessary, or instrumentation (e.g. a mechanism) has to be protected from becoming too cold, then heaters can be used. However, one is reluctant to use heaters if they can be avoided because they use precious electrical power. If temperatures below approximately 150 K are needed, then more complex cooling will be necessary. Closed-cycle sterling cycle coolers or helium cryostats may then be used. For extremely low temperatures (< 1 K) even more exotic technologies, such as dilution or adiabatic demagnetisation refrigerators, are needed. For such situations thermal design is critical and very complex thermal models of the spacecraft have to be developed and verified to support the design process.

Payload. Each scientific payload instrument is a sub-system of the spacecraft. While some coordination of operating modes may occur between instruments, they are normally largely independent (although in some cases instruments will be combined into a package to share common functional elements such as electronics). Typically an instrument will mainly interface with the spacecraft through a communications link (bus), and power ('raw' 28 V, which the instrument further regulates and converts to useful levels). More subtle interfaces include pointing jitter, contamination, thermal and mechanical.

14.4 The Systems Engineering Process

Systems engineering addresses all aspects of the process of system development, from conception through to operation and even disposal. The development of the systems engineering life cycle has been most marked in the areas of space systems, defence systems and IT. NASA divides its systems life cycle into the following phases:

Pre-phase A — The development of user requirements
Phase A — The development of system requirements
Phase B — Architectual design
Phases C, D — Component development
Phase E — Launch and validation
Phase F — Operations

Pre-phase A studies are concerned with identifying potential scientific missions which address important scientific needs. In this case the user is the scientist and the language is that of the scientist. Pre-phase A work also includes strategic, enabling technology development.

Phase A studies are concerned with verifying that a practical solution exists which will meet the scientific need. Viable engineering is the issue and the language is that of the engineer.

Phase B is concerned with the evolution of a design concept. The fundamental functional decomposition is performed and an architectural design of the mission elements (including space and ground segments) is developed. The outcome of the phase (preliminary design review) demonstrates the possibility of a viable engineering solution without performing the detailed engineering design.

Phases C, D are usually combined and represent the development of the system up to launch. They include detailed design, fabrication and ground test. Testing involves both functional testing at various levels (components → sub-systems) and performance testing. Designs are qualified for application in a particular mission largely through the testing of a highly representative instrument model (engineering model or qualification model — see below). Once verified the flight hardware can be constructed and tested to a lower, acceptance level more typical of the environmental stress levels expected during the mission. Environmental tests will include vibration, shock, acceleration, thermal vacuum and EMC.

Phase E represents the launch and commissioning activities. Commissioning in orbit typically takes ~ 1–30 days and involves a progressive turn-on and testing of each spacecraft function. The payload is usually the last item to be checked out and the validation of the payload usually consists of measuring the properties of well-understood astronomical targets, collation with other observatories and/or use of on-board calibration sources.

Phase F is the operations phase and will include in-orbit maintenance where appropriate. Where decommissioning after use is required (for example, a controlled re-entry into the Earth's atmosphere), this comes at the end of the phase.

The listing of the phases above is shown according to a classic 'staircase' model. Individual phases are distinct and sequential. It is often more useful to show the phases in a more two-dimensional format, known as the V diagram (Figure 14.2).

Time generally runs from left to right, down and up the V. Overlap between elements of the V indicates the presence of concurrent engineering, in which issues important in later phases are addressed. Lines across the V represent the verification and validation links. The upper line (labelled validation) represents the link between user needs (from which user requirements can be distilled) and operational

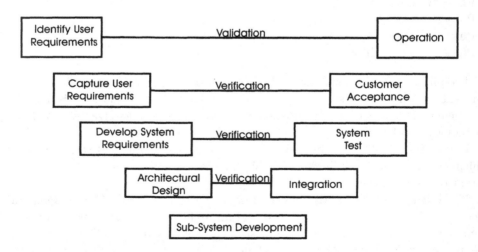

Figure 14.2   The V diagram of systems engineering.

Space System
Development Process

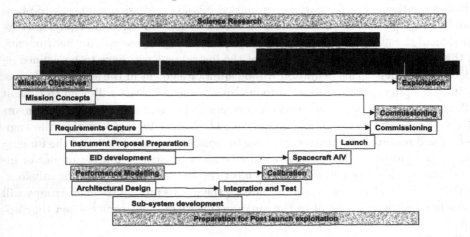

Figure 14.3 A V diagram specific to scientific instrument development. Time runs from left to right. Bars are used to demonstrate typical durations. White bars are primarily engineering activities, black are management activities and mottled are science-based.

performance. Lower lines link the two sides of the V via the process of verification. Verification ensures that the engineering of the instrument is performed with high quality but it does not guarantee that the user needs will be met (they will not be met if the user requirements have not been well captured or if the translation to system requirements has been flawed).

Validation means building the right system. Verification means building the system right.

A more detailed elaboration of the systems engineering process as applied to university-based scientific instrument development for spacecraft is shown in Figure 14.3.

14.5 Requirements Capture

The development of the technical requirement for a scientific instrument is a complex process, which involves iteration and feedback at several levels. Let us first consider how mission opportunities come about. Each of the major space agencies has a declared scientific research programme involving the deployment of scientific satellites. In formulating such programmes the agency will take account of scientific opinion (the opinions of the scientific community) and indeed include members of that community on advisory committees, etc., which help formulate its scientific programme. Initially, missions will be proposed which address a particular scientific need without any confirmation that it is either viable within the likely resources or of sufficient priority to warrant the associated expenditure. Peer review will deal with the relative scientific merits of proposed missions. Mission studies (pre-phase A and phase A) will take mission ideas forward and develop 'strawman' payload concepts, against which competing groups of scientists can propose specific instruments. At this time the technical capabilities of the instrument will have to be defined — indeed the quality of this definition will be a major element in the selection process.

In developing specific, technical requirements the scientists will need to consult with instrument scientists, engineers, managers and others. Technical requirements must be achievable yet highly competitive. They must be viable within the engineering and resource constraints set down by both the space agency and the funding bodies. Individual science groups will be conducting strategic programmes of instrumentation development designed to meet the needs of future scientific missions. By this means and through an awareness of the above process, science groups will be ready to respond in detail to the announcements of opportunity when they appear.

Therefore the specification of technical requirements is a subtle combination of technology push (what can instrumentation offer science objectives?) and market pull (what sort of instrument performance do scientists need?).

Inspection of the V diagram for scientific instrument development will indicate that requirements capture is linked to commissioning via the process of validation (designing the right system). One hopes that when the instrument is proven in orbit it will meet the expectations of those who agreed on the technical requirements. Of course not all the aspirations of the scientists will be achievable (not all of their needs). However, if the requirements capture process is sound, then the agreed aspirations should be met. A sound approach will involve an iteration of requirements, designs and models. From a proposed instrument design (even a high level design) estimates can be made of predicted performance, which can be compared with expectations. Requirements or design may then be modified until an acceptable solution is reached. Naturally, pragmatic issues such as cost, timescale, the use of spacecraft resources and technical risk must also be borne in mind.

Later we will find that this model may well evolve to form the basis of the ground calibration of the instrument.

14.6   Architectural Design, Integration and Test, and Sub-system Development

Figure 14.4 shows how architectural design (AD), the integration and test (I&T) programme and sub-system development (SSD) are related. Note the elaboration of SSD into development models — discussed below.

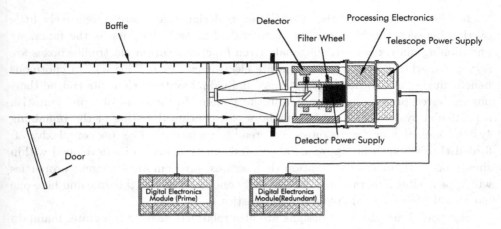

Figure 14.4   *XMM-Newton* — optical monitor architectural design.

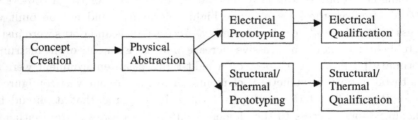

Figure 14.5   Design qualification through the use of physical abstraction.

Architectural design is the process of dividing the system into (usually) segregated sub-systems. For scientific instruments these may consist of detector units, electronics boxes, structure, optics, thermal control, mechanisms, OBDH interface, power conversion, software, etc. During architectural design the functionality of each sub-system is defined, as is its interface with other sub-systems and, where appropriate, the spacecraft and ground test equipment.

Architectural designs (sometimes also called system designs) may be shown diagrammatically in many ways, usually dependent upon the type of abstraction that most naturally describes the system at the time. Function block diagrams lend themselves to electronics/software systems. Optical layouts are naturally good for optical systems while physical layouts are good for structures. Shown in Figure 14.5 is an architectural design for the *XMM-Newton* Observatory Optical/UV Monitor. Here an approximately scaled drawing is found to be useful although some elements are exaggerated in size for clarity. The optical ray path is shown to aid understanding. Of course this simple diagram is supplemented by function and interface definitions. Although simple it provides a conceptual model of the instrument that is shared by the entire development team and is used as a backdrop for discussion both within the team and elsewhere.

It will be noted from the *XMM-Newton* design that there is relatively little overlap between the functionality of individual system elements — the function/ partitioning matrix is fairly diagonal. Even functions within electronics boxes are segregated where possible onto separate electronics cards. This has an enormous benefit during integration and test since individual systems elements can be thoroughly tested prior to inclusion within the system. In the event of a problem with a particular systems element, a simulator, for continued testing of the remaining systems elements or for diagnosis, can readily replace it. The low complexity afforded through such an approach allows full functionality to be developed within simulators without the expectation that serious, problematic emergent properties will appear after integration. (EMC is sadly one exception to this rule and here one has to fall back on good engineering heuristics.)

Segregated architectural designs are also relatively easy to integrate, maintain and understand. While these hold great advantages for spacecraft instrumentation, they do not fully overcome the pressures on spacecraft resources such as mass and power. Some compromise is usually inevitable. In the case of the *XMM-Newton* Optical/UV Monitor, the system, while highly modular, could not be built with easy access to certain modules after assembly. Access panels and doors were just too difficult to include without an excessive increase in mass. Elements of the instrument whose primary purpose was to provide a light-tight, clean environment were also required to be important structural elements. The instrument was configured as a tube-like structure and the strength of the tube required that it should have no weakening doors. Access to the elements within the tube was from either end or only after separation of the tube at one of its bulkheads. The instrument was consequently awkward to assemble optically.

Let us further explore Figure 14.4. The SSD was divided in this case into three overlapping processes: breadboarding, engineering model and flight model.

Breadboarding represents laboratory demonstrations of functionality. These are not designed to show space-worthiness but rather that the sub-system in question is able to meet a given subset of functional performance requirements. Usually these requirements are chosen to be those of highest technical risk (or those necessary to demonstrate another sub-system function). The development of elements of the instrument that can easily be reproduced by off-the-shelf hardware, albeit unsuitable for flight (e.g. electrical power conversion), may be deferred. Such elements can be emulated early in the SSD process with minimal cost and risk. Breadboards offer a very cost-effective way to develop the basic instrument concept. When the programme matures, the usefulness of the breadboard model decreases, as the requirements imposed by the rigours of the space environment and the demands imposed by the limited spacecraft resources take over.

To show that the instrument is capable of meeting the stringent reliability requirements, interface requirements, performance requirements and resource constraints, one or more engineering models are needed. Engineering models are always representative of the flight instrument in at least one area of abstraction.

Table 14.1   Abstractions, tests and models used in space system engineering.

Abstractions	Tests	Mathematical models
Mechanical	Functional	Dynamic load
Thermal	Vibration	Stress
Electronic (digital)	Shock	Thermal equilibrium
On-board software	Thermal	Contamination
Electronic (analog)	Thermal vacuum	Stray light
Electronic (power conditioning)	EMC	Electronic simulation
Contamination	ESD	Radiation dosage
Optics/particule analyser	Contamination	Operations scenarios
Detector physics	Stray light	Reliability
	Lifetime	Payload instrument simulator
	Performance	

For instance, an electrical model will provide representative electrical functionality, albeit only as viewed at the spacecraft interface. Similarly, a structure and thermal model is often produced. Models, like much of the design process, proceed in various forms of abstraction. These too are shown in Table 14.1. The familiar physical abstractions of e.g. mechanical, thermal or electrical are present. But so too are more exotic and specific abstractions, such as contamination (considering the flow of contamination in the system) or detector performance.

The entire design of the instrument may be qualified through the use of a qualification model (QM). A QM is highly representative of the flight model. It is environmentally tested to demonstrate its ability to withstand higher-than-expected stresses (vibration, shock, thermal vacuum, etc.). This margin is essential to give sufficient confidence in the design in the face of uncertainties in both the expected stress levels and the actual build standards. The process of qualification can often be sufficiently arduous that it may degrade the model to the point where it is unfit to be flown in space. A new, flight model is usually then developed and environmentally tested to a lower level to ensure that a good build standard has been achieved.

An alternative process of design qualification is shown in Figure 14.6. Here qualification is performed in a piecemeal fashion following a separation into two complementary physical abstractions. While plausible, such a qualification approach is less desirable since emergent cross-abstraction properties are missed (e.g. vibration can cause spurious electrical signals).

Qualification of mechanisms is a particular issue. Accelerated lifetests are performed and/or complex arrangements are made to simulate weightless conditions in space.

In fact models are not confined to physical items. Mathematical and computer-based models and simulations are commonplace in the SSD process. Table 14.1

Figure 14.6   Conflicting issues in project management.

indicates the range of such models that may be developed during this period. The outputs of these models are an important part of the design process. Accurate simulations allow designs to be developed with every expectation that they will lead to viable solutions when built.

With the advent of high-speed computers, finite element modelling is now a very valuable tool for the instrument modeller. The system in question is modelled as a number of interacting nodes. The interaction of each node with every other node is calculated and an equilibrium model developed by numerically solving the very large number of simultaneous equations on the computer.

Ideally the flight model development does not begin until the QM has been fully tested. However, this will usually compromise the schedule to an unacceptable level. Rather, the design process is typically of sufficiently high quality that some work can begin on the FM before the end of the QM stage. Indeed, the QM might better be viewed as part of the verification process rather than the development process.

14.7   Project Management

One cannot review the process of instrument development without a few words about the role of project management. Project management for instrumentation to be flown on a spacecraft is particularly challenging for a number of important reasons:

- The quality objectives of a project include meeting a required performance standard in the delivered instrument, remaining within a finite budget and delivery to a given schedule.
- Flight hardware provision is very highly competitive. Instrument performance is critical for selection and so inevitably any project will contain a significant element of originality. Even concepts that appear at first glance to be repeats of earlier experience soon unravel to become unique and challenging following the pressure for increased performance (requirements creep), problems with the continued availability of parts and a general predisposition to innovation in the engineering team.
- Resources are invariably very limited. The Particle Physics and Astronomy Research Council usually fund projects within the UK. The Council caps project

budgets, which often include contingencies of less than 10%. Comparable US or industrial projects will typically carry a contingency of more than twice this. Further, the contingency available to many research council projects is limited to specific areas that are outside the control of the project team (e.g. launch delays or currency fluctuations).

- Schedules are constrained by the space agency's master schedule, which is part of a contract with the prime contractor for the spacecraft. Delays in this schedule are likely to be very expensive, especially if declared late. The project teams are under constant pressure to maintain progress.

- Therefore the projects tend to be both highly constrained and very challenging. 'Requirements creep' — the phenomenon of gradually increasing requirements — is commonplace and hard to resist in the presence of competing international science missions and ground-based science facilities. The rigours of space flight instrument development typically lead to quite long development cycles. The need to be conservative in component selection (and the limitation of available components that are space-qualified) means that flight hardware is 'out of date' when flown. It is therefore essential to adopt at least in part a fast-track approach to development to remain competitive. Some flexibility must be retained to prevent obsolescence. While hardware flexibility is challenging (but not impossible), flexible, re-loadable software is a workable approach. Many missions have been rejuvenated by judicious software improvements after launch.

- Flight hardware projects are almost invariably the province of international consortia of participating science groups. Usually no formal, legal arrangement exists between the members of such consortia and each member is separately funded. As with any gathering of organisations, there will be conflicting agendas at some level. Each member will need to prioritise its activities in the context of its own engineering and science programme. An overall project manager will be appointed but will have little or no direct authority over consortium members (with the exception of their own local team). The challenge of project management to coordinate effectively without authority is particularly apposite to scientific instrument development by university/research laboratory research groups.

Project managers must understand the technologies that they are dealing with and adopt a hands-on attitude.

14.8   Underpinning Instrument Development

Earlier, the need for an underpinning, strategic instrumentation development programme was mentioned. Here we will review this in a little more detail.

Technologies for space flight may spend many years in research and development prior to eventual deployment in space. For detector-based technologies this typically begins in a research laboratory (possibly not one engaged directly in space instrumentation development). As new physical phenomena are identified, so the scientific community of detector physicists take up the challenge of development and understanding. Understanding is the key. To be able to develop reliable, predictable hardware that will be able to function unattended in a difficult, uncertain environment, one must appreciate its basic, physical and engineering characteristics. As this understanding improves, the concept of a practical flight instrument might be demonstrated (proof of concept). While this indicates the potential viability of a useful instrument, it falls far short of a full instrument development. Selection for a flight opportunity could be based on only a predicted performance of a new technology but this is a very high-risk strategy (and one that has all too often failed). More likely, a prototype will be needed which can be qualified for space flight. The use of sounding rockets remains one of the most convincing and cost-effective approaches to prototype demonstration. It is cost-effective because sounding rocket instrument developments can cut many expensive corners. Many valuable lessons can be learnt from even a few minutes of operation in space.

Each flight opportunity is different and considerable effort is necessary to prepare for a particular mission proposal. These mission studies seek to optimise the instrument design for a particular application. Space agencies are becoming increasingly demanding about the detail and quality of the instrument proposals. Detailed design definition and interface definition is invariably necessary.

Bibliography

Acton, L.W., et al., 1980, *Solar Physics*, **65**, 53.

Adler, D., Bazin, M., and Schiffer, M., 1975, *Introduction to General Relativity*, International Student Edition, McGraw-Hill, Tokyo.

Alcock, C., et al., 2001, *Astrophys. J.*, **550**, L169.

Alfvén, H., 1943, *Ark. F. Mat. Ast. Fys.*, **29A**, 1.

Aly, J.J., 1984, *Astrophys. J.*, **283**, 349.

Antiochos, S.K., Devore, C.R., and Klimchuk, J.A., 1999, *Astrophys. J.*, **510**, 485.

Bailey, J., Ferrario, L., and Wickramasinghe, D.T., 1991, *MNRAS*, **251**, 37P.

Barnes, W.L., Thomas, S.P., and Salomonson, V.V., 1998, *IEEE Transactions on Geoscience and Remote Sensing*, **36**, 1088.

Barton, I.J., 1995, *J. Geophys. Res.*, **100**, 8777.

Barton, I.J., 1998, *J. Geophys. Res.*, **103**, 8139.

Baumjohann and Treumann, *Basic Space Plasma Physics*, Imperial College Press, 1996.

Benoit, A., et al., in *Sixth European Symposium on Space Environmental Control Systems*, Noordwijk, The Netherlands, 20–22 May 1997, ed. Guyenne, T.D., 1997, ESA SP-400, 497.

Berger, M.A., 1984, *Geophys. Astrophys. Fluid Dynamics*, **30**, 79.

Berger, M.A., and Field, G.B., 1984, *J. Fluid. Mech.*, **147**, 133.

Biesecker, D.A., Myers, D.C., Thompson, B.J., Hammer, D.M., and Vourlidas, A., 2002, *Astrophys. J.*, **569**.

Bittencourt, 1986, *Fundamentals of Plasma Physics*, Pergamon.

Blain, A.W., 2000, astro-ph/0011387.

Bowles, J.A., Patrick, T.J., Sheather, P.H., and Eiband, A.M., 1974, *P. Phys.*, **E7**, 183.

Boyd and Sanderson, *Plasma Dynamics*, Nelson.

Brinchmann, J., et al., 1998, *Astrophys. J.*, **499**, 112.

Brinkmann, A.C., et al., 1997, *Proc. SPIE*, **3113**, 181.

Brown, O.B., Brown, J.W., and Evans, R.H., 1985, *J. Geophys. Res.*, **90**, 11667.

Brown, S.J., Harris, A.R., and Mason, I.M., 1997, *J. Geophys. Res.*, **102**, 27973.

Bruijn, M.P., Hoevers, H.F.C., Mels, W.A., den Herder, J.W., and de Korte, P.A.J., 2000, *Nucl. Inst. Meth.*, **A444**, 260.

Burlaga, L., Sittler, E., Mariani, F., and Schwenn, R., 1981, *J. Geophys. Res.*, **86**, 6673.

Burnight, T.R., 1949, *Phys. Rev.*, **76**, 165.

Butler, D.M., Hartle, R.E., Abbott, M., Ackley, S., Arvidson, R., Chase, R., Delwiche, C.C., Gille, J., Hays, P., Kanemasu, E., Leovy, C., McGoldrick, L., Melack, J.,

Mohnen, V., Moore, B., Phillips, R., Rango, A., Robin, G. deQ., Suomi, V., and Zinke, P., 1984, NASA Technical Memorandum 86129.

Butler, D.M., Hartle, R.E., Abbott, M., Cess, R., Chase, R., Christensen, P., Dutton, J., Fu, L-L., Gautier, C., Gille, J., Gurney, R., Hays, P., Hovermale, J., Melack, J., Miller, D., Mohnen, V., Moore, B., Norman, J., Schneider, S., Sherman, J., Suomi, V., Tapley, B., Watts, R., Wood, E., and Zinke, P., 1987, *From Pattern to Process: The Strategy of the Earth Observing System. Earth Observing System, Volume II, Science Steering Committee Report*, NASA, Washington DC, USA.

Cairns, 1985, *Plasma Physics*, Blackie.

Canfield, R.C., Hudson, H.S., and McKenzie, D.S., 1999, *Geophys. Res. Lett.*, **26**, 627.

Canizares, C.R., Schattenburg, M.L., and Smith, H.I., 1985, *SPIE*, **597**, 253.

Cargill, P.J., 2001, in *Space Storms and Space Weather Hazards*, ed. Daglis, I.A., NATO Science Series II, Mathematic, Physics & Chemistry, **38**, 177.

Catura, R.C., Joki, E.G., Bakke, J.C., Culhane, J.L., and Rapley, C.G., 1974, *Mon. Not. R. Astr. Soc.*, **168**, 217.

Chandrasekhar, S., 1992, *The Mathematical Theory of Black Holes*, Oxford University Press, Oxford.

Chelton, D.B., Ries, J.C., Haines, B.J., Fu, L.-L., and Callahan, P.S., 2001, 'Satellite Altimetry', in *Satellite Altimetry and Earth Sciences*, eds. Fu, L.-L., and Cazeneve, A., Academic Press.

Chen, *Introduction to Plasma physics*, Plenum, 1974; 2nd edition, 1984.

Chervenak, J.A., et al., 2000, *Nucl. Inst. Meth.*, **A444**, 107.

Church, J.A., Gregory, J.M., Huybrechts, P., Kuhn, M., Lambeck, K., Nhuan, M.T., Quin, D., and Woodworth, P.L., and 28 other contributing authors, 2001, 'Changes in Sea Level', in *Climate Change 2001: The Scientific Basis*, eds. Houghton et al., Cambridge University Press.

Clauer, C.R., and McPherron, R.L., 1980, *J. Geophys. Res.*, **85**, 6747.

Clemmow and Dougherty, 1969, *Electrodynamics of Particles and Plasmas*, Addison-Wesley.

Coates, A.J., 1999, *Philos. Trans. R. Soc. Lond. A*, **357**, 3299.

Colless, M., et al., 2001, *MNRAS*, **328**, 1039.

Connolly, A.J., Szalay, A.S., Dickinson, M., Subbarao, M.U., and Brunner, R.J., 1997, *Astrophys. J.*, **486**, L11.

Cowling, T.G., 1953, ed. Kuiper, G.P., Univ. Chicago Press, Chicago, p. 532.

Cracknell, A.P., and Hayes, L.W.B., 1991, *Introduction to Remote Sensing*, Taylor and Francis.

Culhane, J.L., et al., 1991, *Solar Physics*, **136**, 89.

Culhane, J.L., et al., 2002, *Proc. ISSI Workshop on the Calibration of the SOHO Instruments*, in press.

de Bruijne, J.H.J., et al., 2002, *Astron. Astrophys.*, **381**, L57.

de Graauw, Th., Helmich, F.P., and the HIFI Consortium, *Proc. Symp. 'The Promise of the Herschel Observatory'*, 12–15 Dec., Toledo, Spain, eds. Pilbratt, G.L., Cerpicharo, J., Pruste, T., and Harris, R., 2001, ESA SP-460, 45.

DeForest, C.E., and Gurman, J.B., 1998, *Astrophys. J.*, **501**, L217.

Delderfield, J., Llewellyn-Jones, D.T., Bernard, R., de Javel, Y., Williamson, E.J., Mason, I.M., Pick, D.R., and Barton, I.J., 1986, Proceedings of the Meeting, Cannes, France, Nov. 1985, *SPIE International Society for Optical Engineering*, **589**, 114.

Démoulin, P., Mandrini, C.H., van Driel-Gesztelyi, L., López Fuentes, M.C., and Aulanier, G., 2002b, *Solar Physics*, in press.

Démoulin, P., Mandrini, C.H., van Driel-Gesztelyi, L., Thompson, B., Plunkett, S., Kövári, Zs., Aulanier, G., and Young, A., 2002a, *Astron. Astrophys.*, **382**, 650.

den Herder, J.W., et al., 2001, *Astron. Astrophys.*, **365**, L7.

Dendy, *Plasma Dynamics*, Oxford, 1990.

Dere, K.P., Brueckner, G.E., Howard, R.A., Michels, D.J., and Delaboudiniere, J.P., 1999, *Astrophys. J.*, **516**, 465.

Dessler, A.J., and Parker, E.N., 1959, *J. Geophys. Res.*, **64**, 2339.

DeVore, C.R., 2000, *Astrophys. J.*, **539**, 944.

Dressler, A., et al., 1998, *ApJS*, **122**, 51.

Duggin, M.J., and Saunders, R.W., 1984, 'Problems Encountered in Remote Sensing of Land and Ocean Features', in *Satellite Sensing of a Cloudy Atmosphere*, ed. Henderson-Sellers, A., Taylor and Francis.

Dungey, J.W., 1953, *Philos. Mag.*, **44**, 725.

Edwards, T., Browning, R., Delderfield, J., Lee, D.J., Lidiard, K.A., Milbarrow, R.S., McPherson, P.H., Peskett, S.C., Toplis, G.M., Taylor, H.S., Mason, I.M., Mason, G., Smith, A., and Stringer, S., 1990, *Journal of the British Interplanetary Society*, **43**, 160.

Elachi, C., 1987, *Introduction to the Physics and Techniques of Remote Sensing*, Wiley.

Ellis, R.S., Colless, M., Broadhurst, T., Heyl, J., Glazebrook, K., et al., 1996, *MNRAS*, **280**, 235.

Erdélyi, R., Doyle, J.G., Perez, M.E., and Wilhelm, K., 1998, *Astron. Astrophys.*, **337**, 287.

Esaias, W.E., Abbott, M.R., Barton, I.J., Brown, O.B., Campbell, J.W., Carder, K.L., Clark, D.K., Evans, R.H., Hoge, F.E., Gordon, H.R., Balch, W.M., Letelier, R., and Minnett, P.J., 1998, *IEEE Transactions on Geoscience and Remote Sensing*, **36**, 1250.

Fang, L.Z., and Ruffini, R., 1983, *Basic Concepts in Relativistic Astrophysics*, World Scientific, Singapore.

Fisk, L.A., 1996, *J. Geophys. Res.*, **1011**, 5547.

Flores, H., et al., 1999, *Astrophys. J.*, **517**, 148.

Forbes, T.G., and Acton, L.W., 1996, *Astrophys. J.*, **459**, 330.

Forsyth, R., and Gosling, J.T., 2001, 'Co-rotating and Transient Structures in the Heliosphere', in *The Heliosphere Near Solar Minimum: The Ulysses Perspective*, eds. Balogh, A., Marsden, R.G., and Smith, E.J., Springer-Praxis, Chichester, UK.

Fraser, G.W., *X-Ray Detectors in Astronomy*, 1989, Cambridge University Press.

Freedman, W.L., et al., 2001, *Astrophys. J.*, **553**, 47.

Friedman, H., Lichtman, S.W., and Byram, E.T., 1951, *Phys. Rev.*, **83**, 1025.

Fu, L.-L., and Cazeneve, A., eds., 2001, *Satellite Altimetry and Earth Sciences*, Academic Press.

Furth, H.P., Killeen, J., and Rosenbluth, M.N., 1963, *Phys. Fluids*, **6**, 459.

Gaizauskas, V., Mandrini, C.H., Démoulin, P., Luoni, M.L., and Rovira, M.G., 1998, *Astron. Astrophys.*, **332**, 353.

Gallaudet, T.C., and Simpson, J.J., 1991, *Remote Sensing of Environment*, **38**, 77.

Galloway, D.J., and Weiss, N.O., 1981, *Astrophys. J.*, **243**, 945.

Gary, G.A., 1989, *Astrophys. J. Suppl. Ser.*, **69**, 323.

Giacconi, R., et al., 1969, *Space Sci. Rev.*, **9**, 3.

Giacconi, R., and Rossi, B., 1960, *J. Geophys. Res.*, **65**, 773.

Glover, A.G., Ranns, N.D.R., Harra, L.K., and Culhane, J.L., 2000, *Geophys. Res. Lett.*, **27**, 2161.

Gobron, N., Pinty, B., Verstraete, M., and Widlowski, J.-L., 2000, *IEEE Transactions on Geoscience and Remote Sensing*, **38**, 2489.

Goertz, C.K., and Strangeway, R.J., 1995, 'Plasma waves', in Kivelson, M.G., and Russel, C.T., eds., *Introduction to Space Physics*, Cambridge University Press, 356.

Golub, L., et al., 1990, *Nature*, No. 6269, 842.

Gonzales, et al., 1994, *J. Geophys. Res.*, **99**, 5771.

Gonzales, W.D., and Tsurutani, B.T., 1987, *Planet Space Science*, **35**, 1101.

Griffin, M.J., 2000, *Nucl. Inst. Meth.*, **A444**, 397.

Gurney, R.J., Foster, J.L., and Parkinson, C.L., 1993, eds., *Atlas of Satellite Observations Related to Climate Change*, Cambridge University Press.

Haller, E.E., 1985, *Infrared Phys.*, **25**, 257.

Harra, L.K., and Sterling, A.C., 2001, *Astrophys. J. Lett.*, **561**, L215.

Harries, J.E., 1995, *Earthwatch: The Climate From Space*, Wiley.

Harris, A.R., and Mason, I.M., 1992, *International Journal of Remote Sensing*, **13**, 881.

Harrison, R.A., et al., 1995, *Solar Physics*, **162**, 233.

Harrison, R.A., 1990, *Solar Physics*, **126**, 185.

Hasinger, G., 2001, in *ISO Surveys of a Dusty Universe*, Lecture Notes in Physics, 548, 423, eds. Lemke, D., Stickel, M., and Wilke, K., Springer.

Hendrick, J., 1957, *J. Opt. Soc. Ann.*, **47**, 165.

Hepburn, I.D., Smith, A., Duncan, W., and Kerley, N., 1996, Proc. 30th ESLAB Symp., *Submillimetre and Far-Infrared Space Instrumentation*, ESA SP-388.

Huete, A.R., 1988, *Remote Sensing of Environment*, **25**, 295.

Huete, A.R., Liu, H.Q., Batchily, K., and van Leeuwen, W., 1997, *Remote Sensing of Environment*, **59**, 440.

Hundhausen, A.J., 1993, *J. Geophys. Res.*, **98**, 13177.

IGBP, 1990, *The International Geosphere–Biosphere Programme: A Study of Global Change: The Initial Core Projects*, IGBP Report 12 (June 1990), IGBP Secretariat, Royal Swedish Academy of Sciences, Stockholm, Sweden.

Illing, R.M.E., and Hundhausen, A.J., 1985, *J. Geophys. Res.*, **90**, 275.

IPCC, 1990, *Climate Change: The IPCC Scientific Assessment*, eds. Houghton, J.T., Jenkins, G.J., and Ephraums, J.J., Cambridge University Press.

IPCC, 1996, *Climate Change 1995: The Science of Climate Change*, eds. Houghton, J.T., Meira Filho, L.G., Callander, B.A., Harris, N., Kattenberg, A., and Maskell, K., Cambridge University Press.

IPCC, 2001, Climate Change 2001: The Scientific Basis, eds. Houghton, J.T., Ding, Y., Griggs, D.J., Noguer, M., van der Linden, P.J., Dai, X., Maskell, K., and Johnson, C.A., Cambridge University Press.

Irwin, K.D., 1995, *Appl. Phys. Lett.*, **66**, 1998.

Jackson, J.D., 1975, *Classical Electrodynamics*, 2nd edition, Wiley, New York.

Janesick, J., Hynecek, J., and Blouke, M., 1981, *Proc. Soc. Photo-Opt. Eng. (SPIE)*, **290**, 161.

Jones, L.R., et al., 1998, *Astrophys. J.*, **495**, 100.

Jonker, P.G., and van der Klis, M., 2001, *Astrophys. J.*, **553**, L43.

Justice, C.O., Vermote, E., Townshend, J.R.G., Defries, R., Roy, D.P., Hall, D.K., Salomonson, V., Privette, J.L., Riggs, G., Strahler, A., Lucht, W., Myneni, R.B., Knazikhin, Y., Running, S.W., Nemani, R.R., Wan, Z., Huete, A.R., van Leuwen, W., Wolfe, R.E., Giglio, L., Muller, J.-P., Lewis, P., and Barnsley, M.J., 1998, *IEEE Transactions on Geoscience and Remote Sensing*, **36**, 1228.

Karl, T.R., ed., 1996, *Long Term Climate Monitoring by the Global Climate Observing System*, Kluwer Academic Publishers.

Kaufman, Y., and Tanré, D., 1992, *IEEE Transactions on Geoscience and Remote Sensing*, **30**, 261.

Kelley, R.L., et al., 2000, *Nucl. Inst. Meth.*, **A444**, 170.

King, M.D., and Greenstone, R., eds., 1999, *1999 EOS Reference Handbook*. Available from http://eospso.gsfc.nasa.gov/eos_homepage/misc_html/refbook.html. Published by NASA/Goddard Space Flight Centre, Greenbelt, MD 20771, USA.

Kiplinger, A.L., 1995, *Astrophys. J.*, **453**, 973.

Kivelson and Russell, eds., 1995, *Introduction to Space Physics*, Cambridge University Press.

Kivelson, M.G., and Russell, C.T., 1997, *Introduction to Space Physics*, Cambridge University Press.

Kivelson, M.G., 1995, 'Pulsations and Magnetohydrodynamic Waves', in Kivelson, M.G., and Russel, C.T., eds., *Introduction to Space Physics*, Cambridge University Press, 330.

Klein, K.-L., and Trottet, G., 2001, *Space Science Reviews*, **95**, 215.

Klein, L.W., and Burlaga, L.F., 1982, *J. Geophys. Res.*, **87**, 613.

Kosovichev, A.G., et al., 1997, *Solar Physics*, **170**, 43.

Landau, L.D., and Lifshitz, E.M., 1971, *Classical Theory of Field*, Pergamon, Oxford.

Lang, 1995, *Sun, Earth and Sky*, Springer-Verlag.

Leka, K.D., Canfield, R.C., McClymont, A.N., and van Driel-Gesztelyi, L., 1996, *Astrophys. J.*, **462**, 547.

Lemen, J., 2002, private communication.

Li, X., Pichel, W., Maturi, E., Clemente-Colón, P., and Sapper, J., 2001, *International Journal of Remote Sensing*, **22**, 699.

Lilly, S.J., Le Fevre, O., Hammer, F., and Crampton, D., 1996, *Astrophys. J.*, **460**, L1.

Lilly, S.J., Tresse, L., Hammer, F., Crampton, D., and Le Fevre, O., 1995, *Astrophys. J.*, **455**, 108.

Liu, H.Q., and Huete, A.R., 1995, *IEEE Transactions on Geoscience and Remote Sensing*, **33**, 457.

Longair, M.S., *High Energy Astrophysics*, Vol. 1, 2nd edition, 1992, Cambridge University Press.

Longcope, D.W., and Welsch, B.T., 2000, *Astrophys. J.*, **545**, 1089.

Low, B.C., 1996, *Solar Physics*, **167**, 217.

Luhmann, J.G., 1991, in *Cometary Plasma Processes*, ed. Johnstone, A.D., AGU Geophys. Mono., 61, 5.

Lumb, D.H., Berthiaume, G.D., Burrows, D.N., Garmire, G.P., and Nousek, J.A., 1991, *Experimental Astronomy*, **2**, 179.

Luminet, J.P., 1992, *Black Holes*, Cambridge University Press.

MacKenzie, D.E., and Hudson, H.S., 1999, *Astrophys. J.*, **519**, L93.

Madau, P., et al., 1996, *MNRAS*, **283**, 1388.

Magorrian, J., et al., 1998, *AJ*, **115**, 2285.

Mason, G., 1991, *D.Phil. thesis*, University of Oxford.

Mason, I.M., Sheather, P.H., Bowles, J.A., and Davies, G., 1996, *Appl. Opt.*, **35**, 629.

Mauskopfm, P.D., Bock, J.J., del Castillo, H., Holzapfl, W.L., and Lange, A.E., 1997, *Appl. Opt.*, **36**, No. 4, 765.

McCammon, D., Moseley, S.H., Mather, J.C., and Mushotzky, R.F.J, 1986, *Appl. Phys.*, **56**, 1263.

McClain, E.P., Pichel, W.G., and Walton, C.C., 1985, *J. Geophys. Res.*, **90**, 11587.

McComas, D.J., Gosling, J.T., and Skoug, R.M., 2000, *Geophys. Res. Lett.*, **27**, 2437.

McLean, I.S., *Electronic Imaging in Astronomy*, 1997, John Wiley & Sons in association with Praxis Publishing.

McPherron, R.L., 1995, 'Magnetospheric Dynamics', in *Introduction to Space Physics*, eds. Kivelson and Russell, Cambridge University Press.

McPherron, R.L., Russell, C.T., and Aubry, M., 1973, *J. Geophys. Res.*, **78**, 3131.

Merchant, C.J., and Harris, A.R., 1999, *J. Geophys. Res.*, **104**, 23579.

Merchant, C.J., Harris, A.R., Murray, M.J., and Zavody, A.M., 1999, *J. Geophys. Res.*, **104**, 23565.

Mestel, L., 1999, *Stellar Magnetism*, Oxford University Press.

Mewaldt, R.A., Selesnik, R.S., and Cummings, J.R., 1996, 'Anomalous Cosmic Rays: The Principal Source of High Energy Heavy Ions in the Radiation Belts', in *Radiation Belts, Models and Standards*, eds. Lemaire, J.F., Heynderickx, D., and Baker, D.N., AGU Geophys. Mono., 97.

Milman, A.S., and Wilheit, T.T., 1985, *J. Geophys. Res.*, **90**, 11631.

Minnett, P.J., 1990, *J. Geophys. Res.*, **95**, 13497.

Misner, C.M., Thorne, K.S., and Wheeler, J.A., 1973, *Gravitation*, Freeman, San Francisco.

Miura, T., Huete, A.R., and Yoshioka, H., 2000, *IEEE Transactions on Geoscience and Remote Sensing*, **38**, 1399.

Miura, T., Huete, A.R., Yoshioka, H., and Holben, B.N., 2001, *Remote Sensing of Environment*, **78**, 284.

Moseley, S.H., Mather, J.C., and McCammon, D., 1984, *Appl. Phys.*, **56**, 1257.

Mulligan, T.L., Russell, C.T., and Luhmann, J.G., 1998, *Geophys. Res. Lett.*, **25**, No. 15, 2959.

Murray, J., Allen, M.R., Merchant, C.J., and Harris, A.R., 2000, *Geophys. Res. Lett.*, **27**, 1171.

Nicholson, *Introduction to Plasma Theory*, Wiley.

NOAA, 2000, *NOAA KLM User's Guide*. Available from http://www2.ncdc.noaa.gov/docs/klm/index.htm, National Oceanic and Atmospheric Administration, Washington, D.C.

Orlowska, A.H., Bradshaw, T.W., and Hieeth, J., 1990, *Cryogenics*, **30**, 246.

Page, M.J., Mason, K.O., McHardy, I.M., Jones, L.R., and Carrera, F.J., 1997, *MNRAS*, **291**, 324.

Page, M.J., Stevens, J.A.S., Mittaz, J.P.D., and Carrera, F.J., 2001, *Science*, **294**, 2516.

Pandey, P.C., and Kniffen, S., 1991, *International Journal of Remote Sensing*, **12**, 2493.

Parker, *Cosmical Magnetic Fields*, Clarendon Press, 1979.

Parker, E.N., 1957, *J. Geophys. Res.*, **62**, 509.

Parker, E.N., 1963, *Astrophys. J. Supp.*, **8**, 177.

Parkes, I.M., Sheasby, T.N., Llewellyn-Jones, D.T., Nightingale, T.J., Zavody, A.M., Mutlow, C.T., Yokoyma, R., Tamba, S., and Donlon, C.J., 2000, *International Journal of Remote Sensing*, **21**, 3445.

Parks, 1991, *Physics of Space Plasmas: An Introduction*, Addison-Wesley.

Peacock, A., 1998, *Astron. Astrophys. (Suppl.) Series*, **127**, 497.

Perlmutter, S., et al., 1999, *Astrophys. J.*, **517**, 565.

Perryman, M.A.C., Favata, F., Peacock, T., Rando, N., and Taylor, B.G., 1999, *Astron. Astrophys.*, **A346**, L30.

Petschek, H.E., 1964, *Magnetic Field Annihilation*, in *Physics of Solar Flares*, ed. Hess, W.N., NASA-SP 50, Washington, DC, p. 425.

Pettini, M., Kellogg, M., Steidel, C.C., Dickinson, M., Adelberger, K.L., and Giavalisco, M., 1998, *Astrophys. J.*, **508**, 539.

Phillips, 1992, *Guide to the Sun*, Cambridge University Press.

Picaut, J., and Busalacchi, A.J., 2001, 'Tropical Ocean Variability', in *Satellite Altimetry and Earth Sciences*, eds. Fu, L.-L., and Cazeneve, A., Academic Press.

Poglitsch, A., Waelkens, C., and Geis, N., *Proc. Symp. 'The Promise of the Herschel Observatory'*, 12–15 Dec., Toledo, Spain, eds. Pilbratt, G.L., Cerpicharo, J., Pruste, T., and Harris, R., 2001, ESA SP-460, 29.

Press, W., Teukolsky, S., Vetterling, W., and Flannery, B., 1996, *Numerical Recipes*, 2nd edition, Cambridge University Press.

Priest, E.R., and Forbes, T.G., 1986, *J. Geophys. Res.*, **91**, 5579.

Priest, E.R., and Forbes, T., 2000, *Magnetic Reconnection. MHD Theory and Applications*, Cambridge University Press.

Priest, E.R., Titov, V.S., Vekstein, G.E., and Richard, G.J., 1994, *J. Geophys. Res.*, **99**, 21467.

Priest, 1987, *Solar Magnetohydrodynamics*, Reidel.

Puchnarewicz, E.M., Mason, K.O., Murdin, P.G., and Wickramasinghe, D.T., 1990, *MNRAS*, **244**, 20P.

Qi, J., Chehbouni, A., Huete, A.R., Kerr, Y.H., and Sorooshian, S., 1994, *Remote Sensing of Environment*, **48**, 119.

Ramsay, G., Wu, K., Cropper, M., Schmidt, G., Sekiguchi, K., Iwamuro, F., and Maihara, T., 2002, *Monthly Notices of the Royal Astronomical Society*, **333**, 575.

Rapley, C.G., Culhane, J.L., Acton, L.W., Catura, R.C., Joki, E.G., and Bakke, J.C., 1977, *Rev. Sci. Inst.*, **48**, 1123.

Rapley, C.G., Griffiths, H.D., Squire, V.A., Lefebvre, M., Birks, A.R., Brenner, A.C., Brossier, C., Clifford, L.D., Cooper, A.P.R., Cowan, A.M., Drewry, D.J., Gorman, M.R., Huckle, H.E., Lamb, P.A., Martin, T.V., McIntyre, N., Milne, K., Novotny, E., Peckham, G.E., Schgounn, C., Scott, R.F., Thomas, R.H., and Vesecky, J.F., 1983, *A Study of Satellite Radar Altimeter Operation Over Ice-Covered Surfaces*, ESA Contract Report 5182/82/F/CG(SC).

Rapley, C.G., Guzkowska, M.A.J., Cudlip, W., and Mason, I.M., 1987, *An Exploratory Study of Inland Water and Land Altimetry Using Seasat Data*, ESA Contract Report 6483/85/NL/BI.

Reames, D.V., 1999, *Space Science Reviews*, **90**, 413.

Richardson, A.J., and Wiegand, C.L., 1997, *Photogrammetric Engineering and Remote Sensing*, **43**, 1541.

Rieke, G.H., 1994, *Detection of Light from UV to Sub-mm*, Cambridge University Press.

Rieke, M., et al., 2001, *Proc. SPIE*.

Saunders, R.W., and Kriebel, K.T., 1988, *International Journal of Remote Sensing*, **9**, 123.

Saunders, M.A., and Russell, C.T., 1986, *J. Geophys. Res.*, **91**, 5589.

Schmidt, M., 1968, *Astrophys. J.*, **151**, 394.

Schmidt, 1996, *Physics of High Temperature Plasmas*, Academic Press.

Sckopke, N., 1966, *J. Geophys. Res.*, **71**, 3125.

Sckopke, N., Paschmann, S.J., Bame, J.T., Gosling, J.T., and Russell, C.T., 1983, *J. Geophys. Res.*, **88**, 6121.

Sheeley, N.R. Jr., Walters, J.H., Wang, Y.-M., and Howard, R.A., *J. Geophys. Res.*, **104**, 24739.

Shepherd, A., Wingham, D.J., Mansley, J.A.D., and Corr, H.F.J., 2001, *Science*, **291**, 862.

Shing, L., Stern, R., Catura, P., Morrison, M., Eaton, T., Pool, P., and Shivamoggi, B.K., 1985, *Phys. Rep.*, **127**, 99.

Short, N., 1982, *The Landsat Tutorial Workbook: Basics of Satellite Remote Sensing*, NASA Ref. Publ. 1078.

Simpson, J.J., McIntyre, T.J., Stitt, J.R., and Hufford, G.L., 2001, *International Journal of Remote Sensing*, **22**, 2585.

Simpson, J.J., Schmidt, A., and Harris, A.R., 1998, *Remote Sensing of Environment*, **65**, 1.

Slavin, J.A., 1998, in *New Perspectives in Magnetotail Physics*, eds. Nishida, A., Cowley, S.W.H., and Baker, D.N., AGU Monograph, 105, 225.

Smail, I., Ivison, R.J., Blain, A.W., and Kneib, J.-P., 2002, *MNRAS*, **331**, 495.

Smith, E.J., 'Solar Wind Magnetic Fields', in *Cosmic Winds and the Heliosphere*, eds. Jokipii, J.R., Sonett, C.P., and Giampapa, M.S., p. 425, University of Arizona Press, Tucson.

Spjeldvik, W.N., and Rothwell, P.L., 1985, 'The Radiation Belts', in *Handbook of Geophysics and the Space Environment*, ed. Jura, A.S., AFGL, USAF.

St. Cyr, O.C., et al., 2000, *J. Geophys. Res.*, **105**, 18169.

Sterling, A.C., and Moore, R.L., 2001, *Astrophys. J.*, **560**, 1045.

Stevens, G.L., 1994, *Remote Sensing of the Lower Atmosphere: An Introduction*, Oxford University Press.

Struder, L., et al., 2001, *Astron. Astrophys.*, **365**, L18.

Stull, R.B., 2000, *Meteorology for Scientists and Engineers*, 2nd edn., a technical companion book to C. Donald Ahrens' *Meteorology Today* (Brooks/Cole).

Sturrock, 1994, *Plasma Physics*, Cambridge.

Sweet, P.A., 1958, 'The Neutral Point Theory of Solar Flares', in *Electromagnetic Phenomena in Cosmical Physics*, IAU Symp. 6, ed. Lehnert, B., Cambridge University Press, p. 123.

Tarbell, T.D., et al., 1994, Proc. 3rd SOHO Workshop, ESA-SP-373.

Tarpley, J.D., 1991, *The NOAA Global Vegetation Index Product: A Review. Palaeogeography, Palaeoclimatology, Palaeoecology (Global and Planetary Change Section)*, **90**, 189.

Thompson, B.J., et al., 1999, *Geophys. Mon.*, **109**, ed. Burch, J.L., et al., AGU, Washington D.C., 31.

Tinkler, D., Pick, D.R., Stringer, S.J., and Woods, C.G., 1986, 'The Design of the Focal Plane Assembly for the Along Track Scanning Radiometer', in *Instrumentation for Optical Remote Sensing*, Proceedings of the Meeting, Cannes, France, Nov. 1985, *SPIE International Society for Optical Engineering*, **589**, 129.

Townshend, J.R.G., Tucker, C.J., and Goward, S.N., 1993, 'Global Vegetation Mapping', in *Atlas of Satellite Observations Related to Climate Change*, eds. Gurney et al., Cambridge University Press.

Tsuneta, S., 1997, *Astrophys. J.*, **483**, 507.

Tsuneta, S., et al., 1991, *Solar Physics*, **136**, 37.

Tsurutani, B.T., 2001, in *Space Storms and Space Weather Hazards*, ed. Daglis, I.A., NATO Science Series II. Mathematic, Physics & Chemistry, Vol. 38, 103.

Turner, M.J.L., et al., 2001, *Astron. Astrophys.*, **365**, L27.

Underwood, J.H., Bruner, M.E., Haisch, B.M., Brown, W.A., and Acton, L.W., 1987, *Science*, **238**, 61.

Verhoeve, P., 2000, *Nucl. Inst. Meth.*, **A444**, 435.

Walker, A.B.C., Jr., Hoover, R.B., and Barbee, T.W., 1993, in *Physics of Solar and Stellar Coronae*, eds. Linsky, J., and Serio, S., Kluwer Academic Publishers, Dordrecht, Netherlands.

Walton, C.C., Pichel, W.G., and Sapper, J.F., 1998, *J. Geophys. Res.*, **103**, 27999.

Weinberg, S., 1972, *Gravitation and Cosmology*, Wiley, New York.

Werrett, S.T., et al., 1986, *Adv. Cryo. Eng.*, **31**, 791.

Wickramasinghe, D.T., and Martin, B., 1985, *MNRAS*, **212**, 353.

Wilhelm, K., et al., 1995, *Solar Physics*, **162**, 189.

Windt, D., 1998, *Proc. SPIE*, **3448**, 371.

Windt, D., et al., 2002, *Proc. SPIE*, **4851**, in press.

Wingham, D.J., Ridout, A.J., Scharroo, R., Arthern, R.J., and Shum, C.K., 1998, *Science*, **282**, 456.

Yokoyama, T., Akita, K., Morimoto, T., Inoue, K., and Newmark, J., 2001, *Astrophys. J.*, **546**, L69.

Zarnecki, J.C., and Culhane, J.L., 1977, *Mon. Not. R. Astr. Soc.*, **178**, 57.

Zavody, A.M., Gorman, M.R., Lee, D.J., Eccles, D., Mutlow, C.T., and Llewellyn-Jones, D.T., 1994, *International Journal of Remote Sensing*, **15**, 827.

Zavody, A.M., Mutlow, C.T., and Llewellyn-Jones, D.T., 2000, *Journal of Atmospheric and Oceanic Technology*, **17**, 595.

Zombeck, M.V., *Handbook of Space Astronomy and Astrophysics*, 2nd edition, 1990, Cambridge University Press.

Zwally, H.J., and Brenner, A.C., 2001, 'Ice Sheet Dynamics and Mass Balance', in *Satellite Altimetry and Earth Sciences*, eds. Fu and Cazenave, Academic Press.

Glossary

accretion The process by which the mass of an object increases by collecting matter from its surroundings. In the planetary context: coalescence of material from an interstellar medium, solar nebula or environment close to a body of gas, dust or heavier material by gravitational attraction. In the context of X-ray astronomy, it describes the process by which matter falls onto a compact star under the influence of gravity, which can generate substantial luminosity by the release of gravitational potential energy.

active galactic nucleus (AGN) A luminous accreting black hole at the centre of a galaxy.

ADEOS-II Advanced Earth Observing Satellite–II.

Albedo The proportion of (usually solar) electromagnetic radiation reflected by a surface, expressed as a fraction or percentage.

Alfvén speed The speed at which magnetic disturbances travel along the field. It is derived from the momentum equation, equating the change of momentum with the magnetic (Lorenz) force. It is expressed as $v_A \equiv B/(\mu_0 \cdot \rho)^{1/2}$, where B is the magnetic field strength, μ_0 is the magnetic permeability and ρ is the density of the plasma.

Alfvén wave An MHD wave mode in a magnetised plasma, arising as a result of restoring forces associated with the magnetic field. Also referred to as the shear, or transverse, or intermediate wave.

Alfvén wings Electromagnetic interaction of a moving, conducting body in a plasma at sub-sonic and sub-Alfvénic speeds causes wave structures in the surrounding plasma moving with the body at the Alfvén speed.

anthropogenic Produced by humans.

469

asteroid   A minor planetary body. Many such bodies are in the 'main belt' between the orbits of Mars and Venus, some are in closer heliocentric orbits and some are in orbits crossing the Earth's.

atmospheric circulation   Temperature-driven motion in planetary atmospheres. Examples are Hadley circulation (equator–pole flow at high altitudes, return at low); ocean-driven flows; flows disturbed by topography; day–night flow and pole-to-pole sublimation flow (Mars).

auroral electrojet   A current that flows in the ionosphere at a height of ~ 100 km.

auroral oval   An elliptical band around the geomagnetic pole where auroral emission is predominant. It ranges from $75°$ magnetic latitude at local noon to $67°$ at midnight in quiet conditions. During magnetic storms the oval widens and extends to greater latitudes.

baryon   Composite particles made up of three quarks, which include protons and neutrons.

bathymetric   Concerning measurements of the depth of the ocean, i.e. of the seabed topography.

binary X-ray sources   Double star systems in which one component is very compact (a white dwarf, neutron star or black hole) and is accreting matter from its companion, generating high temperatures and emitting in the X-ray band.

biogeochemical   Relating to the Earth's biology and chemistry.

biosphere   The Earth's living organisms, or the regions containing them.

bispherical shell   A particle distribution function of cometary ions, centred (in velocity space) on upstream- and downstream-propagating Alfvén waves.

black body   A hypothetical body that absorbs all electromagnetic radiation that falls on it. An example of a black body would be a closed object with a small aperture.

black hole   A star that is so compact that no light can escape from its gravitational attraction.

black-body spectrum   The idealised spectrum that is radiated by a perfect emitter.

blazar   An active galaxy in which there is a relativistic jet whose direction of motion is close to our line of sight. In these circumstances motion of the jet at near the speed of light results in a substantially boosted emission in that direction, such that the emission observed from the galaxy is dominated by non-thermal emission from the jet.

bolometer   A device that measures the amount of electromagnetic radiation falling on it by registering the heat generated. Bolometers are useful in the infrared band. New devices operating at close to absolute zero ($-273°C$) can measure the heat generated by individual X-ray photons, allowing them to be used as a sensitive spectrometer.

bombardment phase   The first 0.8 billion years of the Solar System following its formation 4.6 billion years ago were characterised by collisions that first formed the planets and then bombarded them with asteroids and comets.

bow shock   A collisionless shock wave standing upstream of a planetary obstacle, which acts to reduce the solar wind flow to subsonic, sub-Alfvénic speeds, to heat the plasma and compress the plasma and magnetic field.

bow shock (heliospheric)   Theoretical collisionless shock ahead of the heliosphere in the local interstellar medium, or LISM, required if the heliosphere moves supersonically with respect to the LISM.

braids Structures seen by *Voyager* in Saturn's rings.

bremsstrahlung   'Braking radiation' produced when the path of an electron in an ionised plasma is accelerated (deflected) by an interaction with an ion.

brown dwarf   A stellar body that contains insufficient matter for ignition of fusion reaction. A failed star.

Cassini division   The gap between Saturn's A and B rings, first noted by Cassini in 1675.

cataclysmic variable   A close binary star containing a white dwarf that is accreting matter from its companion. The term 'cataclysmic' arises because many such systems exhibit spectacular variability in their output on short timescales.

CCD   Charge-coupled device. A silicon chip containing an array of light-sensitive diodes, used for capturing images. These chips are used in devices such as digital cameras, and as highly sensitive astronomical detectors. In the X-ray band, the amount of energy deposited by individual photons can be measured, allowing the device to function as an imaging spectrometer.

chromosphere   The region of the solar atmosphere above the photosphere with a temperature between 5000 and 50,000 K.

closed magnetosphere   A cavity formed in the solar wind flow by virtue of it being completely frozen out of the region occupied by a planetary magnetic field.

clumps   Structures seen by *Voyager* in Saturn's rings.

clusters of galaxies   A gravitationally bound assemblage of galaxies.

CMOS   Complementary metal oxide semiconductor. Electronic components designed for low power consumption.

collective behaviour   The condition for a population of particles to form a plasma, whereby they respond as a collection to long-range electromagnetic forces.

comet   A planetesimal from the outer Solar System, principally water ice but with additional volatiles, dust and crust. When in the inner Solar System, sublimation results in a coma, ion tail and dust tail.

cometary nucleus   The solid heart of a comet; a 'dirty snowball' with additional volatiles, dust and crust. Comet Halley's nucleus was 15×8 km.

Compton reflection   A process whereby a photon stream can be 'reflected' from a cloud of free electrons (for example ionised gas) by the mechanism of Compton scattering.

Compton scattering   An interaction between a photon and an electron in which the direction of the photon is changed and its energy reduced.

configuration space   Three-dimensional coordinates representing a particle position in physical space.

constellation mission   A multi-spacecraft space plasma mission. The first generation of this type is the four-spacecraft *Cluster* mission; future missions may involve 30–50 satellites.

convection zone   The region in the Sun below the surface in which energy is transported predominantly by the bodily movement of plasma.

core   The central part of a planetary structure. Some terrestrial planets have iron-rich cores causing planetary magnetic fields; gas giants are thought to have icy–rocky cores.

Coriolis force   Apparent force in planetary atmospheres associated with rotation: with respect to an observer on the ground, deviation is to the right in the northern hemisphere and to the left in the southern.

corona   The upper region of the solar atmosphere, which has a temperature of > 1 million K.

coronal holes   Regions on the Sun with 'open' magnetic flux (closes via the heliopause), the source of the fast solar wind.

coronal mass ejection (CME)   Explosive transient outflow of plasma and magnetic field from the Sun, seen as a discrete density enhancement in coronagraph images.

co-rotating interaction region (CIR)   Compression in solar wind plasma density associated with a fast solar wind stream running into the back of a slower stream, and often associated with interplanetary shocks. It may persist over many solar rotations, such that a given CIR may be observed to pass the Earth once every ~ 27 days.

co-rotation   Rotation of plasma populations and magnetic flux tubes at the same angular velocity as the associated planetary body.

cosmic strings   A hypothesised line-like defect in the fabric of space–time predicted by some models of the early Universe.

cratering record   Impact crater features on planetary bodies exposed to asteroid or cometary bombardment. The record may be eroded by surface winds, ocean, climate and internal activity (volcanism).

cross-tail current   A sheet of current flowing from dawn to dusk across the equatorial magnetotail. It acts to support the reversal of the field direction between the north and south lobe regions.

cryosphere   The frozen water at or below the Earth's surface (e.g. snow, permafrost, land ice, sea ice).

cryovolcanism   An internally driven process involving warmed icy volatiles, e.g. icy slush, on outer planetary moons.

current sheet   A thin current-carrying plasma layer across which the magnetic field changes either in direction or magnitude or both. Magnetic reconnection typically occurs in current sheets.

cusp   Funnel-shaped features in a planetary magnetic field where shocked solar wind from the magnetosheath can directly penetrate towards the atmosphere.

cycloid   The shape described by cyclotron (circular) motion with a superimposed drift, usually where electric field and magnetic field are non-parallel ($\mathbf{E} \times \mathbf{B}$ drift).

cyclotron frequency   The natural frequency of a plasma associated with gyration motion of a given particle species around the magnetic field direction. Also known as the gyrofrequency.

dark matter   Thus-far-unidentified particles different from 'normal' matter, which are believed to make up a substantial fraction of the mass of the Universe.

dayside   The sunlit side of a planet.

Debye length   The characteristic length scale in a plasma over which the particles in the system are able to shield out any net charge imbalance within that system.

degenerate dwarf   see White dwarf.

differentiation   Attraction of heavier elements towards the planetary core during planetary formation.

discontinuity (MHD)   May be a shock, a tangential discontinuity or a rotational discontinuity in a plasma (e.g. solar wind).

distribution function   The statistical representation of a plasma in six-dimensional phase space. It indicates the number of particles in a given volume of velocity space and a given volume of real space at a given time.

Doppler shift   The change in wavelength of a (sound or light) wave caused by motion of the source with respect to the observer.

Drake equation   A formulation of the number of civilisations in our galaxy, depending on the product of several, mostly unknown, factors.

draping (magnetic field)   As mass is added to a flowing plasma, magnetic field drapes (as in comets) if frozen-in approximation holds.

driver gas   Plasma and magnetic field ejected from the Sun with a speed greater than the ambient solar wind speed which can drive a shock wave ahead of itself.

DST index   A measure of the variation in the geomagnetic field due to the ring current used to define the intensity of a geomagnetic storm.

dynamo   The process that transforms the kinetic energy of a conducting object into magnetic energy.

early type star   A massive star with a hot spectrum.

eccentricity   A measure of the ellipticity of an orbit. A circular orbit has an eccentricity of 0.

ecliptic plane   A plane containing the path of the Sun over a year as seen from the Earth. The orbit of most planets is close to this plane.

electron plasma frequency   The natural frequency of a plasma associated with electron oscillations due to small-scale charge density perturbations within the plasma.

EMC   Electromagnetic compatibility. A set of tests that determine whether an instrument or spacecraft operates correctly in a common electromagnetic environment. A complementary series of EMI (electromagnetic interference) tests measure the electromagnetic radiation properties of a given instrument or spacecraft system to determine whether it is within designed tolerance. These are part of an Integration and Test (I&T) programme routinely applied to instrumentation designed to fly in space.

ERS   European Remote-sensing Satellite.

ESA   European Space Agency.

exosphere   The region in the planetary atmosphere where the probability of escape is 1/2.

fast mode magnetosonic wave An MHD wave mode in a magnetised plasma, arising as a result of restoring forces associated with the magnetic field.

fast mode shocks A shock wave associated with the non-linear steepening of fast mode magnetosonic waves. Most bow shocks and interplanetary shocks are fast mode shocks.

field stop An aperture in the focal plane of a telescope, defining its (instantaneous) field of view.

field-aligned current Regions of current flow that are parallel to the underlying, global magnetic field direction. A vehicle by which stresses imposed from the magnetosphere are transferred down to the ionosphere.

first adiabatic invariant Also known as the magnetic moment, μ — a constant of a charged particles motion associated with its gyromotion around the magnetic field direction. It is invariant provided the field does not change significantly during one gyroperiod.

fixing Dissolution of gas, e.g. carbon dioxide, in the oceans may lead to fixing of carbon in the rocks (e.g. formation of carbonates).

fluid approach The level of description of a plasma, generally involving one electron and one ion fluid. Treating these as a single fluid is MHD.

fluorescent emission Emission from an atom that has been excited by a source of higher-energy radiation.

flux transfer events (FTEs) Open flux tubes on the dayside magnetosphere created by localised, sporadic reconnection, which links the magnetosheath and magnetospheric plasma populations. Observed by spacecraft as they unwind across the magnetopause surface.

force-free field A magnetic field which exerts no force on the surrounding plasma ($j \times B = 0$). This can either be a field with no electric current ($j = 0$; potential field) or a field with currents which all flow along (or opposite to) the magnetic field lines, i.e. parallel to B.

Fraunhofer lines Absorption lines in the spectrum of the Sun discovered by Fraunhofer — the most prominent of which are the Na D lines and the Ca H and K lines.

frozen-in flux approximation A key concept in space plasmas — if the conductivity is very high and the length scales are large, diffusion of the magnetic field through the plasma is negligible and the field convects exactly with the plasma. As a result a field line threading a particular plasma parcel will always thread that parcel, no matter how it moves.

gaps Structures seen by *Voyager* in Saturn's rings.

Gauss   A unit of magnetic field strength, $= 0.0001$ tesla.

geomagnetic activity   A generic term for terrestrial magnetospheric variations, and their ground-based signatures, associated with the variable input of solar wind energy.

geomagnetic storm   A worldwide disturbance of the Earth's magnetic field characterised by a decrease in horizontal magnetic field intensity. Storms have three phases: (1) the initial phase or sudden storm commencement (SSC), where the magnetic field increases over a period of ~ 5 min; (2) the main phase, where the magnetic field decreases by ~ 100 nT over a period of 7–10 h; (3) the recovery phase, where the magnetic field returns to pre-storm ambient values. Often just referred to as a magnetic storm.

GMS   Geostationary Meteorological Satellite.

GOES   Geostationary Operational Environmental Satellite.

GOMS   Geostationary Operational Meteorological Satellite.

granulation   Convective cells, each about 1400 km in diameter, that lie near the surface of the Sun.

grating   Or 'diffraction grating'. A surface on which are ruled a series of closely spaced lines that break light into a spectrum by diffraction. Diffraction gratings can be constructed to transmit the diffracted light ('transmission gratings') or reflect it ('reflection gratings').

gravitational lensing   An effect that results from the bending of the path of light in a gravitational field. If a massive object is interposed in the line of sight to a background scene, it can act in the same way as an optical lens to focus light, magnifying the image scale of the background and enhancing its apparent brightness.

greenhouse effect   Blanketing effect of a planetary atmosphere, which transmits some wavelengths of electromagnetic radiation but not others. It can lead to global warming.

greenhouse effect — runaway   Positive feedback effect in Venus atmosphere where early warming due to initial water in the atmosphere led to more heating, more water in the atmosphere and so on. It currently involves carbon dioxide in the atmosphere, which cannot be 'fixed' as there are no oceans.

grooves   Structures seen by *Voyager* in Saturn's rings.

ground station   A terrestrial station for communicating with spacecraft (i.e. sending commands and receiving data).

gyrofrequency The natural frequency of a plasma associated with gyration motion of a given particle species around the magnetic field direction. Also known as the cyclotron frequency.

gyromotion Motion of charged particles in circular orbits in the plane perpendicular to the magnetic field direction.

gyroradius The radius of the circular gyromotion of a charged particle about the magnetic field direction. Also known as the Larmor radius.

hadron Elementary particles that participate in strong nuclear reactions. These include baryons such as protons and neutrons, and mesons such as pions.

heliocentric A system of coordinates with the origin at the centre of the Sun.

heliopause The electromagnetic edge of the heliosphere, analogous to the Earth's magnetopause. The boundary between solar magnetic field and the local interstellar medium (LISM) magnetic field. Expected at $\sim$ 150 AU on upstream side with respect to the flowing LISM.

helioseismology The study of the global oscillations in the Sun.

heliosphere The region of influence of the solar wind.

HILDCAA High Intensity Long Duration Continuous AE Activity events — periods of continuous sub-storm activity caused by highly fluctuating IMFs.

Hubble expansion The expansion of the Universe, discovered by Edwin Hubble.

ICME Interplanetary manifestation of a CME.

IMF Interplanetary magnetic field. Magnetic field carried with the solar wind.

in situ In place. Used when describing observations, and referring to those made directly, i.e. not by remote sensing.

induction equation A combination of Maxwell's equations and Ohm's law leads to the induction equation, which eliminates the electric field and relates v and B. It is a basic equation to describe the magnetic behaviour in MHD. It expresses that local change of B can be due to advection and diffusion.

inflation An epoch of accelerated expansion in the early Universe.

instantaneous field of view The angular region that can be observed by a telescope at a given instant. In remote sensing, it is usually the smallest region defined by a field stop, detector or detector element in the focal plane.

intercombination line A spectral line produced between two levels that require a spin change. Such transitions have a low probability of occurrence.

intermediate polar   A type of cataclysmic variable in which the white dwarf has a magnetic field strong enough to influence the accretion flow, and funnel material into restricted areas on the surface of the white dwarf. Rotation of the white dwarf star in such systems usually causes the flux observed from it to be modulated with the white dwarf spin period.

intermediate wave   An MHD wave mode in a magnetised plasma, arising as a result of restoring forces associated with the magnetic field. Also referred to as the shear, or transverse, or Alfvén wave. The term 'intermediate' refers to the propagation velocity lying between those of the fast and slow magnetosonic modes.

interplanetary magnetic field (IMF)   Magnetic field of solar origin, which is carried out into interplanetary space by the solar wind flow. On average it forms a spiral structure within the heliosphere.

inverse Compton scattering   A process in which a fast-moving electron imparts energy to a photon when they collide.

Io plasma torus   A torus of plasma within the Jovian magnetosphere formed around the orbit of Io by the outflow of volcanic material from that moon.

ionopause   The boundary separating regions dominated by plasma of ionospheric origin from that dominated by plasma of solar wind origin at an unmagnetised planet such as Venus.

ionosphere   A high altitude layer of a planetary atmosphere in which photo-ionisation creates a (generally collisional) plasma of sufficient density to have significant influence on the dynamics of the layer.

isotopes — primordial   Left over from the beginning of the Solar System.

isotopes — radiogenic   Produced from radioactive decay of planetary materials.

Keplerian orbit   An orbit that obeys Kepler's laws of motion.

kinks   Structures seen by *Voyager* in Saturn's rings.

Kuiper belt   A hypothetical 'storage' zone for comets, near the ecliptic plane and at radial distances beyond Pluto.

Lagrangian point   A point at which a small body can remain in the orbital plane of two massive bodies where the net perturbative force due to the two bodies is zero. The *SOHO* spacecraft is located at one such Lagrangian point.

land ice   Glaciers and ice caps (usually including floating ice shelves at the edges of glaciers/ice caps).

Larmor radius   The radius of the circular gyromotion of a charged particle about the magnetic field direction. Also know as the gyroradius.

late type star   A low-mass star with a cool spectrum.

lepton   Elementary particles such as electrons, muons and neutrinos that do not participate in the strong nuclear interactions.

lobate scarp   Thrust faults on a planetary body resulting from crustal compressional stress.

local interstellar medium (LISM)   The region of the spiral arm of the Milky Way galaxy in which the Solar System is located. It moves with respect to the Solar System.

local thermodynamic equilibrium (LTE)   The balance of processes that create and destroy photons.

Lorenz force   Or magnetic force ($j \times B$) is perpendicular to the magnetic field, so any plasma acceleration along or parallel to the magnetic field must be caused by other forces. It has two components: the magnetic pressure force $-\nabla(B^2/2\mu_0)$, which appears when B^2 changes with position, and is directed from high magnetic pressure to low pressure; the other component is the magnetic tension force, which appears when the field lines are curved, and tries to shorten the field lines. It is inversely proportional to the radius of curvature and proportional to B^2.

loss cone   The part of a particle distribution function occupied, on closed field lines, by particles that will not mirror at altitudes above the ionosphere. As a result, the ionosphere rapidly absorbs such particles, such that the loss cone is usually a relatively empty part of the distribution function.

luminosity function   The number of objects per unit logarithmic luminosity interval per unit volume.

lunar wake   A relative plasma void found downstream of the Moon, created by the absorption of the oncoming solar wind at the dayside surface.

magnetic annihilation   The process whereby oppositely directed magnetic field lines come together and cancel out at a thin current sheet.

magnetic buoyancy   Leads to the rise of flux tubes e.g. from the bottom of the convection zone up to the surface of the Sun and beyond. In magnetic flux tubes the internal pressure, which is the sum of the plasma and magnetic pressures, keeps balance with the external plasma pressure ($p_e = p_i + p_m$;, where $p_m = B^2/2\mu_0$). For magnetic flux tubes, which are in thermal equilibrium with their surroundings ($T_i = T_e$), the lower internal plasma pressure ($p_i < p_e$) implies lower plasma density, leading to a buoyancy force.

magnetic cloud   A coherent magnetic structure, generally associated with a coronal mass ejection, in which the magnetic field exhibits a steady rotation over a period of several tens of hours.

magnetic curvature drift   A drift of charge particles located in a curved field geometry. The drift is in the direction perpendicular to both the local magnetic field direction and the radius of the curvature vector. The drift direction is charge-dependent, so ions and electrons drift in opposite directions and thus constitute a current.

magnetic gradient drift   A drift of charge particles located in a region of magnetic field strength gradient directed perpendicular to the field direction. The drift is in the direction perpendicular to both the local magnetic field direction and the field strength gradient vector. The drift direction is charge dependent, so ions and electrons drift in opposite directions and thus constitute a current.

magnetic helicity   Measures structural properties of the magnetic field such as twist, shear, shearing and braiding. In ideal MHD (with infinite conductivity) it is conserved, and even in coronal plasmas with high conductivity ($R_m \gg 1$), reconnection conserves magnetic helicity with a very good approximation.

magnetic moment μ — the first adiabatic variant — a constant of a charged particle motion associated with its gyromotion around the magnetic field direction. It is invariant provided the field does not change significantly during one gyroperiod.

magnetic monopole   A hypothesised entity predicted by some models of the early Universe that acts like an isolated north or south pole of a magnet.

magnetic pressure   A force associated with the magnetic field that mimics a pressure. This force tends to oppose the compression of the magnetic field.

magnetic reconnection   A process by which oppositely directed field lines are cut, then connect with others to form new magnetic links, allowing topological changes of the magnetic field to occur. Magnetic reconnection is associated with specific patterns of plasma flow (inflows and Alfvénic, jet-like outflows), and results in the conversion of magnetic energy to kinetic and thermal energy of the plasma. Reconnection requires the dissipation of electric currents, and thus the redirection of the field is associated with plasma no longer being tied to the field lines. It happens in a small region (current sheet). Reconnection is invoked to explain the heating of plasmas and acceleration of particles that are observed e.g. in solar flares, magnetic substorms and storms.

magnetic Reynolds number   The ratio of the rate that the magnetic field convects with a plasma to the rate that it diffuses through the plasma. A high magnetic Reynolds number indicates that the magnetic field is frozen into the plasma.

magnetic tension   A force associated with the magnetic field that mimics a tension. This force tends to oppose the bending of the magnetic field.

magnetohydrodynamic waves   Are initiated by small perturbations of an equilibrium situation in a (magnetised) plasma. They can be classified by their restoring forces, which can be driven by e.g. the plasma pressure (sound waves), magnetic pressure (compressional Alfvén waves) or magnetic tension (Alfvén waves), or, in the case of two forces acting together, e.g. the ones driven by plasma and magnetic pressures, the result is magneto-acoustic waves. Alfvén waves are transverse waves, which propagate at the Alfvén speed. The two modes of the magneto-acoustic waves (slow and fast) propagate slower and faster than the Alfvén speed. Propagation characteristics of MHD waves depend on the direction of propagation relative to the magnetic field.

magnetohydrodynamics (MHD)   Fluid mechanics adapted to include the electromagnetic forces that act on a plasma. The fluid flow (v) and the magnetic field (B) influence each other: the Lorenz force, $j \times B$, accelerates the fluid, while the electromotive force, $v \times B$, creates a current, which modifies the electromagnetic field. The basic equations of MHD consist of Maxwell's equation, Ohm's law, three conservation principles (conservation of mass, momentum, and internal energy) as well as the equation of state or gas law.

magnetopause   A boundary current sheet separating the solar wind plasma and IMF from the region occupied by the planetary magnetic field.

magnetosheath   The region between the bow shock and magnetopause occupied by the shocked solar wind plasma.

magnetosonic wave   MHD waves in a magnetised plasma, arising as a result of restoring forces associated with the magnetic field.

magnetosphere   The cavity produced by the interaction of the solar wind with the field of a magnetised planet.

magnetospheric convection   Motion of plasma and fields within the magnetosphere driven by reconnection and the cross-tail electric field.

magnetospheric substorm   A cycle of geomagnetic activity in which energy is first tapped from the solar wind, stored within the magnetotail, and then explosively released. Energy is dissipated in the auroral ionosphere, the ring current, and within plasmoids expelled into the solar wind.

mass-loading   The process in which solar wind field lines are loaded with newly ionised ionospheric or cometary material. This causes a slowing of the solar wind in the vicinity of the associated body and leads to an induced magnetic tail in the downstream region.

Maunder minimum   The period when there was a lack of indication of solar activity (sunspots), between 1645 and 1715.

Maxwell's equations   Are the four basic equations of electromagnetism. Poisson's equation and Faraday's law express that either electric charges or time-varying magnetic fields give rise to electric fields. Ampère's law states that either currents or time-varying electric fields may produce magnetic fields, while Gauss's law for B states that there are no magnetic monopoles.

meridional winds   North–South winds.

metallic hydrogen   The high pressure state of hydrogen, possibly liquid or solid, in which electrical conductivity and other properties are similar to those of a metal.

metastable level   An energy level that has a very low transition probability.

MHD approximations   Usually, plasma velocities are low in MHD ($v \ll c$, where c is the speed of light), and thus second-order terms in v/c are negligible. Furthermore, in the basic computations, the effects of local electric charge densities and displacement currents, and of radiation pressure, are neglected.

microlensing   An effect caused by gravitational lensing, the bending of light around an object due to gravity. The term is usually applied to the small increase in apparent brightness that results when a small body (such as a planet or low mass star) passes in front of a background star.

MIL standard   A quality standard for electronic components used in military programmes.

NASA   National Aeronautics and Space Administration (of the USA).

NASDA   National Space Development Agency (of Japan).

near-Earth neutral line (NENL)   A reconnection X-line that is formed in the near-Earth magnetotail at or near the onset of a magnetospheric substorm.

neutral line   A magnetic null point formed at the point where magnetic reconnection occurs. Also referred to as an X-line.

neutrino   A weakly interacting particle that is emitted during radioactive decay.

neutron star   A very small, dense star whose gravity is sufficiently strong to force electrons and protons to combine to form neutrons. Such an object with a mass similar to that of the Sun would have a diameter of only about 30 km.

nightside   The shadowed (away from the Sun) side of a planet.

NOAA   National Oceanic and Atmospheric Administration (of the USA).

nova   Not literally a 'new' star, but one that has brightened substantially from relative obscurity. Nova explosions occur in binary systems where a white dwarf is accreting matter from its companion. They are caused by the sudden nuclear burning of material that has collected on the surface of the accreting star. The frequency with which this happens depends, among other things, on the rate at which material is accreted, where the nova explosion is triggered when the pressure of overlying material exceeds a threshold. The interval between explosions can range from thousands of years to a few years. Systems that have been observed to explode more than once are known as 'recurrent' novae.

nucleosynthesis The formation of atomic nuclei.

Ohm's law This expresses that in a frame moving with the plasma, the electric field (E') is proportional to the current. The moving plasma in the presence of magnetic field B is subject to an electric field $v \times B$, in addition to its electric field E.

Oort cloud A hypothetical 'storage' region for comets, spherical in shape and peaked at 20,000–50,000 AU.

open magnetosphere A cavity occupied by a planetary magnetic field which is linked to the solar wind by the occurrence of magnetic reconnection at the day-side magnetopause. Solar wind mass, momentum and energy can be imparted to an open magnetosphere by the open field lines threading the magnetopause boundary.

optical depth A measure of how far light will travel through a medium before it is absorbed or scattered. The emergent flux is related to the input flux by $F_{em} = F_{in}e^{-\tau}$, where τ is the optical depth. So, for an optical depth of 1, the emergent light is reduced to 37% of its original level. If the value of τ is much less than 1, the medium is said to be *optically thin*. If it is greater than 1, the medium is said to be *optically thick*.

penumbra The outer region of a sunspot.

phase space Six-dimensional coordinates representing a particle position in both configuration and velocity space.

photoelectric absorption Absorption of photons by means of the photoelectric effect, in which an electron that is bound in an atom absorbs the energy of a photon, causing it to be ejected from the atom. The probability of a photon being absorbed peaks when the photon energy exactly matches the binding energy of the electron. Below this critical energy the electron cannot be ejected, resulting in a pronounced edge in the absorption cross-section when plotted as a function of energy, corresponding to the critical energy of the electron in that atom. Above the energy of each edge, the cross-section falls by about the third power of energy. In X-ray astronomy, photoelectric absorption by the atoms in the interstellar medium is an important process. This attenuates X-ray sources, particularly at low photon energies where the absorption cross-section due to the summed effect of the various atomic species that make up the interstellar medium is highest.

photon A quantum of electromagnetic radiation.

photosphere   The Sun's surface with a temperature of 6000 K.

phytoplankton   Plankton (microscopic living organisms drifting in the ocean and inland waters) that consists of plant rather than animal species.

pick-up   The process whereby newly ionised particles suddenly see the solar wind convection electric field and are thus swept up into the solar wind flow.

Planck's constant $h = 6.6262 \times 10^{-34}$ Js.

planet   A large, differentiated planetary body (nine in our Solar System).

planetesimal   A body in the early Solar System formed by accretion and collisions. The building block of the planets.

plasma beta   A parameter which is the ratio of the thermal plasma pressure and the magnetic pressure, $p/(B^2/2\mu_0)$. In low-beta plasma, like the solar corona, the magnetic forces dominate over the plasma forces.

plasma kinetic theory   Mathematical application for dealing with plasmas at the distribution function level.

plasma sheet   A region of plasma occupying the central region of the magnetotail and containing the cross-tail current sheet.

plasma torus   An ionised region around the orbit of a planetary moon, which is itself a source of neutral particles: either from an atmosphere (Titan, Triton) or volcanism (Io).

plasmasphere   A region of cold plasma of ionospheric origin located close to the Earth and co-rotating with the planet.

plasmoid   A section of a plasma sheet that becomes disconnected from the Earth near substorm onset. It comprises a magnetic bubble of loop-like magnetic fields.

polar cap (space plasma physics)   A region surrounding the magnetic poles of the Earth that contains the footpoints of the open magnetic field lines within the magnetosphere.

polar cusps   The two regions in the magnetosphere near magnetic local noon, about 15° towards the equator from the north and south poles, that mark the transition from geomagnetic field lines on the sunward side to the field lines in the polar cap that are swept back towards the magnetotail.

polar   Or 'AM Her' star. A magnetised accreting white dwarf that is similar to the intermediate polars except that the rotation of the white dwarf is phase-locked to the binary orbit. This means that the spin and orbital periods of the white dwarf are the same and one side of the white dwarf faces the companion star permanently.

precipitation Ice particles or water droplets large enough to fall at least 100 m below cloud level before evaporating (e.g. rain, hail, snow).

primordial nebula A cloud of rotating gas and dust at the beginning of the Solar System.

pristine Unchanged since original formation.

protoplanetary disc The region around a star, containing dust which obscures the star as seen from the Earth. It may contain forming planetesimals, comets or planets.

QPO Quasi-periodic oscillation, referring to oscillations in the electromagnetic flux emitted by a source.

quantum mechanics The study of radiation and matter at the atomic level.

quantum numbers Four numbers that can describe each orbit of an electron.

quasar Or quasi-stellar object (QSO). Quasars are the bright nucleus of an active galaxy seen at high redshift (large distance). The name is a contraction of 'quasi-stellar', referring to its star-like appearance. The term QSO is used historically for quasars that are relatively faint in the radio band, but the two terms are increasingly used interchangeably. It is believed that the enormous energy of a quasar is generated by accretion on to a very massive black hole.

quasi-neutrality The property of a plasma by which the charge density of ions and electrons cancel over all but the smallest of spatial scales.

quasi-parallel shock The portion of the bow shock or interplanetary shock at which the magnetic field is oriented close to parallel to the normal to the shock surface. Such shocks are difficult to identify in data as accelerated particles can easily escape into the upstream region.

quasi-perpendicular shock The portion of the bow shock or interplanetary shock at which the magnetic field is oriented close to perpendicular to the normal to the shock surface. Such shocks are readily identified in data as a sharp jump in magnetic field and plasma parameters.

radian A unit of angle, being the angle subtended at the centre of a circle by an arc equal to the circle's radius.

radiation belts Regions of the magnetosphere in which charged particles are trapped. They are located about 1.2–6 R_E above the equator. The inner belt, at about 5000 km above the Earth's surface, contains high energy (MeV) protons produced by the decay of neutrons created by cosmic ray collisions with neutral particles in the atmosphere. The outer belt, from about 10 R_E on the sunward side to 6 R_E on the night side, has lower energy protons (< 1 MeV) from the solar wind and ionsophere. The radiation belts are also commonly known as the van Allen belts.

radiation pressure The force exerted by photons on matter, for example due to scattering by electrons in an ionised cloud.

radiometer An instrument for measuring the intensity of electromagnetic radiation.

radiometric Concerning the measurement of the intensity of electromagnetic radiation.

radiosonde A balloon-borne meteorological instrument package, released from the Earth's surface so as to ascend freely and measure vertical profiles of atmospheric temperature, humidity and pressure (and sometimes wind speed).

Rayleigh–Jeans tail The low-energy tail of a blackbody spectrum, where the amount of flux emitted as a function of wavelength can be approximated by the inverse fourth power of the wavelength.

red giant The phase during the lifetime of a star (e.g. our Sun) where hydrogen fuel is exhausted. Other nuclear fusion reactions start and the diameter expands. In the Sun's case the likely new radius is near the Earth's orbit.

redshift The amount by which the wavelength of light is increased (shifted redward) by the Doppler effect when the object that emits it is receding from the observer. The converse effect, due to an object approaching the observer, is known as 'blueshift'. The redshift caused by the cosmological expansion of the Universe is generally given the symbol z.

reforming On an icy satellite, thawing and re-freezing of the planetary surface. On a terrestrial type planet it may be deposition of volcanic lava. Both destroy the body's cratering record.

regolith A layer of loose and broken rock and dust on a planetary surface. Grains may contain material from impacts at any point on the surface.

resonances Structures seen by *Voyager* in Saturn's rings.

resurfacing See 'reforming'.

ring current The current that flows near the geomagnetic equator in the outer radiation belt. It is produced by gradient and curvature drift of trapped 10–300 keV charged particles. It is strengthened during geomagnetic storms as a result of hot plasma injected from the magnetotail and oxygen ions from the ionosphere. An increase in the ring current produces depression of the Earth's horizontal magnetic field.

ring distribution   Ion distribution in velocity space formed by a cycloid in real space (e.g. new cometary ions). It is an unstable distribution that creates plasma waves.

Roche lobe   In a binary star system, the surface beyond which, matter is not gravitationally bound to one star or the other.

Scavenging   The process by which solar wind pick-up can remove ionospheric material from the dayside and transport it into the downstream region at an unmagnetised planet such as Venus.

Schwarzschild radius   The boundary around a non-rotating black hole from within which even light cannot escape.

sea ice   The frozen ocean surface (but not including icebergs, which are floating pieces of ice that have broken off glaciers or ice shelves).

second adiabatic invariant J   A constant of a charged particles motion associated with its bounce motion between mirror points in a magnetic bottle configuration, such as the closed field lines in the inner magnetosphere. It is invariant provided the field does not change significantly during one bounce period.

sedimentation   Deposition of material in layers, e.g. on the ocean bed.

selection rules   'Rules' that define what transitions an electron can make within an atom.

SEP   Solar energetic particle.

Seyfert galaxy   A galaxy with a bright, active nucleus, of a kind first catalogued by Carl Seyfert. Such galaxies are believed to be a lower-luminosity example of the quasar phenomenon.

shear wave   An MHD wave mode in a magnetised plasma, arising as a result of restoring forces associated with the magnetic field. Also referred to as the Alfvén wave, or transverse, or intermediate wave.

shell distribution   Ion distribution in velocity space formed (approximately; see 'bispherical shell') by scattered ring distribution.

shell star   Also known as a 'P-Cygni star', an early-type star that has absorption and emission features due to a surrounding envelope of material.

shepherding   Structures driven by gravitational interactions with nearby small moons seen by *Voyager* in Saturn's rings.

shock foot   A region on the upstream side of a quasi-perpendicular shock in which there are relatively small changes of the field and plasma parameters. It appears to result from the confinement of reflected particles to within 1 gyroradius of the shock ramp.

shock ramp   The main discontinuity in field and plasma parameters at a quasi-perpendicular shock.

slow mode magnetosonic wave   An MHD wave mode in a magnetised plasma, arising as a result of restoring forces associated with the magnetic field.

slow mode shocks   A shock wave associated with the non-linear steepening of slow mode magnetosonic waves. There is some evidence that these occur in the regions downstream of a magnetic reconnection neutral line.

solar cycle   The periodicity in solar activity, which is normally measured by the number of sunspots. The period of a cycle lasts for approximately 11 years.

solar flares   Fast releases of energy across the electromagnetic spectrum over a small area within an active region.

solar nebula   See 'primordial nebula'.

solar-terrestrial relations   Interdisciplinary study of the input and controlling influence of the solar wind on the terrestrial system.

solid angle   Three-dimensional angle (e.g. formed by the apex of a cone).

space density   The number density of objects in a given volume.

spectral signature   The characteristic shape of the spectrum of electromagnetic radiation as reflected or emitted by an object or material.

spectral type   Classification of stars based on what atomic transitions are dominant in their spectra. The current classification (OBAFGKM) is based on the surface temperature.

spicules   An elongated type feature seen in the upper part of the chromosphere. They are seen strongly on the red side of the Hα line.

splits   Structures seen by *Voyager* in Saturn's rings.

spokes   Structures, perhaps due to dust–plasma interactions, seen by *Voyager* in Saturn's rings.

stand-off distance   The distance upstream that a boundary, such as a bow shock or magnetopause, stands upstream of the planet.

steradian (sr)   A unit of solid angle, being the solid angle subtended at the centre of a sphere by part of its surface whose area is equal to the square of its radius.

STJ   Superconducting tunnel junctions, or Josephson junction detectors, are devices for measuring the energy of photons. They consist of two thin films of a superconducting metal such as niobium, tantalum or hafnium separated by a thin insulating layer. When operated at a temperature well below the superconductor's critical temperature (typically less than $1°$ K), the equilibrium state of the junction is easily perturbed by any photon striking it. By applying a small bias voltage across the junction, an electric charge proportional to the energy of the perturbing photon can be extracted from the device. STJ detectors are capable of energy discrimination even in the optical band.

storm main phase Enhancement of the ring current associated with the injection of enhanced fluxes of energetic particles during enhanced and prolonged coupling between the solar wind and the magnetosphere.

storm recovery phase Decay of the ring current over several days following a storm main phase.

storm sudden commencement (SSC) Compression of the magnetospheric magnetic field at the beginning of a storm. Often associated with the arrival of a CME at the Earth.

stratosphere The atmospheric layer (typically between about 10 km and 50 km in altitude) where the temperature increases (or is constant) with increased height, thus preventing convection.

sublimation Evaporation from the solid to vapour phase directly.

substorm current wedge The diversion of the cross tail current down magnetospheric field lines and across the auroral ionosphere during the expansion phase of a magnetospheric substorm.

substorm expansion phase The energy release phase of a magnetospheric substorm.

substorm growth phase The energy storage phase of a magnetospheric substorm.

substorm recovery phase The relaxation phase of a magnetospheric substorm.

sunspot Cool areas of the Sun's surface. These have strong magnetic fields.

supernova A cataclysmic explosion due to the collapse of the core in a massive star approaching the end of its life, or the collapse of an accreting white dwarf in a binary system when its mass exceeds the threshold for stability.

super-rotation Rotation of atmosphere faster than that of the host planet (e.g. Venus).

tectonic A plate-like geological process as at the Earth.

terminal shock A hypothetical boundary inside a heliopause where solar wind flow becomes abruptly subsonic.

terraforming   Changing a planetary environment by human intervention.

thermohaline circulation   An ocean circulation that is driven by gradients in temperature and salinity.

third adiabatic invariant   A constant of a charged particle's motion associated with its gradient and curvature drift motion around a magnetised planet. It is invariant provided the field does not change significantly during one drift-orbit period.

Thomson scattering   The scattering of photons off free electrons.

transverse wave   An MHD wave mode in a magnetised plasma, arising as a result of restoring forces associated with the magnetic field. Also referred to as the Alfvén wave, or shear, or intermediate wave.

triple point   A temperature condition where a substance (e.g. water or methane) can occur in solid, liquid and gas state.

T-Tauri   The early phase in the lifecycle of a Sun-like star in which the star heats nearby dust and solar wind intensity is much higher than present.

ultraviolet catastrophe   The prediction from classical physics that the higher the frequency the more radiation there should be.

umbra   The dark centre of sunspots.

validation   Making valid. When relating to measurements made by remote sensing, checking those measurements against *in situ* data, so as to confirm their accuracy (or to enable improvements in accuracy to be made).

velocity space   Three-dimensional coordinates allowing representation of a particles velocity vector.

volatile   Water and other substances, which can evaporate or can be trapped in forming planetary bodies, eventually becoming atmospheric constituents.

warps   Structures seen by *Voyager* in Saturn's rings.

white dwarf   A small, dense remnant of a star whose supply of nuclear fuel has been exhausted. White dwarfs can have a mass up to about 1.4 times that of the Sun. In a higher mass object atomic forces are insufficient to withstand the force of gravity, and the star will collapse still further to form a neutron star or black hole.

X-line   A magnetic null point formed at the point where magnetic reconnection occurs. Also referred to as a neutral line.

X-ray burster   A neutron star which emits occasional flashes of X-rays caused by a nuclear reaction in matter that has collected on its surface due to accretion.

X-ray   Electromagnetic radiation with a wavelength in the approximate range of 0.01–10 nm, lying between extreme ultraviolet and gamma rays.

Zeeman splitting   The splitting of an atom's spectral lines under the influence of a magnetic field.

zonal winds   East–West winds.

Index

3C 273, 225

absorption coefficient, 32, 34
accretion, 223, 227–230, 232, 233,
  235–239, 244, 247–249
accretion disc, 232, 235, 244, 247, 248
accretion power, 223
ACE, 98, 103
active galactic nuclei (AGNs), 243–245,
  248, 344, 345, 377, 382, 439
active galaxies, 263
active regions on the Sun, 196, 197, 202,
  204, 210–212, 217, 218, 220
adiabatic demagnetisation refrigerator
  (ADR), 399, 441, 449
advanced very high resolution radiometer
  (AVHRR), 20, 21, 29, 38, 43–49, 52–60
Alfvén speed, 124, 125, 127, 288, 291,
  293–295, 299, 305
Alfvén wave, 124, 125, 128, 129, 279, 301,
  302, 305, 306
Alfvén waves, 101, 167
  in the solar atmosphere, 203, 209, 211
along track scanning radiometer (ATSR),
  29, 38, 39, 42, 52–58, 60, 61
AM Her stars, 237
American Science and Engineering, Inc.,
  224
Anglo-Australian Telescope (AAT), 339,
  345
angular momentum quantum number, l,
  257
angular resolution, 369, 370, 377, 378, 380,
  381, 384, 416, 418, 421, 426, 437, 439

Antarctica, 70, 72
anthropic principle, 334
apogee, 10, 11
Apollo, 3, 4, 6
Appleton, 95
architectural design, 449, 452, 453
Arecibo telescope, 108
argon, 191
Ariane 5, 98
Aristarchus, 334
Aristotle, 334
ASCA, 226, 227
asteroid mining, 104, 105
asteroids, 73, 76, 79, 81, 82, 88, 94, 95,
  103
Astro E, 9
Astro E2, 398, 399, 405
Atacama Large Millimetre Array
  (ALMA), 345
atmosphere–ocean general circulation
  models, 18
atomic oxygen, 445
attitude control
  of spacecraft, 443, 452
aurorae, 74, 97, 160, 162–165, 213, 214,
  218
auroral oval, 149, 163, 164
autoionisation, 261

Babcock, 197, 202
Balmer lines, 268, 277
Balmer series, 201
batteries
  on spacecraft, 447, 454

COLOUR PLATES

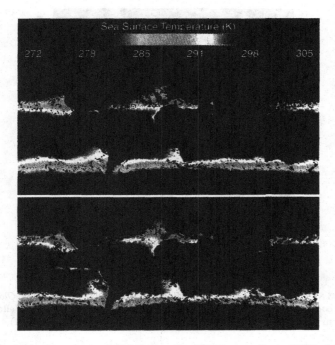

Plate 1 (Figure 2.18)   Examples of global monthly SST images from ATSR data, illustrating SST variations between an El Niño event (December 1997 — upper image) and a La Niña event (December 1998 — lower image). Blank areas have persistent could. Images from NERC/ RAL/ESA.

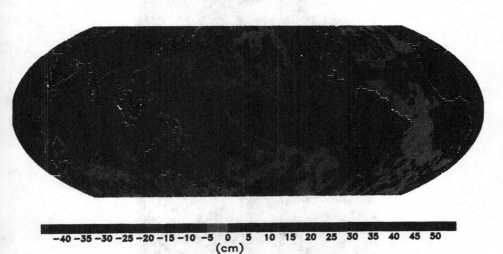

Plate 2 (Figure 2.22)   Global sea level variations due to ENSO, measured by the TOPEX/ POSEIDON radar altimeter. Anomalies are December 1997 (El Niño) minus December 1998 (La Niña) heights. From Picaut and Busalacchi (2001). Copyright 2001, with permission from Elsevier.

507

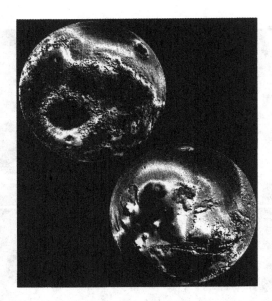

Plate 3 (Figure 3.8)   Mars Orbiter Laser Altimeter (MOLA) data from *Mars Global Surveyor.*
Height is colour-coded and exhibits a range of topography of 30 km on Mars. Hellas, 9 km deep
and some 2100 km across, is seen at the lower left of the topmost image. (NASA)

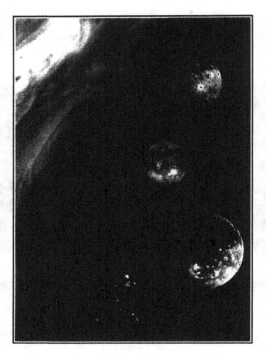

Plate 4 (Figure 3.15)   Size comparison for Jupiter and the Galilean satellites Io, Europa,
Ganymede and Callisto. (NASA)

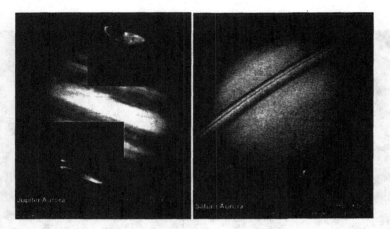

Plate 5 (Figure 3.21)   Aurora, imaged in the ultraviolet by HST, on Jupiter and Saturn. (STSci/NASA)

Plate 6 (Figure 3.23)   The *Cassini–Huygens* spacecraft during ground testing. (NASA)

Plate 7 (Figure 5.5)   The aurora australis observed from the space shuttle *Endeavour* in April 1994. (NASA)

Plate 8 (Figure 6.10)   A model of the magnetic field of the surface of the Sun produced from several instruments on board the *SOHO* spacecraft. The green and white plane is a coronal image from the EUV Imaging Telescope and the magnetic field is seen as black and white spots showing opposite polarities. Photo credit: Standford-Lockheed Institute for Space Research, Palo Alto, CA, and NASA Goddard Space Flight Center, Greenbelt, MD.

Plate 9 (Figure 6.16)   A soft X-ray image from the soft X-ray telescope on board the *Yohkoh* spacecraft. This clearly shows the magnetic complexity of the corona, with bright loop structures present, from small scale to large scale. A coronal hole (where there is no emission) is clearly seen towards the southern pole.

Plate 10 (Figure 6.21)   This is an image of a coronal mass ejection (CME) which occurred on 20 March 2000, obtained using the C2 occulting disc on the *LASCO* instrument. The *SOHO/LASCO* data here are produced by a consortium of the Naval Research Laboratory (USA), Max-Planck Institut für Aeronomie (Germany), Laboratoire d'Astronomie (France) and the University of Birmingham (UK). *SOHO* is a project of international cooperation between ESA and NASA.

Plate 11 (Figure 7.3)   A comparison of the sky in the region of the constellation of Orion as seen in visible light (*left*) and in soft X-rays by the ROSAT telescope (*right*). Picture courtesy of the Max Planck Institute for Extraterrestrial Physics, Garching.

Plate 12 (Figure 7.4)   A mosaic of images taken with the *XMM-Newton* X-ray telescopes showing the Andromeda galaxy (M31) in the 0.3–10 keV energy range. The five *XMM-Netwon* observations are arranged along the major axis of the galaxy, with the central one covering the core of the galaxy. The colours are related to the slope of the X-ray spectrum, with blue sources being hard and red sources soft. Notice the concentration of bright sources towards the central core of the galaxy, surrounded by cool diffuse gas. Courtesy of Sergey P. Trudolyubov, Los Alamos National Laboratory.

X-ray   **Optical**

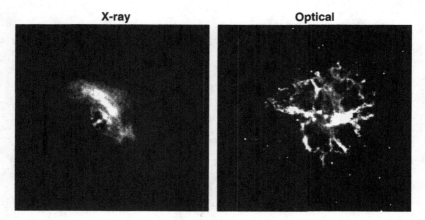

Plate 13 (Figure 7.10)   An example of a filled supernova remnant, the Crab nebula, seen in X-rays by the *Chandra* spacecraft and in the optical. (NASA)

Plate 14 (Figure 7.11)   The shell-like supernova remnant Cas-A pictured in X-rays by the *Chandra* satellite. The spectral slope is represented by colour, with red being the softest spectrum and blue the hardest. (NASA)